STOCHASTIC MODELS IN QUEUEING THEORY

STOCHASTIC MODELS IN QUEUEING THEORY

J. Medhi

Emeritus Professor of Statistics
Gauhati University
Guwahati, India

ACADEMIC PRESS, INC.
Harcourt Brace Jovanovich, Publishers
Boston San Diego New York
London Sydney Tokyo Toronto

This book is printed on acid-free paper. ∞

ACADEMIC PRESS, INC.
1250 Sixth Avenue, San Diego, CA 92101

United Kingdom Edition published by
ACADEMIC PRESS LIMITED
24–28 Oval Road, London NW1 7DX

Library of Congress Cataloging-in-Publication Data

Medhi, J. (Jyotiprasad)
Stochastic models in queueing theory / Jyotiprasad Medhi.
p. cm.
Includes bibliographical references.
ISBN 0-12-487550-5 (alk. paper)
1. Queueing theory. 2. Stochastic processes. I. Title.
T57.9.M45 1990
519.8′2—dc20 90-254
CIP

Printed in the United States of America
91 92 93 94 9 8 7 6 5 4 3 2 1

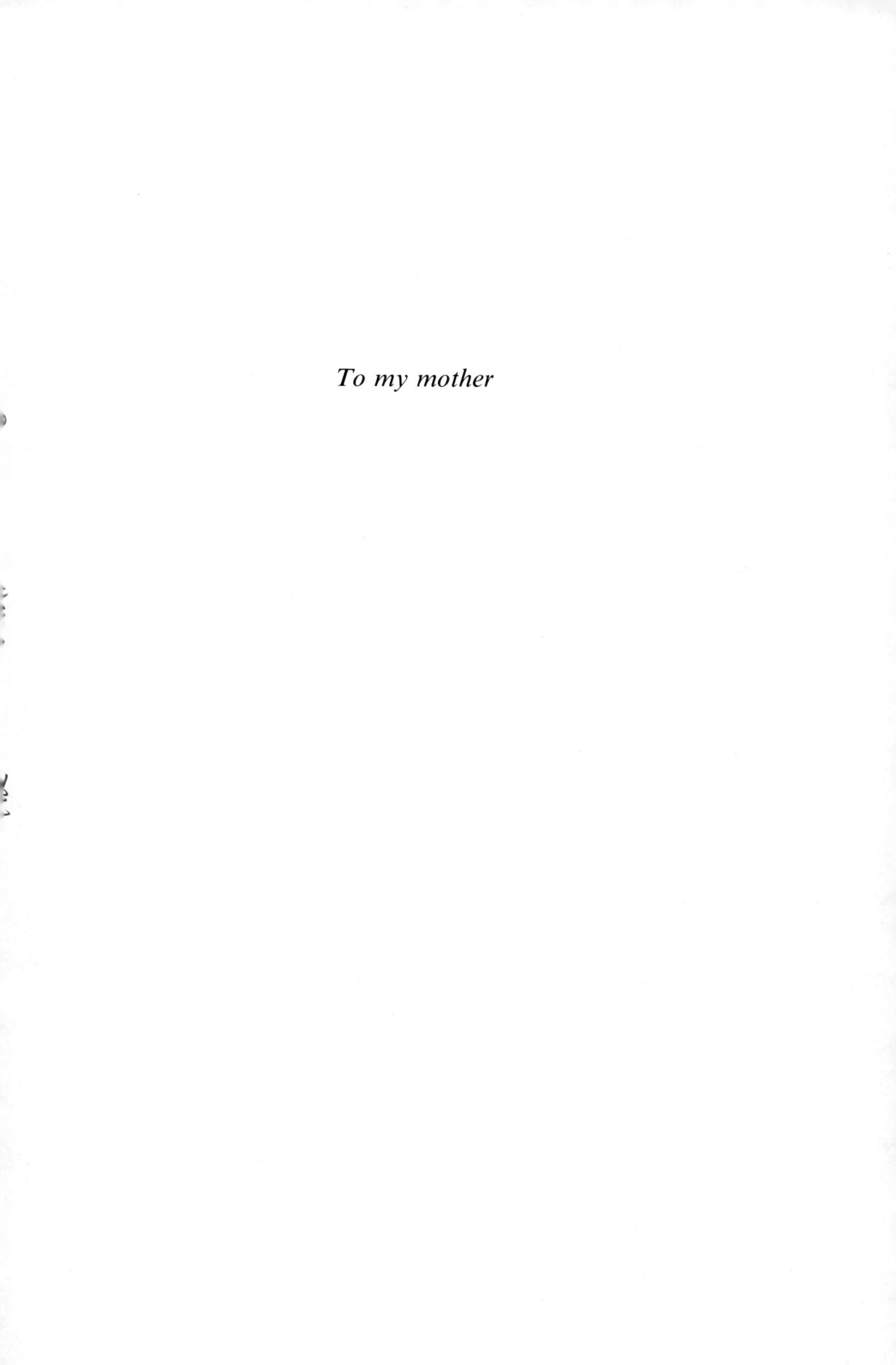

To my mother

karmaṇy evā 'dhikāras te
  mā phaleṣu kadācana
mā karmaphalahetur bhūr
  mā te saṅgo'stv akarmaṇi

You have right only over your own work
but never over the results thereof;
You should not be the cause of the results of your actions
nor should you have the inclination towards inactivity.

Bhagavadgītā, II,47

Contents

Preface

The study of queueing models has been of considerable active interest for the last eight decades. Queueing theory continues to be one of the most extensive theories of stochastic models. Its progress and development, both in methodology and in applications, are ever growing. Innovative analytic treatments toward its theoretical development are being advanced, and newer areas of application are emerging.

There is a large and growing audience interested in the study of queueing models. The level of background and preparation amongst them varies a great deal, as do the requirements for depth of coverage. The audience is comprised of advanced undergraduate and graduate students from a number of disciplines. In addition to students of standard graduate courses, there is a large number of researchers, professionals, and industry analysts who require an in-depth knowledge of the subject.

There are, of course, some excellent advanced works, monographs, and texts on the subject also. The rapid development of the subject demands updated texts, especially for the type of audience indicated above. Further, the style of presentation and the approach of individual authors appeal to different sections of this large and varied audience.

The author feels that there is enough scope as well as material for more texts, especially at a graduate level, in this ever-growing subject area. The book has grown out of the author's long experience of teaching and research. A highly complimentary comment made by a reviewer on the author's earlier book, *Stochastic Processes* (Wiley Eastern, and Halsted, Wiley 1982), has been a source of inspiration to the author to undertake the endeavor of preparing a book on queueing models of a similar readable style.

The prerequisites for using this book are a course on applied probability and a course on advanced calculus. Most of the users will have fulfilled these prerequisites. The book is divided into eight chapters. Chapter 1 is a summary of basic results in stochastic processes. This should be helpful to users in eliminating the need to refer frequently to other books on stochastic processes just for basic results. Chapter 2, which is devoted to general concepts, contains some discussion on concepts such as PASTA, superposition of arrival processes, and customer and time averages. Chapters 3 and 4 deal with birth-and-death queueing models and non-birth-and-death systems, respectively. Transient behavior and busy period analysis have been discussed at some length, and a uniformity of approach is emphasized. Some models of bulk queues have also been included because of their importance in transportation science. Chapter 5 is devoted to network of queues and Chapter 6 to non-Markovian queueing systems. In Chapter 7, systems with general arrival and service patterns are discussed. Chapter 8 covers miscellaneous topics such as asymptotic methods and queues with vacations, with a brief excursion into the design and control of queues. Diffusion approximations, which have emerged as powerful tools, have been discussed in some detail. We believe this chapter will be especially useful to researchers and professionals who wish to have a broad, general idea of the diffusion approximation methods.

Each of the chapters (except Chapter 2) contains a number of worked examples, problems and complements, and somewhat elaborate references, which include some of the most recent (1989) publications. The problems and complements contain some materials that have been discussed, keeping in mind the researchers and those who wish to pursue the subject further.

It is my hope that this book will be found readable and useful by those for whom it is intended. It could be used for a one- or two-semester course at the advanced undergraduate or graduate level on operations research, computer science, systems science, industrial (and other branches of) engineering, telecommunications, economics, management, and business (with programs having orientation in quantitative methods) besides mathematics and statistics.

Teachers would be the best judge of topics to be covered in a course. The following suggestions are for the consideration of the teachers:

For a two-semester course:
  the whole book.
For a one-semester course:
  Sections 1.1 through 1.5;
  Sections 2.1 through 2.7;

Section 3.1 through 3.8 and 3.11;
Sections 5.1 through 5.4;
Sections 6.1 through 6.4, 6.7, and 6.8; and
Sections 7.1 and 8.1

Exercises are to be selected from problems and complements.

Acknowledgments

I am intellectually indebted to all those whose works have stimulated my interest in this subject area. I have drawn freely and widely from the ever-increasing amount of literature.

In preparing this book, I have received encouragement and assistance in various ways from a number of experts, friends, and colleagues from this country and abroad. I am thankful to them all.

I am indeed most grateful to Professor J. G. C. Templeton (University of Toronto) and to Professor David D. Yao (Columbia University, formerly of Harvard University), both of whom painstakingly read portions of the manuscript and offered useful comments and valuable suggestions.

My elder son, Dr. Deepankar Medhi (Computer Science Program, University of Missouri, Kansas City, formerly of AT&T Bell Laboratories), and my elder daughter, Shakuntala Choudhury (AT&T Technology Systems, Bridgewater, NJ.), have rendered invaluable technical assistance. My younger son, Shubhankar, and younger daughter, Alakanandaa, have been of great help; so also have been my granddaughter, Namrata Gargee Choudhury and grandson Neiloy.

Last but not least, it is my wife, Prity, who bore with me, patiently, through the long hours that kept me engaged for months and months and who was seldom tired of waiting!

Assistance received from the University Grants Commission, India, in the form of a financial grant is gratefully acknowledged. Thanks are also due to Mrs. Susan M. Gay and other concerned Editors of Academic Press for their care and cooperation.

Brahmā Kripāhi Kevalam
Lord's Blessing Only (Sought)

Jyotiprasad Medhi
U. N. Bezbarua Road
Silpukhuri West
Guwahati 781 003
Assam, India

1 Stochastic Processes

1.1. Introduction

The theory of stochastic processes mainly originated from the needs of physicists; it began with the study of physical phenomena as random phenomena changing with time. Let t be a parameter assuming values in a set T, and let $X(t)$ represent a random or stochastic variable for every $t \in T$. The family or collection of random variables $\{X(t),\ t \in T\}$ is called a stochastic process. The parameter or index t is generally interpreted as time and the random variable $X(t)$ as the state of the process at time t. The elements of T are time points or epochs and T is a linear set, denumerable or nondenumerable. If T is countable (or denumerable) then the stochastic process $\{X(t),\ t \in T\}$ is said to be a discrete-parameter or discrete-time process, while if T is an interval of the real line, then the stochastic process is said to be a continuous-parameter or continuous-time process. For example $\{X_n,\ n = 0, 1, 2, \ldots\}$ is a discrete-time and $\{X(t),\ t \geq 0\}$ is a continuous-time process. The set of all possible values that the random variable $X(t)$ can assume is called the state space of the process; this set may be countable or noncountable. Thus, stochastic processes may be classified into the four types:

(i) discrete-time and discrete state space,

(ii) discrete-time and continuous state space,

(iii) continuous-time and discrete state space, and

(iv) continuous-time and continuous state space.

A discrete state space process is often referred to as a chain. A process such as (i) is a discrete-time chain and a process such as (iii) is a continuous-time chain.

A stochastic process, that is, a family of random variables thus provides description of the evolution of some physical phenomenon through time. Queueing systems provide many examples of stochastic process. For example, $X(t)$ might be the number of customers that arrive before a service counter by time t; then $\{X(t), t \geq 0\}$ is of the type (iii) above. Again W_n might be the queueing time of the nth arrival, then $\{W_n, n = 0, 1, 2, \ldots\}$ is of the type (ii) above.

Stochastic processes play an important role in modeling queueing systems. Certain stochastic processes are briefly discussed in this chapter.

1.2. Markov† Chains

1.2.1. *Basic Ideas*

Suppose that we observe the state of a system at discrete set of time points $t = 0, 1, 2, \ldots$. The observations at successive time points define a set of random variables (RVs) $X_0, X_1, X_2, \ldots$. The values assumed by the RVs X_n are the states of the system at time n. Assume that X_n assumes the finite set of values $0, 1, \ldots, m$; then $X_n = i$ implies that the state of the system at time n is i. The family of random variables (RVs) $\{X_n, n \geq 0\}$ is a stochastic process with discrete parameter space $n = 0, 1, 2, \ldots$ and discrete state space $S = \{0, 1, \ldots, m\}$.

Definition. A stochastic process $\{X_n, n \geq 0\}$ is called a Markov chain, if, for every $x_i \in S$,

$$Pr\{X_n = x_n | X_{n-1} = x_{n-1}, \ldots, X_0 = x_0\}$$
$$= Pr\{X_n = x_n | X_{n-1} = x_{n-1}\}, \tag{2.1}$$

† A. A. Markov (1856–1922).

provided the first member (LHS) is defined. Equation (2.1) indicates a kind of dependence between the RVs X_n; intuitively, it implies that given the present state of the system, the future is independent of the past. The conditional probability

$$Pr\{X_n = k \mid X_{n-1} = j\}, \qquad j, k \in S$$

is called the *transition probability* from state j to state k. This is denoted by

$$p_{jk}(n) = Pr\{X_n = k \mid X_{n-1} = j\}. \tag{2.2}$$

The Markov chain will be called (temporally) homogeneous if $p_{jk}(n)$ does not depend on n, i.e.,

$$Pr\{X_n = k \mid X_{n-1} = j\} = Pr\{X_{n+m} = k \mid X_{n+m-1} = j\}$$

for $m = -(n-1), \ -(n-2), \ldots, 0, 1, \ldots$. In such cases we denote $p_{jk}(n)$ simply by p_{jk}. The transition probability p_{jk}, which is the probability of transition from state j to state k in one step, i.e., from step $n-1$ to next step n (or from step $n+m-1$ to step $n+m$) is called one-step transition probability. The transition probability

$$Pr\{X_{r+n} = k \mid X_r = j\}$$

from state j to state k in n steps, (from state j in step r to state k in step $r+n$) is called the n-step transition probability. We denote

$$p_{jk}^{(n)} = Pr\{X_{r+n} = k \mid X_r = j\} \tag{2.3}$$

so that $p_{jk}^{(1)} = p_{jk}$. Define

$$\begin{aligned} p_{jk}^{(0)} &= 1, \qquad k = j, \\ &= 0, \qquad k \neq j. \end{aligned}$$

Then (2.3) is defined for $n = 0, 1, 2, \ldots$. Denote

$$\pi_j = Pr\{X_0 = j\} \quad \text{and} \quad \boldsymbol{\pi}(0) = \{\pi_0, \pi_1, \ldots, \pi_m\}.$$

$\boldsymbol{\pi}(0)$ is the initial probability vector. We have

$$\begin{aligned} &Pr\{X_0 = x_0, X_1 = x_1, \ldots, X_n = x_n\} \\ &\quad = Pr\{X_0 = x_0, \ldots, X_{n-1} = x_{n-1}\} \\ &\qquad \times Pr\{X_n = x_n \mid X_0 = x_0, \ldots, X_{n-1} = x_{n-1}\} \\ &\quad = Pr\{X_0 = x_0\} p_{x_0 x_1} p_{x_1 x_2} \cdots p_{x_{n-1} x_n}. \end{aligned} \tag{2.4}$$

Thus, given p_{jk} and $\pi(0)$, the joint probability given by (2.4) can be determined.

The matrix $P = (p_{jk})$, $j, k \in S$ is called the transition matrix or *transition probability matrix* (TPM) of the Markov chain. P is a nonnegative square matrix with unit row sums, i.e., $0 \le p_{jk} \le 1$, $\sum_k p_{jk} = 1$ for every $j \in S$.

A nonnegative square matrix P with unit row sums is called a stochastic matrix.

It can be easily shown that P^n is also a stochastic matrix and that

$$(p_{jk}^{(n)}) = P^n. \tag{2.5}$$

That is,

$$p_{jk}^{(2)} = \sum_{r \in S} p_{jr} p_{rk} \quad \text{for every} \quad j, k \in S$$

$$p_{jk}^{(n)} = \sum p_{jr}^{(n-1)} p_{rk};$$

more generally,

$$\begin{aligned} p_{jk}^{(m+n)} &= \sum_r p_{jr}^{(m)} p_{rk}^{(n)} \\ &= \sum_r p_{jr}^{(n)} p_{rk}^{(m)}, \qquad r \in S. \end{aligned} \tag{2.6}$$

Equation (2.6) is a special case of the Chapman–Kolmogorov equation; it is satisified by transition probabilities of a Markov chain.

Remark. To every stochastic matrix $P = (p_{ij})$, $i, j = 0, 1, \ldots$, there exists a homogeneous Markov chain $\{X_n, n = 0, 1, \ldots\}$ with state space $S = \{0, 1, \ldots\}$ and one-step transition probability p_{ij}, $i, j \in S$.

That is, to every stochastic matrix P, there corresponds a Markov chain $\{X_n\}$ for which P is the unit-step transition matrix.

Then $P^2 = (p_{ij}^{(2)})$ is also stochastic; it is the two-step transition matrix for the chain $\{X_n, n = 0, 1, \ldots\}$. However, not every stochastic matrix is the two-step transition matrix of a Markov chain.

1.2.2. Classification of States and Chains

1.2.2.1. Finite Homogeneous Chain. Let $\{X_n, n \ge 0\}$ be a finite homogeneous Markov chain having TPM $P = (p_{jk})$ and state space S, and let i, j, k be arbitrary states of S.

State i is said to lead to state j (or state j is said to be accessible from state

i) and is denoted by $i \to j$, if there exists an integer m (≥ 1) such that $p_{ij}^{(m)} > 0$. If no such integer exists then we say that i does not lead to j: denote this by $i \not\to j$. Two states are said to communicate with each other if $i \to j$ and $j \to i$; this is denoted by $i \leftrightarrow j$. The relations $i \to j$ and $i \leftrightarrow j$ are transitive.

One way to classify the states of a chain is given below.

If $i \to j$ but $j \not\to i$, then index i is said to be inessential. If $i \to j$ implies $i \leftrightarrow j$ for at least one j, then i is said to be essential. All essential states can be grouped into a number of essential classes such that all states belonging to an essential class communicate with one another, but cannot lead to a state outside the class. An essential class is closed, i.e., if i, j, k belong to an essential class and l is another state outside the class then $i \leftrightarrow j \leftrightarrow k$, but $i \not\to l$, $j \not\to l$, $k \not\to l$ (though $l \to i, j$, or k).

Inessential states, if any, can be grouped into a number of inessential classes such that all states belonging to an inessential class communicate with all states in the class. A finite homogeneous Markov chain has at least one essential class of states, whereas a Markov chain with denumerable numbers of states may not necessarily have any essential class.

A Markov chain is said to be *irreducible* if it contains exactly one essential class of states; in this case every state communicates with every other state of the chain. A nonirreducible (or reducible) chain may have more than one essential class of states as well as some inessential classes.

Suppose that $i \to i$, i.e., there exists some m (≥ 1) such that $p_{ii}^{(m)} > 0$. The greatest common divisor of all such m for which $p_{ii}^{(m)} > 0$ is called the period $d(i)$ of the state i. If $d(i) = 1$, then state i is said to be *aperiodic* (or *acyclic*) and if $d(i) > 1$, state i is said to be *periodic* with period $d(i)$. If $p_{ii} > 0$, then clearly state i is aperiodic.

For an irreducible Markov chain, either all states are aperiodic or they are periodic having the same period $d(i) = d(j) = \cdots$. Thus, irreducible Markov chains can be divided into two classes, aperiodic and periodic. An irreducible Markov chain is said to be *primitive* if it is aperiodic and *imprimitive* if it is periodic.

A Markov chain whose essential states form a single essential class and are aperiodic is said to be *regular*. Such a chain may have some inessential indices as well. Note that the transitions between states of the essential class of a regular chain form a submatrix P_1 that is stochastic. The TPM P of a regular chain can be written in canonical form,

$$P = \begin{pmatrix} P_1 & 0 \\ R_1 & Q \end{pmatrix}, \tag{2.7}$$

where the stochastic submatrix P_1 corresponds to transitions between states of essential class, the square matrix Q to transitions between inessential states, and the rectangular matrix R_1 to transitions between inessential states and essential states (see Seneta (1981) for details.)

1.2.2.2. Ergodicity Property. We shall now discuss an important concept: the ergodicity property. The classification given above is enough for this discussion so far as finite chains are concerned.

Before considering ergodicity, we shall describe another concept: invariant measure.

If $\{X_n\}$ is a Markov chain with TPM P and if there exists a probability vector $\mathbf{V} = (v_1, v_2, \ldots)$ (i.e., $0 \le v_i \le 1, \sum v_i = 1$) such that

$$\mathbf{V}P = \mathbf{V},$$

then $\mathbf{V}$ is called an *invariant measure* (or *stationary distribution*) of the Markov chain $\{X_n\}$ or with respect to the stochastic matrix P.

If there exists $\mathbf{V}$ such that $\mathbf{V}P \le \mathbf{V}$, then $\mathbf{V}$ is called a subinvariant measure of the chain with TPM P. Our interest lies in the types of Markov chains that possess invariant measures.

For a finite, irreducible Markov chain with TPM P, an invariant measure exists and is unique, that is, there is a unique probability vector $\mathbf{V}$ such that

$$\mathbf{V}P = \mathbf{V}, \qquad \mathbf{V}\mathbf{e} = 1 \tag{2.8}$$

where $\mathbf{e} = (1, 1, \ldots, 1)$ is a column vector with all its elements equal to unity.

1.2.2.3. Ergodic Theorems.

Theorem 1.1. *Ergodic theorem for primitive chains.*

Let $\{X_n, n \ge 0\}$ be a finite irreducible aperiodic Markov chain with TPM P and state space S. Then, as $n \to \infty$,

$$P^n \to \mathbf{e}\mathbf{V} \tag{2.9}$$

elementwise, where $\mathbf{V}$ is the unique stationary distribution (or invariant measure) of the chain.

Further, the rate of approach to the limit is geometrically fast, that is, there exist positive constants $a, b, 0 < b < 1$ such that $\varepsilon_{ij}^{(n)} \le ab^n$, where $p_{ij}^{(n)} = v_j + \varepsilon_{ij}^{(n)}$.

The above theorem implies that

$$\lim_{n\to\infty} p_{ij}^{(n)} \to v_j \quad \text{for every} \quad j \in S \tag{2.10}$$

exists and is independent of the initial state i and the quantities v_j's are given by the solution of the matrix equation

$$\mathbf{V}P = \mathbf{V} \quad \text{with} \quad \mathbf{Ve} = 1, \tag{2.11}$$

The limiting probability distribution of the chain tends to an equilibrium distribution that is independent of the initial distribution. This tendency is known as ergodicity or ergodic property.

Another ergodic theorem is stated below.

Theorem 1.2. *Let $\{X_n, n \geq 0\}$ be a finite k state regular Markov chain (i.e., having a single essential class of aperiodic states) and having TPM P. Let $\mathbf{V}_1$ be the stationary distribution corresponding to the primitive submatrix P_1 (corresponding to the transitions between the states of the essential aperiodic class).*

Let $\mathbf{V} = (\mathbf{V}_1, \mathbf{0})$ be a $1 \times k$ vector, Then, as $n \to \infty$,

$$P^n \to \mathbf{eV} \tag{2.9a}$$

elementwise.

$\mathbf{V}$ *is the unique stationary distribution corresponding to the matrix P and the rate of approach to the limit in* (2.9a) *is geometrically fast.*

If C denotes the single essential class with states j, $j = 1, 2, \ldots, m$, $(m < k)$, $j \in C$, and $\mathbf{V}_1 = (v_1, v_2, \ldots, v_j, \ldots, v_m)$ is given by the solution of

$$\mathbf{V}_1 P_1 = \mathbf{V}_1, \qquad \mathbf{V}_1\mathbf{e} = 1, \tag{2.12}$$

then the above result implies that, as $n \to \infty$,

$$\begin{aligned} \lim p_{ij}^{(n)} &\to v_j, \qquad j \in C, \quad \text{and} \\ \lim p_{ij}^{(n)} &\to 0, \qquad j \notin C. \end{aligned} \tag{2.13}$$

The above theorem asserts that regularity of the chain is a sufficient condition for ergodicity. It is also a necessary condition.

Example 1.1. Consider the Markov chain having state space $S = \{0, 1, 2\}$, and TPM P

$$P = \begin{pmatrix} 0 & \frac{1}{3} & \frac{2}{3} \\ \frac{1}{2} & 0 & \frac{1}{2} \\ \frac{3}{4} & \frac{1}{4} & 0 \end{pmatrix}$$

This is a finite irreducible chain. Its invariant measure $\mathbf{V} = (v_1, v_2, v_3)$ is given by the solution of

$$\mathbf{V}P = \mathbf{V},$$

which leads to

$$v_1 = \tfrac{1}{2}v_2 + \tfrac{3}{4}v_3$$
$$v_2 = \tfrac{1}{3}v_1 + \tfrac{1}{4}v_3.$$
$$v_3 = \tfrac{2}{3}v_1 + \tfrac{1}{2}v_2$$

As $\mathbf{Ve} = v_1 + v_2 + v_3 = 1$, one of the above equations is redundant. We get

$$\mathbf{V} = (v_1, v_2, v_3)$$

where

$$v_1 = \tfrac{21}{53}, \qquad v_2 = \tfrac{12}{53}, \qquad v_3 = \tfrac{20}{53}.$$

Thus,

$$P^n \to \mathbf{eV}.$$

That is, as $n \to \infty$,

$$\lim p_{i1}^{(n)} = \tfrac{21}{53}, \qquad \lim p_{i2}^{(n)} = \tfrac{12}{53}, \qquad \lim p_{i3}^{(n)} = \tfrac{20}{53}$$

for all $i = 1, 2, 3$.

Example 1.2. Consider the two-state Markov chain with TPM

$$P = \begin{pmatrix} 1-p & p \\ p & 1-p \end{pmatrix}, \qquad 0 \le p \le 1.$$

The equation $\mathbf{V}P = \mathbf{V}$, ($\mathbf{V} = (v_1, v_2)$) leads to

$$v_1 = v_2 = \tfrac{1}{2}, \quad \text{for all} \quad p,$$

so that the invariant distribution is $(\frac{1}{2}, \frac{1}{2})$ for all p. We have, for $p \neq 0$.

$$P^n = \mathbf{e}(\tfrac{1}{2}, \tfrac{1}{2}) + \tfrac{1}{2}(1 - 2p)^n \begin{pmatrix} 1 & -1 \\ -1 & 1 \end{pmatrix}.$$

For $p = 0$, the chain consists of two absorbing states 0 and 1, i.e., it consists of two essential classes, C_1 and C_2, with members 0 and 1, respectively. The chain is decomposable. For $p = 1$,

$$P^n = \begin{pmatrix} 1 & 0 \\ 0 & 1 \end{pmatrix} \text{ when } n \text{ is even and}$$

$$= \begin{pmatrix} 0 & 1 \\ 1 & 0 \end{pmatrix} \text{ when } n \text{ is odd.}$$

The chain is periodic.

Thus, even though invariant distribution exists for all p, $0 \leq p \leq 1$, as $n \to \infty$,

$$\lim p_{i0}^{(n)} = \tfrac{1}{2} \quad \text{and}$$

$$\lim p_{i1}^{(n)} = \tfrac{1}{2}$$

exist only when $p \neq 0, 1$.

Remarks. We can make two deductions: (1) Existence of stationary distribution (i.e., existence of a solution of $\mathbf{V}P = \mathbf{V}$, $\mathbf{Ve} = 1$) does not necessarily imply existence of the limiting distribution

$$\left\{ \lim_{n \to \infty} p_{ij}^{(n)} \right\}.$$

(2) The example also shows how much the transient behavior can vary over the class of transient matrices for a given equilibrium distribution.

See Whitt (1983) for a discussion on this topic.

1.2.2.4. Markov Chain Having a Denumerably Infinite Number of States. So far we have considered homogeneous Markov chains with a finite number of states. Now we shall discuss homogeneous chains having a denumerably infinite number of states. We shall denote the state space by $S = \{0, 1, 2, \ldots\}$ instead of the more general

$$S = \{\ldots, -3, -2, -1, 0, 1, 2, \ldots\}.$$

Notions already defined (e.g., accessibility, communication, and periodicity) and subsequent definitions of essential and inessential states, essential classes, inessential classes, irreducible chains, and primitive chains will remain valid for a chain with a denumerable number of states. Only the definition of a regular chain cannot be carried over to this case. For, whereas in a finite chain, there is at least one essential class (and so the states of a finite chain may constitute exactly one essential class or more than one essential class), there may not be any essential class in the case of a chain with denumerably infinite number of states. For example, consider the chain $\{X_n, n \geq 0\}$ with $S = \{0, 1, 2, \ldots\}$ and TPM $P = (p_{ij})$ where

$$\begin{aligned} p_{ij} &= 1 \qquad j = i + 1, \\ &= 0 \quad \text{otherwise.} \end{aligned}$$

This denumerable chain does not possess any essential class at all. Each state is inessential and no two states communicate. Whereas consideration of classification of states into essential and inessential classes was adequate for dealing with limiting distribution for finite chains, a more sensitive classification of states would be required in the present case of chains with denumerable infinity of states.

1.2.2.5. *Transience and Recurrence.*

Define

$$\begin{aligned} &\{f_{ij}^{(n)}\}, i, j = 1, 2, \ldots, n, \\ &f_{ij}^{(0)} = 0, \qquad f_{ij}^{(1)} = p_{ij}, \quad \text{and} \\ &f_{ij}^{(k+1)} = \sum_{\substack{r \\ r \neq j}} p_{ir} f_{rj}^{(k)}, \qquad k \geq 1. \end{aligned} \tag{2.14}$$

The quantity $f_{ij}^{(k)}$ is the probability of transition from state i to state j in k steps, without revisiting the state j in the meantime. (It is called *taboo* probability, j being the taboo state.) Here $\{f_{ij}^{(k)}\}$ gives the distribution of the first passage time from state i to state j. We can write

$$f_{ij}^{(n)} = Pr\{X_n = j, X_r \neq j, \qquad r = 1, 2, \ldots, n-1 | X_0 = i\}.$$

The relation (2.14) can also be written as

$$\begin{aligned} p_{ij}^{(n)} &= \sum_{r=0}^{n} f_{ij}^{(r)} p_{jj}^{(n-r)}, \qquad n \geq 1 \\ &= \sum_{r=0}^{n} f_{ij}^{(n-r)} p_{jj}^{(r)}. \end{aligned} \tag{2.15}$$

The relations (2.14) and (2.15) are known as *first entrance formulas.*

Let

$$P_{ij}(s) = \sum_n p_{ij}^{(n)} s^n, \qquad F_{ij}(s) = \sum_n f_{ij}^{(n)} s^n, \qquad |s| < 1.$$

Then from the convolution structure,

$$\begin{aligned} P_{ij}(s) &= P_{ii}(s)F_{ij}(s), \qquad j \neq i \\ P_{ii}(s) - 1 &= P_{ii}(s)F_{ii}(s). \end{aligned} \tag{2.16}$$

Definition. A state i is said to be *persistent* if $F_{ii} = F_{ii}(1-0) = 1$ and is said to be *transient* if $F_{ii}(1-0) < 1$.

A persistent state is null or nonnull based on whether $\mu_{ii} = F'_{ii}(1) = \infty$ or $< \infty$, respectively.

Equivalent criteria of persistence and recurrence are as follows.

An index i is persistent *iff* (if and only if)

$$\sum_n p_{ii}^{(n)} = \infty$$

and is transient *iff*

$$\sum_n p_{ii}^{(n)} < \infty.$$

The relationship between these two types of classification of states and chain can be given as follows.

An inessential state is transient and a persistent state is essential. In the case of a finite chain, i is transient *iff* it is inessential, otherwise it is nonnull persistent.

All the states of an irreducible chain, whether finite or denumerable, are of same type: all transient, all null persistent, or all nonnull persistent.

A finite Markov chain contains at least one persistent state. Further, a finite irreducible Markov chain is nonnull persistent. The ergodic theorem for a Markov chain with a denumerable infinite number of states is stated below.

Theorem 1.3. General Ergodic Theorem. *Let P be the TPM of an irreducible aperiodic (i.e., primitive) Markov chain with a countable state space S (which may have a finite or a denumerably infinite number of states).*

If the Markov chain is transient or null persistent, then for each $i, j \in S$,

$$\lim_{n \to \infty} p_{ij}^{(n)} \to 0. \tag{2.17a}$$

If the chain is nonnull persistent, then for each $i, j \in S$,

$$\lim_{n \to \infty} p_{ij}^{(n)} = v_j \tag{2.17b}$$

exists and is independent of i. *The probability vector* $\mathbf{V} = (v_1, v_2, \ldots)$ *is the unique invariant measure of* P, *that is,*

$$\mathbf{V}P = \mathbf{V}, \qquad \mathbf{Ve} = 1; \tag{2.18a}$$

and further

$$v_j = (\mu_{jj})^{-1} \tag{2.18b}$$

where μ_{jj} *is the mean recurrence time of state* j.

The result is general and holds for a chain with a countable state space S. *In case the chain is finite, irreducibility ensures nonnull persistence, so that irreducibility and aperiodicity (i.e., primitivity) constitute a set of sufficient conditions for ergodicity of a finite chain. The sufficient conditions for ergodicity* ($\lim\ p_{ij}^{(n)} = v_j$) *for a chain with a denumerably infinite number of states involve, besides irreducibility and aperiodicity, nonnull persistence of the chain. For a chain with a denumerably infinite number of states, the number of equations given by* (2.18a) *will be infinite. It would sometimes be more convenient to find* $\mathbf{V}$ *in terms of the generating function of* $\{v_j\}$ *than to attempt solve Eqs.* (2.18a) *as such. We shall consider two such Markov chains that arise in queueing theory. See the Note below.*

Example 1.3. Consider a Markov chain with state space $S = \{0, 1, 2, \ldots\}$ having a denumerable number of states and having TPM.

$$P = \begin{bmatrix} p_0 & p_1 & p_2 & p_3 & \cdots \\ p_0 & p_1 & p_2 & p_3 & \cdots \\ 0 & p_0 & p_1 & p_2 & \cdots \\ 0 & 0 & p_0 & p_1 & \cdots \\ \cdots & \cdots & \cdots & \cdots & \cdots \end{bmatrix} \tag{2.19}$$

where $\sum_k p_k = 1$. Let

$$P(s) = \sum_k p_k s^k \quad \text{and} \quad \mathbf{V}(s) = \sum_k v_k s^k, \qquad |s| < 1$$

be the probability-generating functions (PGF) of $\{p_k\}$ and $\{v_k\}$, respectively. Clearly, the chain is irreducible and aperiodic; since it is a denumerable chain, we need to consider transience and persistence of the chain to study its ergodic property.

It can be shown that the states of the chain (which are all of the same type because of the irreducibility of the chain) are transient, persistent null or persistent nonnull according to

$$P'(1) > 1, \qquad P'(1) = 1, \qquad P'(1) < 1,$$

respectively. (See Prabhu, 1965.) Assume that $P'(1) < 1$, so that the states are persistent nonnull; then from (2.18a), we get

$$v_k = p_k v_0 + p_k v_1 + p_{k-1} v_2 + \cdots + p_0 v_{k+1}, \qquad k \geq 0. \tag{2.20}$$

Multiplying both sides of (2.20) by s^k and adding over $k = 0, 1, 2, \ldots$, we get

$$\begin{aligned}\mathbf{V}(s) &= v_0 P(s) + v_1 P(s) + v_2 sP(s) + v_{k+1} s^k P(s) + \cdots \\ &= P(s)[v_0 + (V(s) - v_0)/s];\end{aligned}$$

whence

$$\mathbf{V}(s) = \frac{v_0(1-s)P(s)}{P(s) - s}.$$

Since $\mathbf{V}(1) = 1$, we have

$$\begin{aligned}1 = \lim_{s \to 1} \mathbf{V}(s) &= v_0 \left[\lim_{s \to 1} \frac{(1-s)P(s)}{P(s) - s} \right] \\ &= v_0 \frac{1}{1 - P'(1)}.\end{aligned}$$

Thus,

$$\mathbf{V}(s) = \frac{(1 - P'(1))(1-s)P(s)}{P(s) - s} \tag{2.21}$$

is the generating function of $\{v_j\}$.

Example 1.4. Consider a Markov chain with state space $S = \{0, 1, 2, \ldots\}$ and having TPM

$$P = \begin{bmatrix} h_0 & g_0 & 0 & 0 & 0 & \cdots \\ h_1 & g_1 & g_0 & 0 & 0 & \cdots \\ h_2 & g_2 & g_1 & g_0 & 0 & \cdots \\ \cdots & \cdots & \cdots & \cdots & \cdots & \cdots \\ \cdots & \cdots & \cdots & \cdots & \cdots & \cdots \end{bmatrix} \tag{2.22}$$

where $h_i = g_{i+1} + g_{i+2} + \cdots$, $i \geq 0$, $g_i > 0$, and $\sum_{i=0} g_i = 1$. Here $p_{i0} = h_i$, $i \geq 0$,

$$\begin{aligned} p_{ij} &= g_{i+1-j}, \ i+1 \geq j \geq 1, \ i \geq 0 \\ &= 0, \qquad i+1 < j. \end{aligned}$$

The chain is irreducible and aperiodic. It can be shown that it is persistent nonnull when $\alpha = \sum jg_j > 1$. Then the chain is ergodic and $v_j = \lim_{n\to\infty} p_{ij}^{(n)}$ exist and are given as a solution of (2.18a); these lead to

$$v_0 = \sum v_r h_r \tag{2.23a}$$

$$v_j = \sum v_{r+j-1} g_r, \qquad j \geq 1 \tag{2.23b}$$

$$\sum_{j=0} v_j = 1. \tag{2.23c}$$

Let $G(s) = \sum_r g_r s^r$ be the PGF of $\{g_r\}$. Denote the displacement operator by E so that

$$E^r(v_k) = v_{k+r}, \qquad r=0, 1, 2, \ldots.$$

Then we can write (2.23b) in symbols as

$$\begin{aligned} E(v_{j-1}) = v_j &= \sum g_r E^r(v_{j-1}), \qquad j \geq 1 \quad \text{or} \\ \{E - \sum g_r E^r\} v_j &= 0, \qquad j \geq 0 \quad \text{or} \\ \{E - G(E)\} v_j &= 0, \qquad j \geq 0. \end{aligned} \tag{2.24}$$

The characteristic equation of the above difference equation is given by

$$r(z) \equiv z - G(z) = 0. \tag{2.25}$$

It can be shown that when $\alpha = G'(1) > 1$, there is exactly one real root of $r(z) = 0$ between 0 and 1. Denote this root by r_0 and the other roots by r_1, $r_2, \ldots$, $|r_i| > 1$, $i \geq 1$. The solution of (2.24) can thus be put as

$$v_j = c_0 r_0^j + \sum_{i=1} c_i r_i^j, \qquad j \geq 0$$

where the c's are constants. Since $\sum v_j = 1$,

$$\begin{aligned} c_i &\equiv 0 \quad \text{for} \quad i \geq 1, \\ v_j &= c_0 r_0^j, \qquad j \geq 0, \quad \text{and} \\ c_0 &= 1 - r_0 \end{aligned}$$

so that

$$v_j = (1 - r_0)r_0^j, \qquad j \geq 0, \tag{2.26}$$

r_0 being the root lying between 0 and 1 of (2.25) (provided $\alpha = G'(1) > 1$). The distribution is geometric.

Notes.

(1) The equation $\mathbf{V}P = \mathbf{V}$ is quite well known in the matrix theory. It follows from the well-known Perron–Frobenius theorem of matrix theory, that there exists a solution $\mathbf{V} = (v_1, v_2, \ldots)$ of the matrix equation $\mathbf{V}P = \mathbf{V}$ subject to the constraints $v_i \geq 0, \sum v_i = 1$.

(2) When the order of P is not large, the equations can be solved fairly easily to get $\mathbf{V} = (v_1, v_2, \ldots)$. When the order of P is large (infinite), the number of equations is also large (infinite) and the solution of the equations becomes troublesome. In Example 1.3 we considered and obtained the solution in terms of the generating function $V(s) = \sum v_j s^j$. This method may not always be applicable. Now suitable computer programs are available to solve the equations, however large the number of equations may be. This enables one to do away with the necessity of solving the equations in the usual way, which (though it appears straightforward) may be very cumbersome in practice.

1.3. Continuous-Time Markov Chains

We shall now consider continuous-time Markov chains, that is, Markov processes with discrete state space. Let $\{X(t), 0 \leq t < \infty\}$ be a Markov process with countable state space $S = \{0, 1, 2, \ldots\}$. We assume that the chain is temporally homogeneous. The transition probability function given by

$$\begin{aligned} p_{ij}(t) &= Pr\{X(t+u) = j \mid X(u) = i\} \\ t &> 0, \qquad i, j \in S, \end{aligned} \tag{3.1}$$

is then independent of $u \geq 0$. We have for all t,

$$0 \leq p_{ij}(t) \leq 1, \qquad \sum_j p_{ij}(t) = 1, \quad \text{for all} \quad j \in S.$$

Denote the matrix of transition probabilities by

$$P(t) = [p_{ij}(t)], \qquad i, j \in S.$$

Setting $p_{ij}(0) = \delta_{ij}$, the initial condition can be put as

$$P(0) = I.$$

Denote the probability that the system is at state j at time t by

$$\pi_j(t) = Pr\{X(t) = j\};$$

the vector $\boldsymbol{\pi}(t) = \{\pi_1(t), \pi_2(t), \ldots\}$ is the probability vector of the state of the system at time t. $\boldsymbol{\pi}(0)$ is the initial probability vector. Now

$$\begin{aligned} \pi_j(t) &= \sum_i Pr\{X(t+u) = j \mid X(u) = i\} Pr\{X(u) = i\} \\ &= \sum_i p_{ij}(t) Pr\{X(0) = i\} \\ &= \sum_i p_{ij}(t) \pi_i(0). \end{aligned} \tag{3.2}$$

Thus, given initial probability vector $\boldsymbol{\pi}(0)$ and the transition functions $p_{ij}(t)$, the state probabilities can be calculated and the probabilistic behavior of the system can be completely determined. The matrix form of (3.2) is

$$\boldsymbol{\pi}(t) = \boldsymbol{\pi}(0) P(t). \tag{3.3}$$

1.3.1. Sojourn Time

The time taken (or the waiting time) for change of state from state i is a random variable, say, τ_i, that is, the sojourn time at state i is τ_i. Then

$$\begin{aligned} &Pr\{\tau_i > s + t \mid X(0) = i\} \\ &\quad = Pr\{\tau_i > s + t \mid X(0) = i, \quad \tau_i > s\} \times Pr\{\tau_i > s \mid X(0) = i\}, \qquad t \geq 0. \end{aligned} \tag{3.4}$$

Denote

$$\bar{F}_i(u) = Pr\{\tau_i > u \mid X(0) = i\}, \qquad u \geq 0.$$

Then (3.4) can be written as

$$\bar{F}_i(t + s) = \bar{F}_i(t) \bar{F}_i(s), \qquad s, t \geq 0;$$

whence it follows that

$$\bar{F}_i(u) = e^{-a_i u}, \qquad u \geq 0, \qquad a_i > 0 \quad \text{is a constant.} \tag{3.5}$$

That is, sojourn time τ_i at state i is exponential with parameter a_i. Further, the sojourn times τ_i and τ_j are independent.

We have, for $t \geq 0$, $T \geq 0$,

$$p_{ij}(T+t) = \sum_k p_{ik}(T)p_{kj}(t), \qquad i, j, k \in S \tag{3.6}$$

or, in matrix form,

$$P(T+t) = P(T)P(t). \tag{3.7}$$

which is called the *Chapman–Kolmogorov equation.*

1.3.2. Transition Density Matrix or Infinitesimal Generator

Denote the right-hand derivative at $t = 0$ by

$$\begin{aligned} q_{ij} &= \lim_{h\to 0} \frac{p_{ij}(h) - p_{ij}(0)}{h} = \lim_{h\to 0} \frac{p_{ij}(h)}{h}, \qquad i \neq j \quad \text{and} \\ q_{ii} &= \lim_{h\to 0} \frac{p_{ii}(h) - p_{ii}(0)}{h} = \lim_{h\to 0} \frac{p_{ii}(h) - 1}{h}; \end{aligned} \tag{3.8}$$

write $-q_{ii} = q_i$. It is to be noted that q_{ij}, $i \neq j$ is always finite. While q_i (≥ 0) always exists and is finite when S is finite, q_i may be infinite when S is denumerably infinite. Writing $Q = (q_{ij})$, we can denote (3.8) in matrix notation as

$$Q = \lim_{h\to 0} \frac{P(h) - I}{h}.$$

From (3.8) it follows that, for small h,

$$\begin{aligned} p_{ij}(h) &= hq_{ij} + o(h), \qquad i \neq j, \\ p_{ii}(h) &= hq_i + o(h), \end{aligned} \tag{3.9}$$

where $o(h)$ is used as a symbol to denote a function of h that tends to zero more rapidly than h, i.e., $o(h)/h \to 0$ as $h \to 0$. Again,

$$\sum_j p_{ij}(h) = 1, \quad \text{or}$$

$$\sum_{j\neq i} p_{ij}(h) + p_{ii}(h) - 1 = 0;$$

whence we get

$$\sum_{j \neq i} q_{ij} + q_{ii} = 0 \quad \text{or} \\ \sum_{j \neq i} q_{ij} = q_i. \tag{3.10}$$

The matrix $Q = (q_{ij})$ is called the *transition density matrix* or *infinitesimal generator* or *rate matrix* or simply Q-matrix. The Q-matrix is such that (i) its diagnonal elements are negative and off-diagonal elements are positive, and (ii) each row sum is zero. Let $S = \{0, 1, 2, \ldots, m\}$ be a finite set, then

$$Q = \begin{bmatrix} -q_0 & q_{01} & \cdots & q_{0m} \\ q_{10} & -q_1 & \cdots & q_{1m} \\ \cdots & \cdots & \cdots & \cdots \\ q_{m0} & q_{m1} & \cdots & -q_m \end{bmatrix}.$$

1.3.3. Limiting Behavior: Ergodicity

The states of a continuous-time Markov chain admit of a classification similar to those of a discrete-time chain. A state j is said to be accessible or reachable from state i $(i \to j)$ if, for some $t > 0$, $p_{ij}(t) > 0$. States i and j communicate if $i \to j$ and $j \to i$. A continuous-time Markov chain is said to be *irreducible* if every state can be reached from every other state (or if each pair of states communicates).

Let α_{ij} denote the (first) entrance time from state i to state j without visiting j in the meantime and let $F(.)$ denote its DF, i.e.,

$$\begin{aligned} F_{ij}(t) &= Pr\{\alpha_{ij} < t\}, & t > 0, \\ &= 0, & t \leq 0. \end{aligned}$$

A state i is called *persistent* if

$$\lim_{t \to \infty} F_{ii}(t) = 1$$

and *transient* otherwise.

Criteria of transience and persistence can be expressed in terms of $p_{ij}(t)$ as follows:

State i is transient *iff*

$$\int_0^\infty p_{ii}(t)dt < \infty.$$

If state i, is null-persistent, then

$$\lim_{t\to\infty} p_{ii}(t) = 0,$$

and if state i is non–null-persistent, then

$$\lim_{t\to\infty} p_{ii}(t) > 0.$$

From Chapman–Kolmogorov Eq. (3.7) we get

$$\begin{aligned} p_{ij}(h+t) &= \sum_k p_{ik}(h)p_{kj}(t) \\ &= \sum_{k\neq i} p_{ik}(h)p_{kj}(t) + p_{ii}(h)p_{ij}(t) \end{aligned}$$

so that

$$\frac{p_{ij}(h+t) - p_{ij}(t)}{h} = \sum_{k\neq i} \frac{p_{ik}(h)}{h} p_{kj}(t) + \left(\frac{p_{ii}(h)-1}{h}\right)p_{ij}(t).$$

Taking the limit as $h \to 0$ and assuming that the order of the operations of taking the limit and summation can be interchanged, we get

$$\lim_{h\to 0} \frac{p_{ij}(h+t) - p_{ij}(t)}{h} = \sum_{k\neq i}\left[\lim_{h\to 0}\frac{p_{ik}(h)}{h}\right]p_{kj}(t) + \left[\lim_{h\to 0}\frac{p_{ii}(h)-1}{h}\right]p_{ik}(t) \quad \text{or}$$

$$p'_{ij}(t) = \sum_{k\neq i} q_{ik}p_{kj}(t) + q_i p_{ij}(t), \tag{3.11}$$

which is another form of the Chapman–Kolmogorov (backward) equation; it is in terms of the elements of the Q-matrix. In matrix notation, we get

$$P'(t) = QP(t). \tag{3.11a}$$

Again from (3.7) we get

$$\begin{aligned} p_{ij}(t+h) &= \sum_k p_{ik}(t)p_{kj}(h) \\ &= \sum_{k\neq j} p_{ik}(t)p_{kj}(h) + p_{ij}(t)p_{ij}(h). \end{aligned}$$

Assuming that the operations of limit and summation are interchangeable and proceeding as above, we get

$$p'_{ij}(t) = \sum_{k\neq i} p_{ik}(t)q_{kj} + q_j p_{ij}(t), \tag{3.12}$$

which is Chapman–Kolmogorov *forward* equation. In matrix notation,

$$P'(t) = P(t)Q. \tag{3.12a}$$

Using (3.3), we can also put (3.11a) and (3.12a) in the form

$$\frac{d}{dt}\{\pi(t)\} = Q\pi(t) = \pi(t)Q. \tag{3.13}$$

1.3.4. Transient Solution

Consider a finite $(m + 1)$ state chain, with given rate matrix Q. Solving (3.11a) or (3.12a) we get $P(t) = P(0)e^{Q(t)}$; with $P(0) = I$, we have

$$P(t) = e^{Qt} = I + \sum_{n=1}^{\infty} \frac{Q^n t^n}{n!} \quad \text{or} \tag{3.14}$$

$$\pi(t) = \pi(0)\left(I + \sum_{n=1}^{\infty} \frac{Q^n t^n}{n!}\right). \tag{3.15}$$

Assume that the eigenvalues d_i of Q are all distinct, $d_i \neq d_j$, $i, j = 0, 1, \ldots, m$. Let D be the diagonal matrix having $d_0, d_1, \ldots, d_m$ as its diagonal elements. Then there exists a nonsingular matrix H (whose column vectors are right eigenvectors of Q) such that Q can be written in the canonical form

$$Q = HDH^{-1}.$$

Then

$$Q^n = HD^nH^{-1}$$

and, substituting in (3.14), we get

$$\begin{aligned} P(t) &= H\Lambda(t)H^{-1} \quad \text{and} \\ \pi(t) &= \pi(0)P(t), \end{aligned} \tag{3.16}$$

where $\Lambda(t)$ is the diagonal matrix with diagonal elements $e^{d_i t}$, $i = 0, 1, \ldots, m$.

It may be noted that in the general case when the eigenvalues of the matrix Q are not necessarily distinct, Q can still be expressed in the canonical form $Q = SZS^{-1}$ and $P(t)$ can be obtained as above.

The transient solution can be obtained as given above. While an analytical solution can be obtained, especially when m is small, it becomes difficult when m is large. For such cases, numerical methods have been put forward (Grassman, 1977; Gross and Miller, 1984a,b). See Section 1.6.3.

For many stochastic systems such as queueing systems and reliability systems, computation of the vector $\pi(t)$ transient probabilities is useful. It is specially important when convergence to steady state is slow.

1.3.5. Alternative Definition

A continuous-time Markov chain with state space $S = \{0, 1, 2, \ldots\}$ can be defined in another way as follows (Ross, 1980). It is a stochastic process such that (i) each time it enters state i, the time it spends in that state before making a transition to another state $j\ (\neq i) \in S$, i.e., sojourn time in state i is an exponential RV with mean $1/a_i$ (a_i depends on i but not on j); and (ii) when the process leaves state i, it enters another state $j\ (\neq i)$, with some probability say, p_{ij} (which depends on both i and j), such that, for all i,

$$p_{ii} = 0, \qquad 0 \leq p_{ij} \leq 1,$$

$$\sum_j p_{ij} = 1, \qquad j \in S.$$

Thus, a continuous-time Markov chain is a stochastic process such that (i) its transition from one state to another state of the state space S is as in a discrete-time Markov chain and (ii) the sojourn in a state i (holding time in state i before moving to another state) is an exponential RV whose parameter depends on i but not on the state next visited. The sojourn times in different states must be independent random variables with exponential distribution.

1.3.5.1. Relationship between p_{ij} and $p_{ij}(t)$.

We have

$$p_{ij}(h) = ha_i p_{ij} + o(h)$$

since $p_{ij}(h)$ is the probability that the state of the process changes from i to j in an infinitesimal interval h. Thus,

$$\lim_{h \to 0} \frac{p_{ij}(h)}{h} = a_i p_{ij},$$

but by definition LHS equals q_{ij}, so that

$$q_{ij} = a_i p_{ij}. \tag{3.17}$$

Again, $1 - p_{ii}(h)$ is the probability that the state of the system changes from state i to some other state in the interval h, so that

$$\begin{aligned} 1 - p_{ii}(h) &= a_i h \sum_j p_{ij} + o(h) \\ &= a_i h + o(h). \end{aligned}$$

Thus,

$$\lim_{h \to 0} \frac{1 - p_{ii}(h)}{h} = a_i;$$

but by definition LHS equals q_i so that

$$a_i = q_i. \tag{3.18}$$

Thus, the Q matrix can also be written as

$$Q = \begin{bmatrix} -a_0 & a_0 p_{01} & \cdots & a_0 p_{0m} \\ a_1 p_{10} & -a_1 & \cdots & a_1 p_{1m} \\ \cdots & & \cdots & \\ a_m p_{m0} & a_m p_{m1} & \cdots & -a_m \end{bmatrix}. \tag{3.19}$$

We have the corresponding ergodicity property.

Theorem 1.4. Ergodic Theorem. *If a Markov chain $\{X(t), t \in T\}$ is irreducible, then all the states are of the same type.*

In case they are all transient or null-persistent, then

$$\lim_{t \to \infty} p_{ij}(t) = 0, \qquad i, j \in S.$$

In case they are nonnull persistent, then

$$\lim_{t \to \infty} p_{ij}(t) = u_j \tag{3.20}$$

exists and is independent of the initial state i. Further, $\mathbf{U} = (u_1, u_2, \ldots)$, $(\mathbf{Ue} = 1)$ *is a probability distribution and is given by the solution of*

$$q_j u_j + \sum_{i \neq j} u_i q_{ij} = 0, \qquad i, j \in S, \quad \text{or}$$

$$\sum_i u_i q_{ij} = 0, \quad \text{or} \tag{3.21}$$

$$\mathbf{UQ} = \mathbf{0}, \qquad \mathbf{Ue} = 1. \tag{3.22}$$

We now consider the alternative definition of the continuous time chain. Using (3.17) *and* (3.19) *we get from* (3.21)

$$a_j u_j = \sum_{i \neq j} a_i p_{ij} u_i, \qquad j \in S, \quad \text{with} \quad \mathbf{Ue} = 1 \tag{3.23}$$

from which $\mathbf{U} = (u_1, u_2, \ldots)$ *can be obtained.*

Remarks.

(1) If an irreducible chain is finite, then

$$\lim_{t\to\infty} p_{ij}(t) = u_j, \qquad i, j \in S$$

exists. If the chain has a denumerable state space and if it is nonnull persistent, then u_j exists.

(2) When u_j exists, it can be interpreted as the long-run proportion of time the system is in state j.

(3) Equations (3.23) have an interesting interpretation.

When the process is in state j, it leaves that state at rate a_j, and u_j is the long-run proportion of time it is in state j, so that $a_j u_j$ = rate at which the process *leaves* state j. Again, when the process is in state i, the rate of transition into state j is $a_i p_{ij} = q_{ij}$, so that

$$\sum_{i\neq j} a_i p_{ij} u_i = \text{rate at which the process } \textit{enters} \text{ state } j.$$

Thus, Eqs. (3.23) can be interpreted as follows: in the long run, the two rates are equal, that is, the rate at which the process enters a state j equals the rate at which it leaves the state j (for each $j \in S$). As the two rates balance each other for every state, Eqs. (3.23) are also known as *balance equations.*

The balance equations, as interpreted above, have very useful applications in queueing systems, in particular, and in stochastic systems, in general.

Example 1.5. Two-state process. Suppose that a system can be in two states: operating and nonoperating or under repair (denoted by 0 and 1, respectively). Suppose that the lengths of the operating and nonoperating periods are independent exponential RVs with parameters a and b, respectively. Let $X(t)$ be the state of the process at time t. $\{X(t), t \geq 0\}$ is a Markov process with state space $S = \{0, 1\}$.

We have, because of exponential distribution,

$$\begin{aligned} p_{01}(h) &= Pr\{\text{change of state from operating to nonoperating} \\ &\qquad \text{in an infinitesimal interval } h\} \\ &= ah + o(h), \end{aligned}$$

and so

$$p_{00}(h) = 1 - ah + o(h),$$

$$p_{10}(h) = bh + o(h), \quad \text{and}$$

$$p_{11}(h) = 1 - bh + o(h)$$

so that the Q-matrix is

$$Q = \begin{pmatrix} -a & a \\ b & -b \end{pmatrix}.$$

The Chapman–Kolmogorov forward equation $P'(t) = P(t)Q$ gives, for $i = 0, 1$,

$$p'_{i0}(t) = -ap_{i0}(t) + bp_{i1}(t)$$

$$p'_{i1}(t) = ap_{i0} - bp_{i1}(t).$$

Again,

$$p_{i0}(t) = 1 - p_{i1}(t).$$

Assume that $p_{00}(0) = 1$. Solving, we get

$$p_{00}(t) = \frac{b}{a+b} + \frac{a}{a+b} e^{-(a+b)t},$$

$$p_{01}(t) = \frac{a}{a+b} - \frac{a}{a+b} e^{-(a+b)t},$$

$$p_{11}(t) = \frac{a}{a+b} + \frac{b}{a+b} e^{-(a+b)t}, \quad \text{and}$$

$$p_{10}(t) = \frac{b}{a+b} - \frac{b}{a+b} e^{-(a+b)t}.$$

As $t \to \infty$,

$$p_{00}(t) \to \frac{b}{a+b}, \qquad p_{01}(t) \to \frac{a}{a+b} \quad \text{and}$$

$$p_{10}(t) \to \frac{b}{a+b}, \qquad p_{11}(t) \to \frac{a}{a+b}.$$

These limiting probabilities can also be obtained by using (3.21). We have, as $t \to \infty$, $\lim p_{ij}(t) = u_j$; then from (3.21) we get

$$q_0 u_0 = u_1 q_{10} \Rightarrow a u_0 = b u_1$$

$$\Rightarrow u_1 = +\frac{a}{b} u_0$$

and since $u_0 + u_1 = 1$,

$$u_0 = \frac{b}{a+b}, \; u_1 = \frac{a}{a+b}.$$

That is,

$$\lim_{t \to \infty} p_{i0}(t) = u_0 = \frac{b}{a+b}, \qquad i = 0, 1 \quad \text{and}$$

$$\lim_{t \to \infty} p_{i1}(t) = \frac{a}{a+b}, \qquad i = 0, 1.$$

1.4. Birth and Death Processes

The class of all continuous-time Markov chains has an important subclass formed by the birth and death processes. These processes are characterized by the property that whenever a transition occurs from one state to another, then this transition can be to a neighboring state only. Suppose that the state space is $S = \{0, 1, 2, \ldots, i, \ldots\}$, then transition, whenever it occurs from state i, can be only to a neighboring state $(i - 1)$ or $(i + 1)$.

A continuous-time Markov chain $\{X(t), t \in T\}$ with state space $S = \{0, 1, 2, \ldots\}$ and with rates

$$q_{i,i+1} = \lambda_i(\text{say}), \qquad i = 0, 1, \ldots,$$

$$q_{i,1-1} = \mu_i(\text{say}), \qquad i = 1, 2, \ldots,$$

$$q_{i,j} = 0, \qquad j = i \pm 1, \qquad j \neq i, \qquad i = 0, 1, \ldots, \quad \text{and}$$

$$q_i = (\lambda_i + \mu_i), \qquad i = 0, 1, \ldots, \qquad \mu_0 = 0,$$

is called

(i) a *pure birth process*, if $\mu_i = 0$ for $i = 1, 2, \ldots,$

(ii) a *pure death process*, if $\lambda_i = 0$, $i = 0, 1, \ldots,$ and

(iii) a *birth-and-death-process* if some of the λ_i's and some of the μ_i's are positive.

Using (3.12) we get the Chapman–Kolmogorov forward equations for birth-and-death process.

For $i, j = 1, 2, \ldots,$

$$p'_{ij}(t) = -(\lambda_j + \mu_j)p_{ij}(t) + \lambda_{j-1}p_{i,j-1}(t) + \mu_{j+1}p_{i,j+1}(t) \quad \text{and} \tag{4.1}$$

$$p'_{i0}(t) = -\lambda_0 p_{i0}(t) + \mu_1 p_{i,1}(t). \tag{4.2}$$

The boundary conditions are

$$p_{ij}(0+) = \delta_{ij}, \qquad i, j = 0, 1, \ldots. \tag{4.3}$$

Denote

$$P_j(t) = Pr\{X(t) = j\}, \qquad j = 0, 1, \ldots, t > 0,$$

and assume that at time $t = 0$, the system starts at state i, so that

$$P_j(0) = Pr\{X(0) = j\} = \delta_{ij},$$

then

$$P_j(t) = p_{ij}(t) \tag{4.4}$$

and the forward equations can be written as

$$P'_j(t) = -(\lambda_j + \mu_j)P_j(t) + \lambda_{j-1}P_{j-1}(t) + \mu_{j+1}P_{j+1}(t), \qquad j = 1, 2, \ldots, \tag{4.5}$$

$$P'_0(t) = -\lambda_0 P_0(t) + \mu_1 P_1(t). \tag{4.6}$$

Suppose that all the λ_{is} and μ_is are nonzero. Then the Markov chain is irreducible. It can be shown that such a chain is non-null persistent and that the limits

$$\lim_{t \to \infty} p_{ij}(t) = p_j$$

exist and are independent of the initial state i. Then Eqs. (4.5) and (4.6) become

$$0 = -(\lambda_j + \mu_j)p_j + \lambda_{j-1}p_{j-1} + \mu_{j+1}p_{j+1}, \qquad j \geq 1 \tag{4.7}$$

$$0 = -\lambda_0 p_0 + \mu_1 p_1. \tag{4.8}$$

Define

$$\pi_j = \frac{\lambda_0 \lambda_1 \cdots \lambda_{j-1}}{\mu_1 \mu_2 \cdots \mu_j}, \qquad j \geq 1, \quad \text{and}$$

$$\pi_0 = 1; \tag{4.9}$$

then the solution of the above can be obtained by induction. We have from (4.8)

$$p_1 = \left(\frac{\lambda_0}{\mu_1}\right)p_0 = \pi_1 p_0$$

and assuming $p_k = \pi_k p_0$, $k = 1, 2, \ldots, j$, we get from (4.7)

$$p_{j+1}\mu_{j+1} = \lambda_i \pi_j p_0 \quad \text{or}$$

$$p_{j+1} = \pi_{j+1} p_0.$$

Thus, if $\sum_{k=0}^{\infty} \pi_k < \infty$, then

$$p_j = \frac{\pi_j}{\sum \pi_k}, \qquad j \geq 0. \tag{4.10}$$

Incidentally, $\sum \pi_k < \infty$ is a sufficient condition for the birth-and-death process to have all the states non-null persistent (and therefore for the process to be ergodic).

This process is of particular interest in queueing theory as several queueing systems can be modeled as birth-and-death processes. As an example, we consider the simple queue.

1.4.1. Special Case: M/M/1 Queue

For this queueing model

$$\lambda_i = \lambda, \qquad i = 0, 1, 2, \ldots \quad \text{and}$$

$$\mu_i = \mu, \qquad i = 1, 2, \ldots,$$

$$\mu_0 = 0;$$

then $\pi_j = (\lambda/\mu)^j$ and $\sum \pi_k < \infty$ *iff* $(\lambda/\mu) < 1$ and then

$$\begin{aligned} \sum \pi_k &= 1/[1 - (\lambda/\mu)] \\ p_j &= [1 - (\lambda/\mu)](\lambda/\mu)^j, \qquad j = 0, 1, 2, \ldots. \end{aligned} \tag{4.11}$$

1.4.2. Pure Birth Process: Yule–Furry Process

If $\mu_i = 0$, $i \geq 0$, then we get a pure birth process; further, if $\lambda_i = i\lambda$, for all i, we get the Yule–Furry process for which

$$\begin{aligned} P_j'(t) &= -j\lambda P_j(t) + (j-1)\lambda P_{j-1}(t), \qquad j \geq 1, \quad \text{and} \\ P_0'(t) &= 0. \end{aligned} \tag{4.12}$$

1.5. Poisson Process

If $\mu_i = 0, i \geq 0, \lambda_i = \lambda$ for all i, then we get what is known as the homogeneous Poisson process with parameter λ. It is a pure birth process with constant rate λ.

The Poisson process can be used as a model of a large class of stochastic phenomena and is thus extremely useful from the point of view of application. We shall discuss this process in some detail.

The Chapman–Kolmogorov forward equations are

$$P'_j(t) = -\lambda[P_j(t) - P_{j-1}(t)], \qquad j \geq 1, \tag{5.1a}$$

$$P'_0(t) = -\lambda P_0(t). \tag{5.1b}$$

Let the boundary condition be

$$P_j(0) = \delta_{ij}.$$

The Eqs. (5.1) can be solved in a number of ways. Let us consider the method of generating function. Define

$$P(s, t) = \sum_{j=0}^{\infty} P_j(t)s^j; \tag{5.2}$$

then

$$P(s, 0) = s^i. \tag{5.3}$$

Assuming the validity of term-by-term differentiation, we get from (5.2)

$$\begin{aligned}\frac{\partial}{\partial t} P(s, t) &= \sum_{j=0}^{\infty} \frac{\partial}{\partial t} \{P_j(t)\} s^j \\ &= P'_0(t) + \sum_{j=1}^{\infty} P'_j(t)s^j.\end{aligned}$$

Multiplying (5.1a) by s^j and adding over $j = 1, 2, 3, \ldots$, we get

$$\frac{\partial}{\partial t} P(s, t) - P'_0(t) = -\lambda[P(s, t) - P_0(t) - sP(s, t)].$$

Using (5.1b), we have

$$\frac{\partial}{\partial t} P(s, t) = P(s, t)[\lambda(s - 1)].$$

Solving we get

$$P(s, t) = Ce^{\lambda(s-1)t} \\ = s^i e^{\lambda(s-1)t}; \tag{5.3a}$$

whence

$$\begin{aligned} P_j(t) &= \text{coeff. of } s^j \quad \text{in} \quad P(s, t) \\ &= e^{-\lambda t}\frac{(\lambda t)^{j-i}}{(j-i)!}, \qquad j = i, i+1, \ldots, \\ &= 0, \qquad j = 0, 1, \ldots, i-1. \end{aligned} \tag{5.4}$$

Since the Poisson process is a Markov chain $\{X(t), t \in (0, \infty)\}$ with stationary transition probabilities, we have

$$\begin{aligned} Pr\{X(t+s) - X(s) = k \mid X(s) = i\} &= Pr\{X(t+s) = i + k \mid X(s) = i\} \\ &= \frac{(\lambda t)^k}{k!} e^{-\lambda t}, \qquad i, k = 0, 1, \ldots; t, s \geq 0. \end{aligned} \tag{5.5}$$

We have defined the Poisson process as a birth process with constant birth rate. It can be introduced as a renewal process (as we shall see later in Section 1.7). A third way of defining the Poisson process is given below.

Let $N(t)$ denote the number of occurrences of a specified event in an interval of length t (i.e., during the time period, say, from 0 to t). Let

$$P_n(t) = Pr\{N(t) = n\}, \qquad n = 0, 1, 2, \ldots.$$

We make the following postulates:

(1) *Independence.* The number of events occurring in two disjoint intervals of time are independent, i.e., if $t_0 < t_1 < t_2, \ldots$, then the increments $N(t_1) - N(t_0)$, $N(t_2) - N(t_1), \ldots$ are independent RVs.

(2) *Homogeneity in time.* The RV $\{N(t+s) - N(s)\}$ depends on the length of the interval $(t+s) - s = t$ and not on s or on the value of $N(s)$.

(3) *Regularity* or *orderliness.* In an interval of infinitesimal length h, the probability of *exactly one* occurrence is

$$P_1(h) = \lambda h + o(h)$$

and the probability of two or more occurrences is

$$\sum_{k=2}^{\infty} P_k(h) = o(h).$$

It follows that $P_0(h) = 1 - \lambda h + o(h)$. From the assumption of independence, we get

$$P_0(t + h) = P_0(t)P_0(h) = P_0(t)[1 - \lambda h + o(h)]$$

so that

$$\lim_{h \to 0} \frac{P_0(t + h) - P_0(t)}{h} = -\lambda P_0(t) + \lim_{h \to 0} \frac{o(h)}{h} \quad \text{or}$$

$$P_0'(t) = -\lambda P_0(t).$$

We have

$$\begin{aligned} P_j(t + h) &= P_j(t)P_0(h) + P_{j-1}(t)P_1(h) + \sum_{r=2}^{\infty} P_{j-r}(t)P_r(h) \\ &= P_j(t)[1 - \lambda h + o(h)] + P_{j-1}(t)[\lambda h + o(h)] + o(h) \end{aligned}$$

so that

$$\lim_{h \to 0} \frac{P_j(t + h) - P_j(t)}{h} = -\lambda P_j(t) + \lambda P_{j-1}(t) + \lim_{h \to 0} \frac{o(h)}{h} \quad \text{or}$$

$$P_j'(t) = -\lambda[P_j(t) - P_{j-1}(t)], \qquad j \geq 1.$$

Thus, we get the same Chapman–Kolmogorov equations as given in (5.1).

If $P_j(0) = \delta_{ij}$, then $P_j(t)$ is given by (5.4). When $P_j(0) = \delta_{0j}$ we get

$$P_j(t) = e^{-\lambda t} \frac{(\lambda t)^j}{j!}, \qquad j = 0, 1, 2, \ldots. \tag{5.6}$$

Thus, $N(t)$ follows the Poisson distribution with parameter λt, that is, $\{N(t), t \geq 0\}$ is a Poisson process with parameter λ (or rate λ). We have $E[N(t)] = \lambda t$ and var $[N(t)] = \lambda t$. We shall state below some important properties of the Poisson process. For proof see works on stochastic processes such as Karlin and Taylor (1975), Medhi (1982) and Ross (1980, 1983).

1.5.1. Properties of the Poisson Process

(1) *Additive Property.* Sum of n independent Poisson processes with parameter λ_i, $i = 1, 2, \ldots, n$ is a Poisson process with parameter $\lambda_1 + \lambda_2 + \cdots + \lambda_n$.

(2) *Decomposition Property.* Suppose that $N(t)$ is the number of occurrences of a specified event and that $\{N(t), t \geq 0\}$ is a Poisson process with parameter λ. Suppose further that each occurrence of the event has a probability p of being recorded, and that recording of an occurrence is independent of other occurrences and also of $N(t)$. If $M(t)$ is the number of occurrences so recorded, then $\{M(t), t \geq 0\}$ is also a Poisson process with parameter λp; if $M_1(t)$ is the number of occurrences not recorded, then $\{M_1(t), t \geq 0\}$ is a Poisson process with parameter $\lambda(1 - p)$. Further $\{M(t), t \geq 0\}$, and $\{M_1(t), t \geq 0\}$ are independent.

The above implies that a *random selection* of a Poisson process yields a Poisson process.

(3) *Interarrival Times.* The interarrival times (i.e., the intervals) between two successive occurrences of a Poisson process with parameter λ are IID RVs that are exponential with mean $1/\lambda$.

(4) *Memoryless Property of Exponential Distribution.* Exponential distribution possesses what is known as a *memoryless* or *Markovian* or *nonaging* property and is the only continuous distribution to possess this property. It may be stated as follows. Suppose that X has exponential distribution with mean $1/\lambda$; then

$$Pr\{X \geq x + y | X \geq x\} = Pr\{X \geq y\}$$

is independent of x, for the LHS equals

$$\begin{aligned}\frac{Pr\{X \geq x + y \text{ and } X \geq x\}}{Pr\{X \geq x\}} &= \frac{Pr\{X \geq x + y\}}{Pr\{X \geq x\}} \\ &= \frac{e^{-\lambda(x+y)}}{e^{-\lambda x}} \\ &= e^{-\lambda y} = Pr\{X \geq y\}.\end{aligned}$$

If the interval between two occurrences is exponentially distributed, then the memoryless property implies that the interval to the next occurrence is statistically independent of the time from the last occurrence and has exponential distribution with the same mean.

If τ is an arbitrary epoch in the interval (t_i, t_{i+1}) between the ith and $(i+1)$th occurrences of a Poisson process with parameter λ, then the distribution of the interval $(t_{i+1} - \tau)$ is independent of the elapsed time $(\tau - t_i)$ since the last occurrence and is exponential with mean $1/\lambda$.

In the queueing context, arrivals (or service completions) may be taken as occurrences; so in the case of the Poisson–exponential process, the above remarkable property leads to easily tractable and mathematically agreeable results.

(5) *Randomness Property.* Given that exactly one event of a Poisson process $\{N(t), t \geq 0\}$ has occurred by epoch T, then the time interval γ in $[0, T]$ in which the event occurred has uniform distribution in $[0, T]$. In other words,

$$Pr\{t < \gamma \leq t + dt \mid N(T) = 1\} = \frac{dt}{T}, \qquad 0 < t < T.$$

This is also expressed by saying that an event of a Poisson process is a purely *random* event. The Poisson process is sometimes called a random process.

The preceding result holds in a more general case. This is stated below.

If an interval of length T contains exactly m occurrences of a Poisson process, then the joint distribution of the epochs at which these events occurred is that of m points uniformly distributed over an interval of length T. The result holds in case of (more general) birth process of which the Poisson process forms a special class. (See Problem 1.21.)

1.5.2. Generalization of the Poisson Process

There are several directions in which the classical Poisson process can be generalized.

1.5.2.1. Poisson Cluster Process (Compound Poisson Process). One of the postulates of the Poisson process is that at most one event can occur at a time. Now suppose that several events (i.e., a cluster of events) can occur simultaneously at an epoch of occurrence of a Poisson process $N(t)$ and that the number of events X_i in the ith cluster is a RV, X_is having independent and identical distribution

$$Pr\{X_i = j\} = p_j, \qquad j = 1, 2, \ldots.$$

Then $M(t)$, the total number of events in an interval of length t, is given by

$$M(t) = \sum_{i=1}^{N(t)} X_i.$$

The stochastic process $\{M(t), t \geq 0\}$ is called a *compound Poisson process.* Its PGF is given by

$$G[P(s)] = \exp\{\lambda t[P(s) - 1]\}$$

where $P(s)$ is the PGF of X_i. We have

$$\begin{aligned} Pr\{M(t) = m\} &= \sum_{k=0}^{m} [Pr\{N(t) = k\} Pr\{X_i = m\}] \\ &= \sum_{k=0}^{m} e^{-\lambda t} \frac{(\lambda t)^k}{k!} p_m^{k*} \end{aligned}$$

where p_m^{k*} is the probability associated with a k-fold convolution of X_i with itself.

We have

$$\begin{aligned} E\{M(t)\} &= \lambda t E\{X_i\} \quad \text{and} \\ \text{var}\{M(t)\} &= \lambda t E\{X_i^2\}. \end{aligned}$$

The compound Poisson process is useful in modeling queueing systems with batch arrival/batch service, with exponential interarrival/service time and independent and identical batch-sized distribution.

1.5.2.2. Nonhomogeneous Poisson Process. The parameter λ in the classical Poisson process is assumed to be a constant, independent of time. Generalizations of the Poisson process arise when λ is assumed to be (i) a nonrandom function of time $\lambda(t)$ and (ii) a random variable.

Here it is assumed that the probability that arrival occurs between time t and time $t + \Delta t$, given that n arrivals occurred by time t, is equal to $\lambda(t)\Delta t + o(\Delta t)$, while the probability that more than one arrival occurs is $o(\Delta t)$. The resulting process is the so-called nonhomogeneous Poisson process $\{N(t), t \geq 0\}$. It can be shown that

$$\begin{aligned} p_n(t) &= Pr\{N(t) = n\} \\ &= \exp\left\{-\int_0^t \lambda(x)dx\right\} \frac{[\int_0^t \lambda(x)dx]^n}{n!}, \; n \geq 0. \end{aligned}$$

1.5.2.3. Random Variation of Parameter. Here we assume that λ is a random variable having PDF $f(\lambda)$, $0 \leq \lambda \leq \infty$. Thus,

$$p_n(t) = Pr\{N(t) = n\} = \int_0^\infty e^{-\lambda t} \frac{(\lambda t)^n}{n!} f(\lambda) d\lambda.$$

The case when the parameter λ of a Poisson process is a random function of time $\lambda(t)$ (and so is itself a stochastic process) leads to a doubly stochastic Poisson process. There are several situations where such generalizations of Poisson process may be realistic.

1.5.2.4. Truncated Process. A simple generalization is truncation of the infinite domain of the Poisson process. This case arises in modeling a queueing system with waiting space limited to n, so that arrivals that occur when the waiting space is full are not permitted and are lost to the system. This will be involved only in scaling the Poisson probabilities by a suitable scale factor.

1.5.3. Role of the Poisson Process in Probability Models

The Poisson process and its associated exponential distribution possess many agreeable properties that lead to mathematically tractable results when used in probability models. Its importance is also due to the fact that occurrences of events in many real-life situations do obey the postulates of the Poisson process, and thus its use in probability modeling is considered realistic. An arrival process to a queueing system is often taken to be Poisson.

Consider an event and an interval of time during which the occurrences of the event happen. Suppose that the interval is subdivided into a large number of subintervals (say, n), and that p_i is the probability of occurrence of the event in the ith subinterval. Suppose further that the events occur independently of one another and that $\lambda = p_1 + \cdots + p_n$, while the largest of p_i tends to 0. Then the number of occurrences of the event in the interval tends in the limit to a Poisson distribution with mean λ. The Poisson distribution thus gives an adequate description of the cumulative effect of a large number of events, such that occurrence of an event in a small subinterval is improbable.

There are other contexts arising out of extreme value theory as well as information theory that provide justification of using the Poisson process in modeling.

Note. Rego and Szpankowski (1989) show that there is an equivalence between using entropy maximization with a two-moment constraint and assumption of exponential distribution in a certain queueing context.

1.6. Randomization: Derived Markov Chains

Let $\{X(t), t \geq 0\}$ be a continuous-time Markov chain with transition matrix Q and countable state space S. Assume that $X(t)$ is *uniformizable*, that is, the diagonal elements of Q are uniformly bounded. Let

$$\alpha = \sup_i q_i < \infty.$$

Then there exists a discrete-time Markov chain $\{Y_n, n = 0, 1, \ldots\}$ with state space S and TPM $P = (p_{ij})$ such that

$$P = \frac{Q}{\lambda} + I, \tag{6.1}$$

where λ is any real number not less than α. Since $P(t) = e^{Qt}$ (Eq. 3.14), we have

$$\begin{aligned} P(t) &= e^{Qt} \\ &= e^{\lambda(P-I)t} \\ &= e^{-\lambda t} e^{\lambda P t} \\ &= e^{-\lambda t} \sum_{n=0}^{\infty} \frac{\lambda^n t^n}{n!} P^n \end{aligned}$$

so that elementwise

$$p_{ij}(t) = e^{-\lambda t} \sum_{n=0}^{\infty} \frac{\lambda^n t^n}{n!} p_{ij}^{(n)}, \qquad t \geq 0, \qquad i, j \in S. \tag{6.2}$$

We shall have from the above

$$\pi(t) = \pi(0)P(t) = \pi(0)e^{-\lambda t} \sum_{n=0}^{\infty} \frac{\lambda^n t^n}{n!} P^n \tag{6.3a}$$

or elementwise

$$\pi_j(t) = \pi(0)e^{-\lambda t} \sum_{n=0}^{\infty} \frac{\lambda^n t^n}{n!} p_{ij}^{(n)}. \tag{6.3b}$$

Another interesting fact is that there exists a Poisson process $\{N(t), t \geq 0\}$ with parameter λ such that Y_n and $N(t)$ are independent and that $\{X(t), t \geq 0\}$ and $\{Y_{N(t)}, t \geq 0\}$ are probabilistically identical, i.e., we can write

$$X(t) \equiv Y_{N(t)}.$$

The converse also holds: if $X(t) = Y_{N(t)}$, then $Q = \lambda(P - I)$.

1.6.1. Markov Chain on an Underlying Poisson Process (or Subordinated to a Poisson Process)

The above method of construction leads from a Markov process $\{X(t), t \geq 0\}$ to a derived Markov chain $\{Y_{N(t)}, t \geq 0\}$ by randomization of operational time through events of a Poisson process. For, we can obtain $p_{ij}(t)$ in terms of $p_{ij}^{(n)}$ by conditioning over the number of occurrences of the Poisson process $N(t)$ in $(0, t)$. Conditioning over the number of occurrences of the Poisson process with parameter λ over $[0, 1]$, we get

$$\begin{aligned} p_{ij}(t) &= Pr\{X(t) = j \mid X(0) = i\} \\ &= \sum_{n=0}^{\infty} Pr\{X(t) = j \mid X(0) = i, N(t) = n\} \\ &\quad \times Pr\{N(t) = n \mid X(0) = i\}. \end{aligned}$$

Now

$$Pr\{N(t) = n \mid X(0) = i\} = e^{-\lambda t}\left[\frac{(\lambda t)^n}{n!}\right]$$

and $Pr\{X(t) = j \mid X(0) = i, N(t) = n\}$ is the probability that the system goes from state i to state j in time t during which n Poisson occurrences took place. (That is, n transitions took place. Here the time interval t is replaced by number of transitions.) Thus,

$$Pr\{X(t) = j \mid X(0) = i, N(t) = n\} = p_{ij}^{(n)}.$$

Hence, we have

$$p_{ij}(t) = e^{-\lambda t} \sum_{n=0}^{\infty} \frac{(\lambda t)^n}{n!} p_{ij}^{(n)}. \tag{6.4}$$

1.6.2. Equivalence of the Two Limiting Forms

Let $\{Y_n, n \geq 0\}$ be an irreducible and aperiodic chain with finite state space S and TPM P. Then from the ergodic theorem (Theorem 1.1) we get that

$$\lim_{n \to \infty} p_{ij}^{(n)} = v_j, \qquad i, j \in S,$$

exists and is independent of i, and $\mathbf{V} = \{v_1, v_2, \ldots\}$ is the invariant distribution given by

$$\mathbf{V}P = \mathbf{V}, \qquad \mathbf{Ve} = 1. \tag{6.5}$$

The Markov process $\{X(t) = Y_{N(t)}, t \geq 0\}$ is also aperiodic and irreducible and has the same state space S. From the ergodic theorem (Theorem 1.4), we get that

$$\lim_{t \to \infty} p_{ij}(t) = u_j$$

exists and is independent of i. Further, $\mathbf{U} = \{u_1, u_2, \ldots\}$ is a probability vector and $\mathbf{U}$ is given as the solution of

$$\begin{aligned} \mathbf{UQ} &= \mathbf{0}, \qquad \mathbf{Ue} = 1, \\ \mathbf{UQ} &= \mathbf{0} \Leftrightarrow \mathbf{U}[\lambda(P - I)] = \mathbf{0} \\ &\Leftrightarrow \mathbf{U}P = \mathbf{U}. \end{aligned} \tag{6.6}$$

Thus, from (6.5) and (6.6), we get

$$\mathbf{U} \equiv \mathbf{V}, \quad \text{elementwise.}$$

In other words,

$$\lim_{t \to \infty} p_{ij}(t) = \lim_{n \to \infty} p_{ij}^{(n)}, \qquad i, j \in S. \tag{6.7}$$

1.6.3. Numerical Method

The numerical method is a subject in itself. We discuss the importance of the randomization technique in numerical analysis. This method of construction gives very useful formulas for computation of $p_{ij}(t)$ or $\pi_j(t)$, that is, transient probabilities of a uniformizable Markov process $\{X(t), t \geq 0\}$. Ross (1980) calls this method *uniformization*, though *randomization* appears to be a more generally used term. What is generally done in computational work is to choose a truncation point N and to set N to bound the error of truncation ε as follows. From (6.2),

$$p_{ij}(t) = \sum_{n=0}^{N} e^{-\lambda t} \frac{(\lambda t)^n}{n!} p_{ij}^{(n)} + \sum_{n=N+1}^{\infty} e^{-\lambda t} \frac{(\lambda t)^n}{n!} p_{ij}^{(n)} \tag{6.8}$$

where N is so chosen such that the second term is less than or equal to the desired control error ε. It would ensure that $p_{ij}(t)$ would be accurate to within ε. The same holds for the computation of $\pi_j(t) = Pr\{X(t) = j\}$.

Algorithms for computation have been developed by Grassman (1977) and Gross and Miller (1984a,b). These algorithms have been shown to be useful for computation of transient probabilities of many stochastic systems such as queueing, inventory, reliability, and maintenance systems. (Refer to Gross and Miller, 1984a), for their SERT algorithm.

1.7. Renewal Processes

1.7.1. *Introduction*

We noted that the interarrival (or interoccurrence) times between successive events of a Poisson process are IID exponential random variables. A possible generalization is obtained by removing the restriction of exponential distribution and by considering that the interarrival times are independent and identical random variables with an arbitrary distribution. The resulting process is called a renewal process.

Definition. Let X_n be the interval between the $(n-1)$th and nth events of a counting process $\{N(t), t > 0\}$. Let $\{X_n, n = 1, 2, \ldots\}$ be a sequence of nonnegative IID random variables having distribution function F. Then $\{N(t), t \geq 0\}$ is said to be a *renewal process* generated or induced by the distribution F.

The discrete time process $\{X_n, n = 1, 2, \ldots\}$ also represents the same renewal process. Let

$$S_0 = 0,\ S_n = X_1 + \cdots + X_n,\ n \geq 1.$$

Then

$$N(t) = \sup\{n: S_n \leq t\}. \tag{7.1}$$

If $S_n = t$ for some n, then a renewal is said to occur at time t. Thus, S_n gives the epoch of nth renewal. We have $F_n(x) = Pr\{S_n \leq x\}$, and $F_n = F^{n*}$ where

F^{n*} is the n-fold convolution of F with itself. Assume that $E\{X_i\} = \mu$ exists and is finite. The function $M(t) = E\{N(t)\}$ is called the renewal function (which is a nonrandom function of t); when it exists, the derivative $M'(t) = m(t)$ is called the renewal density (*not* a PDF). The distribution of $N(t)$ is given by

$$\begin{aligned} p_n(t) &= Pr\{N(t) = n\} \\ &= Pr\{N(t) \geq n\} - Pr\{N(t) \geq (n+1)\} \\ &= Pr\{S_n \leq t\} - Pr\{S_{n+1} \leq t\} \\ &= F_n(t) - F_{n+1}(t). \end{aligned} \tag{7.2}$$

It can be easily verified that for X_n exponential, $\{N(t), t \geq 0\}$ is a Poisson process. The average number of renewals by time t equals

$$\begin{aligned} M(t) &= \sum_{n=0}^{\infty} np_n(t) \\ &= \sum_{n=1}^{\infty} F_n(t) = \sum_{n=1}^{\infty} F^{n*}(t) \\ &= F(t) + \sum_{n=1}^{\infty} F^{(n+1)*}(t) \end{aligned} \tag{7.2a}$$

Now,

$$\begin{aligned} \sum_{n=1}^{\infty} F^{(n+1)*}(t) &= \sum_{n=1}^{\infty} \int_0^t F^{n*}(t-x)dF(x) \\ &= \int_0^t \left\{ \sum_{n=1}^{\infty} F^{n*}(t-x) \right\} dF(x) \end{aligned}$$

assuming the validity of interchange of summation and integration operations. Thus,

$$M(t) = F(t) + \int_0^t M(t-x)dF(x). \tag{7.3}$$

The above is known as the *fundamental equation of renewal theory*.

Renewal theorems involving limiting behavior of $M(t)$ are interesting as well as important from the point of view of applications. (For details refer to any work on stochastic processes, such as Karlin and Taylor (1975), Çinlar (1975), Medhi (1982), and Ross (1983).)

1.7.2. Residual and Excess Lifetimes

We discuss below two RVs that arise in several situations. To a given $t > 0$, there corresponds a unique $N(t)$ such that

$$S_{N(t)} \le t < S_{N(t)+1}, \tag{7.4}$$

i.e., t lies in the interval $X_{N(t)+1}$ between $\{N(t)\}$th and $\{N(t)+1\}$th renewals.

The RV $Y(t) = S_{N(t)+1} - t$ (which is the interval between t and the renewal epoch after t) is called the *residual lifetime* or *forward-recurrence time* at t.

The RV $Z(t) = t - S_{N(t)}$ (which is the interval between t and the last renewal epoch before t) is called the *past lifetime* or *spent lifetime* or *backward-recurrence time* at t.

Note that

$$Y(t) + Z(t) = S_{N(t)+1} - S_{N(t)} = X_{N(t)+1} \tag{7.5}$$

is the total life.

These RVs arise in various queueing contexts. The RV $Z(t)$ denotes the elapsed time between t and the last arrival before t or between t and the commencement of the last service before t depending on whether X_i denotes the interarrival or service time. Similarly, $Y(t)$ can be interpreted. We consider now the distribution of $Y(t)$ and $Z(t)$. We have

$$\begin{aligned} Pr\{Y(t) \le x\} &= F(t+x) - \int_0^t [1 - F(t+x-y)]dM(y), \qquad x > 0 \\ &= 0, \qquad x \le 0. \end{aligned} \tag{7.6}$$

If F is not a lattice distribution, then the limiting distribution Y of $Y(t)$ is given by

$$Pr\{Y \le x\} = \lim_{t\to\infty} Pr\{Y(t) \le x\} = \frac{1}{\mu}\int_0^x [1 - F(y)]dy. \tag{7.7}$$

Again,

$$Pr\{Z(t) \le x\} = \begin{cases} 0, & x \le 0, \\ \displaystyle\int_{t-x}^{t} [1 - F(t-y)]dM(y), & 0 < x \le t, \\ 1, & x > t, \end{cases} \tag{7.8}$$

and if F is not a lattice distribution, then the limiting distribution Z of $Z(t)$ is given by

$$\begin{aligned} Pr\{Z \le x\} &= \lim_{t \to \infty} Pr\{Z(t) \le x\} \\ &= \frac{1}{\mu} \int_0^x [1 - F(y)] dy, \qquad x \ge 0 \qquad (7.9) \\ &= 0, \qquad x < 0. \end{aligned}$$

When these exist, the two limiting distributions Y and Z are identical. It can be easily verified that for exponential X_i, the distributions of $Y(t)$ and $Z(t)$ are again exponential with the same mean $\mu = E(X_i)$.

Suppose that $m_r = E(X_i^r)$ exist for $r = 1, 2$. Then

$$E\{Y\} = E\{Z\} = \frac{m_2}{2\mu} = \frac{m_2}{2m_1}. \qquad (7.10)$$

If F is a lattice distribution, then the distributions of $Y(t)$ and $Z(t)$ have no limits for $t \to \infty$ except in some special cases.

1.8. Regenerative Processes

Let $\{X(t), t \ge 0\}$ be a stochastic process with countable state space $S = \{0, 1, 2, \ldots\}$. Suppose that there exists an epoch t_1 such that the continuation of the process beyond t_1 is a probabilistic replica of the whole process starting at $0(=t_0)$. Then this implies the existence of epochs $t_2, t_3, \ldots$ $(t_i > t_{i-1})$ having the same property. Such a process is known as a regenerative process. If $T_n = t_n - t_{n-1}, n = 1, 2, \ldots$, then $\{T_n, n = 1, 2, \ldots\}$ is a renewal process.

A renewal process is regenerative with T_i representing the time of the ith renewal.

Another example of a regenerative process is provided by what is known as an *alternating renewal process*. Such a process can be envisaged by considering that a system can be in one of two possible states, say 0 and 1, i.e., having $S = \{0, 1\}$. Initially, it is at state 0 and remains at that state for a time Y_1, and then a change of state to state 1 occurs in which it remains for a time Z_1, after which it again goes to state 0 for a time Y_2 and then goes to state 1 for a time Z_2 and so on. That is, its movement could be denoted by $0 \to 1 \to 0 \to 1 \ldots$. The initial state could be 1, in which case the movement could be denoted by $1 \to 0 \to 1 \to 0 \ldots$

Suppose that $\{Y_n\}$, $\{Z_n\}$ are two sequences of IID random variables and that Y_n and Z_n need not be independent. Let

$$T_n = Y_n + Z_n, \qquad n = 1, 2, \ldots.$$

Then at time T_1 the process restarts itself, and so also at times $T_2, T_3, \ldots$. The interval T_n denotes a complete cycle and the process restarts itself after each complete cycle. Let

$$E\{Y_n\} = E\{Y\}, \qquad E\{Z_n\} = E\{Z\}.$$

Then the long-run proportions of time that the system is at states 0 and 1 are given, respectively, by

$$p_0 = \lim_{t\to\infty} Pr\{X(t) = 0\} = \frac{E\{Y\}}{E\{Y\} + E\{Z\}} \quad \text{and} \tag{8.1}$$

$$p_1 = \lim_{t\to\infty} Pr\{X(t) = 1\} = \frac{E\{Z\}}{E\{Y\} + E\{Z\}} \tag{8.2}$$

$$= 1 - p_0.$$

1.8.1. Application in Queueing Theory

The results (8.1) and (8.2) have an important application in queueing theory.

Consider a single-server queueing system such that an arriving customer is immediately taken for service if the server is free, but joins a waiting line if the server is busy. The system can be considered to be in two states (idle or busy) according to whether the server is idle or busy. The idle and busy states alternate and together constitute a cycle of an alternating renewal process. A busy period starts as soon as a customer arrives before an idle server and ends at the instant when the server becomes free for the first time. The epochs of commencement of busy periods are regeneration points. Let I_n and B_n denote the lengths of nth idle and busy periods, respectively, and let

$$E\{I_n\} = E\{I\} \quad \text{and}$$
$$E\{B_n\} = E\{B\}. \tag{8.3}$$

Then the long-run proportion of time that the server is idle equals

$$p_0 = \frac{E\{I\}}{E\{I\} + E\{B\}} \tag{8.4}$$

and the long-run proportion of time that the server is busy equals

$$p_1 = \frac{E\{B\}}{E\{I\} + E\{B\}}. \tag{8.5}$$

In particular, if the arrival process is Poisson with mean λt, then it follows (from its lack of memory property) that an idle period is exponentially distributed with mean $1/\lambda$, i.e., $E(I) = 1/\lambda$. Then when p_0 or p_1 is known, $E(B)$ can be found.

The case of the alternating renewal process can be generalized to cover cyclical movement of more than two states. Suppose that the state space of the process $\{X(t), t \geq 0\}$ is $S = \{0, 1, \ldots, m\}$ and its movement from initial state 0 is cyclic as $0 \to 1 \to 2 \cdots \to m \to 0 \ldots$, and that τ_k is the duration of sojourn at state k, having mean $\mu_k = E\{\tau_k\}$, $k = 0, 1, \ldots, m$. Then we shall have

$$p_k = \lim_{t \to \infty} Pr\{X(t) = k\} = \frac{\mu_k}{\sum_{i=0}^{m} \mu_i}, \qquad k = 0, 1, \ldots, m. \tag{8.6}$$

1.9. Markov Renewal Processes and Semi-Markov Processes

We shall now consider a kind of generalization of a Markov process as well as a renewal process. Let $\{X(t), t \geq 0\}$ be a Markov process with discrete countable state space $S = \{0, 1, 2, \ldots\}$, and let $t_0 = 0, t_1, t_2, \ldots$ $(t_i < t_{i+1})$ be the epochs at which transitions occur. The sequence $\{X_n = X(t_n + 0), n \geq 0\}$ forms a Markov chain and the transition intervals $T_n = t_n - t_{n-1}$, $n = 1, 2, \ldots$, are distributed as independent exponential variables having means that may depend on the state of X_n.

We generalize the situation as follows: Suppose that the transitions $\{X_n, n \geq 0\}$ of the process $\{X(t), t \geq 0\}$ constitute a Markov chain but the transition intervals T_n, $n = 0, 1, \ldots$, have an independent arbitrary distribution and that the mean may depend not only on the state of X_n but also on the state of X_{n+1}. The process $\{X(t), t \geq 0\}$ is then no longer Markovian. The two-dimensional process $\{X_n, t_n, n \geq 0\}$ is called a Markov renewal process with state space S. Here

$$\begin{aligned} Pr\{X_{n+1} = j, T_{n+1} \leq t \mid X_0 = x_0, \ldots, X_n = i, T_0, T_1, \ldots, T_n\} \\ = Pr\{X_{n+1} = j, T_{n+1} \leq t \mid X_n = i\} \qquad (9.1) \\ = Q_{ij}(t), \quad \text{say}, \quad i, j \in S. \end{aligned}$$

Let

$$p_{ij} = \lim_{t \to \infty} Q_{ij}(t) \quad \text{and} \quad F_{ij}(t) = \frac{Q_{ij}(t)}{p_{ij}} \quad \text{and} \tag{9.2}$$

$$Y(t) = X_n \quad \text{on} \quad t_n \le t < t_{n+1}. \tag{9.3}$$

Then $\{Y(t), t \ge 0\}$ is called a *semi-Markov process* and the Markov chain $\{X_n, n \ge 0\}$ is called the *embedded Markov* chain of $\{X(t), t \ge 0\}$. $Y(t)$ gives the state of the process at its most recent transition. The chain $\{X_n, n \ge 0\}$ has TPM (p_{ij}). $F_{ij}(t) = Pr\{T_{ij} \le t\}$ is the distribution function of T_{ij}, the conditional transition time (or sojourn time) at state i given that the next transition is to state j. If τ_k is the unconditional waiting time at state k, then $\tau_k = \sum_j p_{ij} T_{ij}$.

For example, a pure birth process is a special type of Markov renewal process having

$$\begin{aligned} Q_{ij}(t) &= 1 - e^{a_i t}, \qquad j = i + 1, \\ &= 0, \quad \text{otherwise.} \end{aligned}$$

Then

$$\begin{aligned} p_{ij} &= 1, \qquad j = i + 1, \\ &= 0 \quad \text{otherwise, and} \\ F_{ij}(t) &= Q_{ij}(t), \qquad T_i = T_{ij}, \qquad j = i + 1. \end{aligned}$$

A Markov renewal process becomes a Markov process when the transition times are independent exponential, independent of the next state visited. It becomes a Markov chain when the transition times are all identically equal to 1. It reduces to a renewal process if there is only one state and then only transition times become relevant. Semi-Markov processes are used in the study of certain queueing systems. Let $p_k = \lim_{t \to \infty} Pr\{Y(t) = k\}$ be the long-run proportion of time the semi-Markov process is at state k. Suppose that the embedded Markov chain $\{X_n, n = 0, 1, 2\}$ is irreducible, aperiodic, and, if denumerable, recurrent nonnull. Then the limiting probabilities

$$v_j = \lim_{n \to \infty} p_{ij}^{(n)}$$

exist and are given as the unique nonnegative solution of

$$v_j = \sum_{k \in S} v_k p_{kj}, \qquad j \in S.$$

Then we shall have

$$p_k = \frac{v_k \mu_k}{\sum_{j \in S} v_j \mu_j} \tag{9.4}$$

where $\mu_k = E\{\tau_k\}$ is the expected sojourn time in state k. One can get this result by extending the result (8.6) through an intuitive argument. For a formal proof, see Medhi (1982).

Problems

1.1. The transition probability matrix of a Markov chain with three states 0, 1, 2 is given by

$$\begin{pmatrix} 0.4 & 0.5 & 0.1 \\ 0.2 & 0.6 & 0.2 \\ 0.3 & 0.3 & 0.4 \end{pmatrix}$$

and the initial distribution is (0.6, 0.3, 0.1). Find (i) $Pr(X_2 = 3)$, and (ii) $Pr\{X_3 = 1, X_2 = 0, X_1 = 2, X_0 = 0\}$. Find the invariant measure of the chain.

1.2. A chain with $S = \{1, 2, 3\}$ has TPM

$$P = \begin{pmatrix} 1 & 0 & 0 \\ 0 & 1 & 0 \\ p_1 & p_2 & p_3 \end{pmatrix}, \qquad p_i > 0, \qquad \sum p_i = 1.$$

Examine the nature of the states. Find P^n.

1.3. Find the invariant measure of a chain with $S = \{0, 1, 2, \ldots, m-1\}$ and a doubly stochastic transition probability matrix.

1.4. Consider a Markov chain with $S = \{1, 2, 3, 4\}$ and TPM

$$\begin{bmatrix} \frac{1}{3} & \frac{2}{3} & 0 & 0 \\ 1 & 0 & 0 & 0 \\ \frac{1}{2} & 0 & \frac{1}{2} & 0 \\ 0 & 0 & \frac{1}{2} & \frac{1}{2} \end{bmatrix}.$$

Is the chain irreducible?

Verify that states 1 and 2 are recurrent. Find μ_1 and μ_2 ($\mu_1 = 5/3$, $\mu_2 = 23/6$).

1.5. Show that for a Markov chain with a finite state space S, the probability of staying forever among the transient states is zero.

1.6. Show that if state j is transient, then

$$\sum_{n=1}^{\infty} p_{ij}^{(n)} < \infty \quad \text{for all} \quad i \in S.$$

1.7. Show that a transient state cannot be reached from a persistent state.

1.8. Consider a service facility having a limited waiting space for m customers, including the one being serviced. The server serves one customer, if any, at epochs 0, 1, 2, ... Assume that the number of arrivals in the intervals $(k, k + 1)$ is given by an IID random variable A with $Pr(A = n) = p_n$, $\sum p_n = 1$. Assume further that arrivals that occur when the waiting space is full leave the system and do not return. Denote by X_n the number of customers present at time n, including the one being served, if any. Show that $\{X_n, n \geq 0\}$ is a Markov chain and find its TPM.

1.9. In what is called a *gambler's ruin problem*, consider a gambler who with capital totaling a agrees to play a series of games with an adversary having a capital $b(a + b = c$, the total capital). The probability of the gambler winning one game (and with it, one unit of money) is p and that of losing one unit is $q = 1 - p$. (There is no draw.) Suppose that successive games are independent. If X_n is the gambler's fortune at time n (at the time of the nth game), show that $\{X_n, n = 0, 1, 2, \ldots\}$ is a Markov chain. Write down its TPM P. Is the chain irreducible? Examine the nature of the states of the chain.

1.10. Consider the Markov chain with $S = \{0, 1, \ldots, m\}$, such that

$$\begin{aligned} p_{0,0} &= q, & p_{01} &= p, & & \\ p_{i,i-1} &= q, & p_{i,i+1} &= p, & i &= 1, 2, \ldots, m-1 \\ p_{m,m-1} &= q, & p_{m,m} &= p, & & \end{aligned}$$

where $p + q = 1$. (Each of the transitions between other pairs of states has probability 0.) Show that the chain is irreducible and aperiodic. Find the limiting distribution $\mathbf{V}$.

1.11. Consider the Markov chain of Example 1.3

(a) Use (2.21) to find $\sum_j j v_j$.

(b) For $p_n = q^n p$, $n = 0, 1, 2, \ldots$, $p + q = 1$, examine the nonnull persistence of the chain. Find $V(s)$ in this case. Consider the particular case $q = \rho/(1 + \rho)$, $\rho < 1$.

1.12. Suppose that customers arrive at a bank in accordance with a Poisson process at the rate of two per minute. Find the probability that the number of customers that arrive during a 10-minute period is

(i) exactly 20,

(ii) greater than 20, and

(iii) between 10 and 20.

1.13. Suppose that customers arrive at a certain service facility center in accordance with a Poisson process $\{N(t), t \geq 0\}$ having parameter λ. A customer can make a preliminary enquiry as to whether he or she actually needs the facility. Suppose that the proportion of customers actually needing the service facility is p $(0 < p < 1)$. If $M(t)$ gives the number of customers actually needing service, show that $\{M(t), t \geq 0\}$ is again a Poisson process with parameter λp.

1.14. Suppose that customers arrive at a service counter in accordance with a Poisson process at a rate of three per minute. Find the probability that the interval between two successive arrivals is (i) between one and three minutes and (ii) less than one-third minute.

1.15. Suppose that messages arrive at a telephone switch board, the interarrival time of messages being exponential with mean 10 minutes. Find the probability that the number of messages received during the five afternoon hours (1–6 P.M.) is (i) exactly 24, (ii) more than 24, and (iii) nil.

1.16. If X_i, $i = 1, \ldots, n$ are IID exponential RVs with parameter a, then show that $S_n = X_1 + \cdots X_n$ has gamma distribution with PDF

$$f_{a,n}(x) = \frac{a^n x^{n-1} e^{-ax}}{\Gamma(x)}, \qquad x > 0$$

$$= 0, \qquad x \leq 0.$$

Find $E\{S_n\}$ and $\text{var}\{S_n\}$.

1.17. Suppose that X and Y are independent exponential RVs with parameters a and b, respectively. Show that:

(a) $W = X + Y$ has PDF

$$f(x) = \frac{ab(e^{-ax} - e^{-bx})}{b - a}, \qquad x > 0, \qquad a \neq b$$

$$= 0, \qquad x \leq 0.$$

(b) $Z = \min(X, Y)$ is exponential with parameter $a + b$.

(c) $M = \max(X, Y)$ has PDF

$$f(x) = ae^{-ax} + be^{-bx} - (a+b)e^{-(a+b)x}, \qquad x > 0$$
$$= 0, \qquad x \leq 0.$$

(d) $Pr\{X \leq Y\} = \dfrac{a}{(a+b)}.$

1.18. If X is an exponential RV, then show that

$$E\{X \mid X > y\} = y + E\{X\} \quad \text{for all } y > 0;$$

that is, $E\{X - y \mid X > y\} = E(X)$, independent of y.

1.19. A piece of equipment is subject to random shocks that occur in accordance with a Poisson process with rate λ. The equipment fails due to the cumulative effect of k shocks. Show that the duration of the lifetime T of the equipment has gamma distribution with PDF $f_{\lambda, k}(x)$. Note that T is the interval between k occurrences of a Poisson process.

1.20. Consider that two independent series of events A and B occur in accordance with Poisson processes with parameters a and b, respectively. Show that the number N of occurrences of the event A between two *successive* occurrences of the event B has geometric distribution with mass function

$$Pr\{N = n\} = (1-q)q^n, \; q = \frac{a}{(a+b)}, \qquad n = 0, 1, 2, \ldots.$$

1.21. Consider a Poisson process with parameter λ. Given that n events happen by time t, show that the PDF of the time of occurrence T_k of the kth event $(k < n)$ is given by

$$f(x) = \frac{n!}{(k-1)!(n-k)!} \frac{x^{k-1}}{t^k} \left(1 - \frac{x}{t}\right)^{n-k}, \qquad 0 < x < t,$$
$$= 0, \qquad x \geq t.$$

Show that $E(T_k) = kt/(n+1)$.

1.22. Suppose that a queueing system has m service channels. The demand for service arises in accordance with a Poisson process with rate a and the service time distribution has exponential distribution with parameter b. Suppose that the service system has no storage facility, that is, a demand that arises when all m channels are busy is rejected and is lost to the system. Let $X(t)$ be the number of busy service channels (number of demands) at time t. Show that $\{X(t), t \geq 0\}$ is a continuous-time Markov chain. Determine the infinitesimal generator. (See Problem 1.8 for the same queue with discrete service time.)

1.23. **The three-state process.** Suppose that an automatic machine can be in three states: working (state 0), failed in mode 1 (state 1), or failed in mode 2 (state 2). Suppose that a failed machine in either mode cannot go to another failed mode. (That is, transitions from state 1 to 2 and from state 2 to 1 are not possible.) Suppose that $X(t)$ denotes the state (condition) of the machine at time t and that the failure times are IID exponential with rates a_i, $i = 1, 2$, and the repair times are IID exponential with rates b_i, $i = 1, 2$. Show that $\{X(t), t \geq 0\}$ is a continuous-time Markov chain. Find the Q-matrix. (State any assumption that you make.)

1.24. Assume that the lifetime X of a device is random having DF $F(x)$. Show that the expected remaining life of a device aged y (which has already attained age y) is given by

$$E\{X - y | X > y\} = \frac{\int_y^\infty [1 - F(t)]dt}{1 - F(y)}.$$

In particular, for the exponential lifetime, this is equal to $E(X)$, independent of y. (See Problem 1.18.)

1.25. Let $\{N(t), t \geq 0\}$ be a renewal process induced by a RV X. Show that for large t

$$E\{N(t)\} \simeq \frac{t}{E(X)} \quad \text{and}$$

$$\text{var}\{N(t)\} \simeq \frac{\text{var}(X)}{[E(X)]^3}\, t.$$

1.26. Let Y be the stationary forward-recurrence time of a random variable X. Suppose that $m_n = E(X^n)$ exists for $n = 1, 2, \ldots$. Then show that

$$E\{Y_n\} = \frac{m_{n+1}}{(n + 1)m_1}.$$

1.27. Prove that the limiting joint distribution of the residual lifetime $Y(t)$ and spent lifetime $Z(t)$ of a random variable with DF $F(x)$ and finite mean μ is given by

$$Pr\{Y(t) > y, Z(t) > z\} = \frac{1}{\mu} \int_{y+z}^\infty [1 - F(u)]du, \qquad y > 0, \qquad z > 0.$$

1.28. **Renewal-reward process.** Consider a renewal process $\{X_n, n = 1, 2, \ldots\}$. Suppose that renewal epochs are $t_0 = 0, t_1, \ldots$, and that $N(t)$ is the number of renewals by time t; associate a RV Y_i $(i = 1, 2, \ldots)$ with renewal epoch t_i $(i = 1, 2, \ldots)$. (A reward or cost is associated, with each renewal, the amount being given by a RV Y_i for ith renewal.)

Let

$$Y(t) = \sum_{i=1}^{N(t)} Y_i.$$

Then the stochastic process $\{Y(t), t \geq 0\}$ is called a renewal-reward process. Suppose that $E(X_n) = E(X)$ and $E(Y_n) = E(Y)$ are finite. Then show that (a) with probability 1,

$$\lim_{t \to \infty} \frac{Y(t)}{t} \to \frac{E(Y)}{E(X)};$$

and that (b)

$$\lim_{t \to \infty} \frac{E\{Y(t)\}}{t} \to \frac{E(Y)}{E(X)}.$$

The relation (b) gives the long-run average reward (cost) per unit time in terms of $E(X)$ and $E(Y)$.

References

Bhat, U. N. (1984). *Elements of Applied Stochastic Processes*, 2nd ed., Wiley, New York.

Çinlar, E. (1975). *Introduction to Stochastic Processes*, Prentice-Hall, Englewood Cliffs, N.J.

Grassman, W. K. (1977). Transient solutions on Markovian queueing systems. *Comp. Opns. Res.* **4**, 47–53.

Gross, D., and Miller, D. R. (1984a). The randomization technique as a modeling tool and solution procedures for transient Markov processes. *Opns. Res.* **32**, 343–361.

Gross, D., and Miller, D. R. (1984b). Multiechelon repairable-item provisioning in a time-varying environment using a randomization technique. *Naval Res. Log. Qrly.* **31**, 347–361.

Karlin, S., and Taylor, H. M. (1975). *A First Course in Stochastic Processes*, 2nd ed., Academic Press, New York.

Lal, R., and Bhat, U. N. (1987). Reduced systems in Markov chains and their applications in queueing theory. *Queueing systems* **2**, 147–172; Correction **4**, 93.

Medhi, J. (1982). *Stochastic Processes*, Wiley Eastern, N. Delhi, & Halsted Press, Wiley, New York.

Prabhu, N. U. (1965). *Stochastic Processes*, Macmillan, New York.

Rego, V., and Szpankowski, W. (1989). The presence of exponentiality in entropy maximized $M/GI/1$ queues. *Comp. Opns. Res.* **16**, 441–449.

Ross, S. M. (1980). *Introduction to Probability Models*, 2nd ed., Academic Press, New York.

Ross, S. M. (1983). *Stochastic Processes*, Wiley, New York.

Seneta, E. (1981). *Non-negative Matrices and Markov Chains*, 2nd ed., Springer, New York.

Whitt, W. (1982). Approximating a point process by a renewal process. Two basic methods. *Opns. Res.* **30**, 125–147.

Whitt, W. (1983). Untold horrors of the waiting time: what the equilibrium distribution will never tell about the queue length process. *Mgmt. Sci.* **29**, 395–408.

2 Queueing Systems: General Concepts

2.1. Introduction

The origin of queueing theory dates back to 1909, when A. K. Erlang (1878–1929) published his fundamental paper on congestion in telephone traffic [for a brief account, see Saaty (1957), and for details on his life and work, see Brockmeyer *et al.* (1948)].* Kendall (1951, 1953) was the pioneer who viewed and developed queueing theory from the perspective of stochastic processes. The literature on queueing theory and the diverse areas of its applications has grown tremendously over the years. For a bibliography of books and survey papers, refer to Prabhu (1987), and for some directions of current analytical innovation, refer to Prabhu (1986).

Queueing theory is the mathematical study of "queues" or "waiting lines." A queue is formed whenever the demand for service exceeds the capacity to provide service at that point in time. A queueing system is comprised of customers or units needing some kind of service who arrive at a service facility where such service is provided, and who join a queue, if service is

* Erlang was a pioneer in the applications of analytical methods to operational problem; his studies appear to mark the beginning of the study of operations research.

not immediately available, and eventually leave after receiving service; there also arise cases where customers leave the system without joining the queue or leave without receiving service even after waiting for some time.

The terms customer and server are generic ones. Customers are those who need some kind of service and arrive at a facility where such service is available. A mechanism that performs the kind of service on customers or units fed into it is called a server or a service channel, for example, customers at a bank or reservation counter, calls arriving at a telephone keyboard, machines with a repairman needing repair, merchandise for shipment at a yard, and so on. Jobs arriving at a component in a computer center are also regarded as customers and the component of the computing system (such as CPU, drum, disk, line printer, etc.) where such facility is provided is considered the server. A customer receiving service is said to be in service. If, upon arrival, a "customer" finds the server busy, it forms or joins a queue.

2.1.1. Basic Characteristics

The basic characteristics of a queueing system are as follows:

(i) The Input or arrival pattern of customers;
(ii) The pattern of service;
(iii) The number of servers or service channels;
(iv) The capacity of the system; and
(v) The queue discipline.

We describe below the above characteristics.

2.1.2. The Input or Arrival Pattern of Customers

The *input pattern* means the manner in which the arrivals occur. It is specified by the interarrival time between any two consecutive arrivals; a measure usually considered is the average length of the interarrival time or its reciprocal, the average number of arrivals per some unit of time. The input pattern also indicates whether the arrivals occur singly or in groups or batches; if in batches, the manner in which these batches are constituted is also to be covered in the input pattern. The interarrival time may be deterministic, so that it is the same between any two consecutive arrivals, or it may be stochastic; when stochastic its distribution is also to be specified. Sometimes an arrival may not join the queue being discouraged by the length of the queue or being debarred from joining the system because the waiting space, when limited, is filled to maximum capacity. Again, the arrivals may occur from an infinite source or sometimes from a finite source, with the same units circulating in the system, e.g., machines coming for repair whenever they fail. There may also be several classes of customers with different arrival rates.

2.1.3. The Pattern of Service

By the *pattern of service*, we mean the manner in which the service is rendered; it is specified by the time taken to complete a service. The time may be constant (*deterministic*) or it may be stochastic. If it is stochastic, the pattern specification involves the distribution of service time of a unit. A measure typically considered is provided by the average time required to serve a unit or by the average number of units served per some unit of time. Sometimes service may be rendered in bulks or batches, as in the case of an elevator, instead of personalized service of one at a time. In this case, the manner of formation of batches for service also has to be specified.

2.1.4. The Number of Servers

A system may have a single server or a number of parallel servers. An arrival who finds more than one free server may choose at random any one of them for receiving service. If he finds all the servers busy, he joins a queue common to all the servers; the first customer from the common queue goes to the server who becomes free first. This kind of situation is common, for example, in a bank or at a ticket counter.

There may also be situations where there is a separate queue in front of each service facility as in the case of a supermarket. There also arise cases of ordered entry when an arrival has to try to find a free server in the order the servers are arranged.

Unless otherwise stated, we shall mean by a *multiple server system* (with a number of parallel channels) a system having a common queue—the head of the queue going to the first free server.

2.1.5. The Capacity of the System

A system may have an infinite capacity, that is, the queue in front of the server(s) may grow to any length; against this there may be limitation of space, so that when the space is filled to capacity, an arrival will not be able to join the system and will be lost to the system. The system is called a *delay system* or a *loss system* according to whether the capacity is infinite or finite.

If finite, it will have to be specified by the number of places available for the queue as well as for the one(s) being served, if any.

2.1.6. The Queue Discipline

The queue discipline indicates the manner in which the units are taken for service. The usual queue discipline is first come, first served, FCFS (first in first out, FIFO), though sometimes there are other service disciplines, such as last come first served (which happens sometimes in case of messages), or service in random order.

When arrivals occur in batches and service is offered individually, then the manner in which customers arriving in a batch are arranged for service is also to be indicated.

There are also such disciplines as processor-sharing, usually adopted in computer systems with a number of terminals.

Sometimes customers may be of several kinds—having a definite order of priority for receiving service with preemptive (or non-preemptive) service discipline.

Three important measures are the average rate of arrivals (denoted by λ), the average rate of service (denoted by μ), and the number ($c \geq 1$) of parallel servers with a single queue. The quantity

$$\rho = \frac{\lambda}{\mu}$$

in the case of a single server system and $\rho = \lambda/c\mu$ in the case of a c-server system, is called the *traffic intensity* or *offered load* of the system. Though dimensionless, it is expressed in *erlangs* as a mark of honor to A. K. Erlang.

Traffic intensity has to be suitably defined in case of batch arrivals and/or bulk service.

Unless otherwise stated, by system we shall mean a single server system with arrivals from an infinite source, with unlimited waiting space, and with FIFO queue discipline. The interarrival (as well as service times) will be assumed to be mutually independent and also to be independent of one another.

2.2. Queueing Processes

The analysis of a queueing system with fixed (deterministic) interarrival and service times does not present much difficulty. We shall be concerned with models or systems where one or both (interarrival and service times) are stochastic. Their analyses will involve stochastic description of the system and related performance measures, as discussed below.

(1) Distribution of the number $N(t)$ in the system at time t (the number in the queue and the one being served, if any); $N(t)$ is also called the queuelength of the system at time t.

(2) Distribution of the waiting time in the queue (in the system). The time that an arrival has to wait in the queue (remain in the system). If W_n denotes the waiting time of the nth arrival, then, of interest is the distribution of W_n.

(3) Distribution of the virtual wairing time $W(t)$—the length of time an arrival has to wait had he arrived at time t.

(4) Distribution of the busy period being the length of time during which the server remains busy. The busy period is the interval from the moment of arrival of a unit at an empty system to the moment that the channel becomes free for the first time. The busy period is a random variable. From a complete description of the above distributions, various performance measures of interest are obtained.

The problems studied in queueing theory may be grouped as:

(i) Stochastic behavior of various random variables, or stochastic processes that arise, and evaluation of the related performance measures;
(ii) Method of solution—exact, transform, algorithmic, asymptotic, numerical, approximations, etc.;
(iii) Nature of solution—time dependent, limiting form, etc.;
(iv) Control and design of queues—comparison of behavior and performances under various situations, as well as, queue disciplines, service rules, strategies, etc.; and
(v) Optimization of specific objective functions involving performance measures, associated cost functions, etc.

Analysts and operations researchers generally will be involved with these types of problems. But in order to study such problems, one will have to study first the types of problems enumerated under (i)–(iii).

2.3. Notation

The notation introduced by Kendall (1951) is generally adopted to denote a queueing model. It consists of the specifications of three basic characteristics: the input, the service time, and the number of (parallel) servers. Symbols used to denote some of the common formulations are as follows:

M: for exponential (having Markov property) interarrival (Poisson input) and service time distribution
E_k: for Erlang-k distribution
H: for hyperexponential distribution
PH: for phase type distribution
D: for deterministic (constant) (interarrival or service time)
G: for arbitrary (general) distribution.

Thus the notation $M/G/1$ denotes a queue or model with Poisson input, general service time distribution, and a single server. Two more descriptors are added, when needed; the fourth one to denote the capacity of the system and the fifth one to denote the size of the (finite) source from which the arrivals occur. Thus, $M/G/1$ model stands for an $M/G/1/\infty$ model; $G/G/c/K/N$ refers to the c-server model with (arbitrary) general interarrival and service time distributions, the space before the server being limited to K (including the ones, being served, if any) and the N being the size of the finite source from which the arrivals occur.

The description is suitably modified to cover more complicated models or models with other characteristics.

2.4. Transient and Steady-State Behavior

Denote by $N(t)$ the number in the system (the number in the queue plus the number being served, if any) at time t measured from a fixed initial moment ($t = 0$) and its probability distribution by

$$p_n(t) = Pr\{N(t) = n\}, \qquad n = 0, 1, 2, \ldots.$$

Then

$$p_i(0) = 1, \ (p_j(0) = 0, j \neq i)$$

implies that the number of customers at the initial moment was i (where i could be 0, 1, 2, ...). For a complete description of the stochastic behavior of the queuelength processes $\{N(t), t \geq 0\}$, we need to find a time-dependent solution $p_n(t)$, $n \geq 0$. It is often difficult to obtain such solutions. Or even when found, these may be too complicated to handle. For many practical situations, however, one needs the equilibrium behavior, that is, the behavior when the system reaches an equilibrium state after being in operation for a sufficiently long time. In other words, one is often interested in the limiting behavior of $p_n(t)$ as $t \to \infty$. Denote

$$p_n = \lim_{t \to \infty} p_n(t), \qquad n = 0, 1, 2, \ldots$$

whenever the limit exists. Thus, p_n is the limiting probability that there are n in the system, irrespective of the number at time 0. Whenever the limit exists, the system is said to reach a steady (or equilibrium) state and p_n is called the steady-state probability that there are n in the system. It is independent of time.

It usually turns out that p_n is equal to the long-run proportion of time that the system contains exactly n customers. In particular, p_0 denotes the proportion of time that the system is empty. It follows that

$$\sum_{n=0}^{\infty} p_n = 1;$$

this is called the normalizing condition.

It is necessary to know the condition(s) under which the limit exists. This will be discussed at appropriate places.

We consider two other sets of *limiting* probabilities $\{a_n, n \geq 0\}$ and $\{d_n, n \geq 0\}$ defined as follows:

a_n = probability that arrivals (arriving customers) find n in the system when they arrive;

d_n = probability that departures (departing customers) find n in the system when they depart.

It turns out that a_n is the long-run proportion of customers who, when they arrive, find n in the system, and d_n is the long-run proportion of customers who, when they depart, find n in the system. The three quantities p_n, a_n, and d_n need not always be all equal.

We prove the following theorem.

Theorem 2.1. *In any queueing system in which arrivals occur one by one and that has reached equilibrium state,*

$$a_n = d_n \qquad \text{for all } n \geq 0.$$

Proof. Consider that an arrival will see, on arrival, n in the system; then the number in the system will increase by 1 and will go from n to $n + 1$. Again, a departure will leave n in the system, implying that the number in the system will decrease by 1 and will go from $n + 1$ to n. In any interval of time T, the number of transitions A from n to $n + 1$ and the number of transitions B between $n + 1$ to n will differ at most by 1; in other words, either $A = B$ or $A \sim B = 1$. Then for large T, the rates of transitions A/T and B/T will be equal. Thus, on the average, arrivals and departures always see the same number of customers, which means that $a_n = d_n$ always and for every $n \geq 0$.

Note. It may be noted that a_n (and d_n) is in general different from p_n; that is, the long-run proportion that arrivals find n in the system is not, in general, equal to the long-run proportion of time that there are n customers in the system. This is stated by saying that arrivals, in general, do not see time averages. In one important case, however, $a_n = p_n$; that is when the arrivals are from a Poisson process. This is discussed in Section 2.7.1.

2.5. Limitations of the Steady-State Distribution

In certain situations, the conditions for the existence of stationary distribution exist and a stationary or equilibrium distribution of the behavior of the queueing process can be obtained. However, even when this happens, there can be, in general, several stochastic processes having the same stationary distribution. Whitt (1983) points to the danger of using only the stationary distribution to describe the stochastic behavior of a queue with detailed study of a $GI/M/1$ queue in that context.

Consider that the stationary distribution is known or given. The problem of finding all queue-length processes or all arrival processes associated with a given stationary distribution is an *inverse* problem. Such inverse problems arise in several other situations (Keller, 1976; Karr and Pittenger, 1978). In the study of Markov chains, the inverse problem is the one of characterizing the class of Markov transition matrices with a given stationary distribution of the chain (Karr, 1978).

In order to be specific about the behavior of queueing processes, Whitt (1983) points to the necessity of examining transient behavior of some characteristics associated with the queueing process in addition to the stationary distribution, which by itself is not enough. The transient behavior that can be considered may be of various first-passage times such as the busy period. That is, in addition to stationary distribution, such other distributions may be considered. The transient behavior is also to be studied because of its importance in the cost–benefit anaysis of an operating system. An example considered by Whitt is the allocation of buffer by a central processor where the stationary distribution of buffer content may be used for determination of the number of buffers required, but the stochastic fluctuations will indicate the load on the central processor for buffer allocation. Cohen (1982) considers relaxation time to describe and measure the rate at which a stochastic process approaches steady state.

2.6. Some General Relationships in Queueing Theory

There are certain useful statements and relationships in queueing theory that hold under fairly general conditions. Though rigorous mathematical proofs of such relations are somewhat complicated, intuitive and heuristic proofs are simple enough and have been known for a long time (Morse 1958). It has been argued (Krakowski, 1973; Stidham 1974) that *conservation methods* could very well be applied to supply simple proofs of some of these relations. Conservation principles have played a fundamental role in physical and engineering sciences as well as in economics and so on. Similar principles may perhaps be applied in obtaining relations for queueing systems in steady state. Some such relations that hold for systems in *steady state* are given below. The most important one is

$$L = \lambda W \tag{6.1}$$

where λ is the mean arrival rate, L is the expected number of units in the system, and W is the expected waiting time in the system in steady state. Denote the expected number in the queue and the expected waiting time in the queue in steady state by L_Q and W_Q, respectively. These are related by a similar formula:

$$L_Q = \lambda W_Q. \tag{6.2}$$

A rigorous proof of the relation has been given by Little (1961) and so the relation is known as Little's formula. Jewell (1967) gave a proof based on renewal theory; Eilon (1969) gave the following simple proof.

Eilon's proof of $L = \lambda W$. This proof is simple and straightforward. It does not depend (i) on the arrival or service-time distributions, (ii) on the number of servers in the system, or (iii) on the queue discipline.

Consider Fig. 2.1. The top line gives the cumulative number of arrivals and the bottom line the cumulative number of departures from the system. The vertical distances between the two lines gives the number of customers present in the system at that instant, while the horizontal distance denotes the waiting time in the system (waiting time in the queue plus service time).

Suppose that the system has been in operation for some time and that it has settled down to a steady-state condition. Consider a time interval T that may include none or more than one busy period. Let

$$
\begin{aligned}
A(T) &= \text{total number of arrivals during the period } T \\
B(T) &= \text{area between the two lines} \\
&= \text{total waiting time (in the system) of all the customers who arrive during } T. \\
&= w_1 + w_2 + \cdots
\end{aligned}
$$

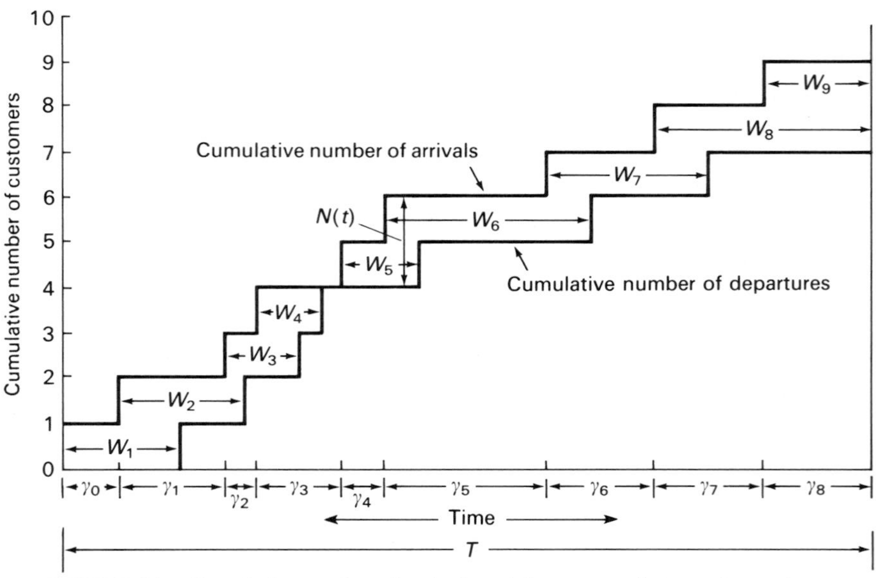

FIGURE 2.1. Cumulative number of arrivals and departures of a queueing system.

$$\lambda(T) = \text{mean arrival rate during } T$$
$$= \frac{A(T)}{T}$$

$$W(T) = \text{mean waiting time in the system during } T$$
$$= \frac{B(T)}{A(T)}$$

$$L(T) = \text{mean number of customers in system during } T$$
$$= \frac{B(T)}{T}.$$

We have

$$L(T) = \frac{B(T)}{T} = \frac{B(T)}{A(T)} \frac{A(T)}{T} = W(T)\lambda(T).$$

Suppose that limits exist as $T \to \infty$ and are given by

$$\lim_{T \to \infty} \lambda(T) = \lambda \quad \text{and}$$

$$\lim_{T \to \infty} W(T) = W.$$

Then a limit for $L(T)$ as $T \to \infty$ also exists and is given by

$$L = \lim_{T \to \infty} L(T)$$

and the three limits satisfy the relation

$$L = \lambda W,$$

which is Little's formula for corresponding L and W in the system.

To get $L = \lambda W$ for L and W in the queue, we refer to the bottom line as representing departures from the *queue* and not from the system; then $L(T)$ and $W(T)$ refer to those in the queue. Proceeding as before we get

$$L_Q = \lambda W_Q.$$

This completes the proof.

This result, of great generality, holds irrespective of the form of interarrival and service time distributions and any discrepancy that may be within the system: it holds under some very general conditions for any system, provided it is in steady-state. Ramalhoto *et al.* (1983) give a survey of contributions on this subject together with applications to various situations besides an exhaustive list of references.

The principle of customer conservation is: the frequency of entries into a service channel equals the frequency of departures from the channel. The frequency of arrivals is λ and the frequency of departures is $\mu(1 - p_0)$. Thus, for a $G/G/1$ queue, we have

$$\lambda = (1 - p_0)\mu \quad \text{or} \quad p_0 = 1 - \rho. \tag{6.3}$$

The average number of busy servers in a $G/G/c$ queue is $c\rho = \lambda/\mu$.

There are other relations that hold under somewhat restrictive conditions. For example, for a system with Poisson input, the Pollaczek–Khinchine formula holds. Some of these results are derived later.

From (6.1) and (6.2) it follows that for a $G/G/c$ queue

$$L - L_Q = \lambda(W - W_Q); \tag{6.4}$$

we know that $W - W_Q$ is the expected service time, which equals $1/\mu$; and $L - L_Q$ is the expected number in service or the expected number of busy servers. Thus, the expected number of busy servers equals $c\rho$ and the expected number of idle servers equals $c - c\rho = c(1 - \rho)$.

Remarks.

(1) The formula $L = \lambda W$ is valid in great generality. It relates customer-average waiting time W to time-average queue length L, given an arrival rate λ.

(2) Heyman and Stidham show that similar relationships exist between more general customer and time averages, which are represented by the formula $H = \lambda G$. For motivation, consider an example provided by Glynn and Whitt (1989) relating to migration of salmon fish up river. As the river narrows, a queue of salmon fish (salmon ladder) is created. The amount of food consumed by *each* fish in the queue can be modeled as a stochastic process. Then H may be taken as the average amount of food consumed *per fish* in the queue among the first n fish (throughout all time) and G as the average amount of food consumed *per time* in the queue by time t (by all fish).

(3) Continuous analogues of the relationships have been obtained by Rolski and Stidham (1983). More general extensions of the relationhips are obtained by Glynn and Whitt (1989).

(4) Attempts to generalize Little's formula has been mainly in two directions: (i) to weaken the sufficient conditions for the validity of the formula and (ii) to relate higher-order moments of the number in the system and sojourn time.

(5) For an $M/G/c$ queue

$$E\{L^r\} = \lambda^r E\{W^r\}/r! \qquad r = 1, 2, \ldots$$

and for certain systems

$$E\{L_{(r)}\} = \lambda^r E\{W^r\}$$

where

$$L_{(r)} = L(L-1)\cdots(L-r+1), \qquad r = 1, 2, \ldots.$$

(See Brumelle, 1972).

2.7. Poisson Arrival Process and Its Characteristics

In Chapter 1, we discussed the role of the Poisson process in modeling of stochastic phenomena. Arrival processes in several situations can be taken as Poisson processes. For example, telephone calls received at a switch board, arrival of customers in a bank, arrival of jobs at a CPU in a computer center, arrival of telecom messages, breakdown of machines in a machine shop having a large number of machines, and many other types of arrivals to a service center usually can be modeled as Poisson processes, that is, arrivals can be considered as events that occur in accordance with a Poisson process.

We mention here yet another interesting type of study. A service facility in a queue may receive input from a number of different sources. It would be reasonable therefore to postulate that the input or arrival process to the service facility is the superposition of a number of component processes that are nearly independent. Albin (1982), Whitt (1982), and Newell (1984) have studied queue-length behavior of systems of the type $\sum_{i=1}^{n} GI_i/G/1$ where the input process is the superposition of n independent renewal processes, Albin has done extensive similation studies of queueing systems of the type $\sum GI_i/M/1$ with exponentially distributed service time and with an input process that is the superposition of n independent renewal processes each with rate λ/n. She observes that as n increases, the average queue length for such a system approaches that of an $M/M/1$ system; however for fixed n the difference between the corresponding characteristic of the two systems increases as ρ increases from 0.5 to 0.9.

Newell (1984) studied in more detail some of the qualitative properties of the $\sum GI_i/G/1$ system. He observes that, as $n \to \infty$, under certain conditions, superposition of renewal processes behaves like a Poisson process. He further shows that for the average queue length the approach to the limiting $M/G/1$ behavior requires that $n(1-\rho)^2 \gg 1$.

These studies further strengthen the basis of assumption of the Poisson process as an arrival process.

2.7.1. A Characteristic of the Poisson Arrival Process: PASTA

Poisson arrivals have a very unique property in that such arrivals behave like *random* arrivals. That is, an observer from a Poisson arrival stream sees or finds the same system state distribution as a *random* observer having nothing to do with the system. For Poisson arrivals $a_n = p_n$, all $n \geq 0$.

If $N(t)(t \geq 0)$ denotes the state of the system (here number of occurrences or arrivals), if t_0 is an arbitrary instant, and if the arrivals are in accordance with a Poisson process, then the distribution of the random variable $N(t_0)$ is independent of whether or not there is an arrival at t_0 (Strauch, 1970.) Since the number of occurrences or arrivals are in accordance with a Poisson process, the interarrival time is exponential. It is memoryless and as such the distribution of $N(t_0)$ will be influenced only by the past history of the process, but not by what happens at t_0. This remarkable property of equality of outside observer's distribution and the arriving customer's distribution for systems with Poisson arrival is not followed by any other type of arrival process.

This property can also be stated as "**Poisson arrivals see time averages**" (**PASTA**). In a queue with Poisson arrivals, the limiting proportion of arrivals that find the system in some state n is equal to the limiting proportion of *time* the system spends in that state n. By PASTA is meant the equality of these two limiting fractions. Thus, for Poisson arrivals, $a_n = p_n$ for all $n \geq 0$.

A simple proof of PASTA is due to Wolff (1982) under the lack of anticipation assumption (LAA).

Let $N(t)$ be the number in the system in a queue where the arrival process $A(t)$ is Poisson. Let B be an arbitrary set in the value space of $N(t)$. Denote

$$\begin{aligned} U(t) &= 1, \text{ if } N(t) \in B \\ &= 0, \text{ otherwise} \end{aligned}$$

$$\mathrm{V}(\mathrm{t}) = \left(\frac{1}{t}\right) \int_0^t U(s)ds$$

$$Y(t) = \int_0^t U(s)dA(s) \quad \text{and}$$

$$Z(t) = \frac{Y(t)}{A(t)}.$$

Then $V(t)$ is the fraction of time during $[0, t]$ that $N(t)$ is in state B and $Z(t)$ is the fraction of arrivals in $[0, t]$ that find $N(t)$ in state B. Then Wolff shows that, under the lack of anticipation assumption, as $t \to \infty$,

$$V(t) \to V(\infty) \text{ with probability (w.p.) } 1$$

iff $$Z(t) \to V(\infty) \text{ w.p. } 1.$$

Notes.

(1) The result may be generalized to the case of a nonstationary Poisson arrival process with rate $\lambda(t)$.

(2) It is the independent increments property of the Poisson process that really accounts for PASTA.

(3) Instances do occur when non-Poisson arrivals also see time averages (Burke, 1976).

(4) PASTA does not hold in case of quasi-random input. When the arrivals occur from a *finite* source, though the interarrivals times are exponential RV, $a_n \neq p_n$. An example is considered in Section 3.8.

(5) Niu (1984) studies the interesting question of when do we have inequalities between the two limiting proportions, and considers three types of interarrival-time distributions in this connection. This problem has been earlier studied by König and Schmidt, Marshall and Wolff, Mori, Rolski, Stoyan, Whitt, and many others.

(6) There has been further ongoing study on this topic.

References

Albin, S. L. (1982). On Poisson approximations for superposition of arrival processes in queues. *Mgmt. Sci.* **28**, 127–137.

Brockmeyer, E., Halstorm, H. L., and Johnson, A. (1948). *The Life and Works of A. K. Erlang* (translation of the Danish Academy of Sciences, no. 2). The Copenhagen Telephone Company, Copenhagen, Denmark.

Brumelle, S. L. (1972). A generalization of $L = \lambda W$ to moments of queue lengths and waiting times. *Opns. Res.* **20**, 1127–1136.

Burke, P. J. (1976). Proof of a conjecture on the interarrival time distribution in an $M/M/1$ queue with feedback. *IEEE Trans. Com.* **24**, 575–576.

Cohen, J. W. (1982). *The Single Server Queue*, 2nd ed., North-Holland, Amsterdam.

Eilon, S. (1969). A simpler proof of $L = \lambda W$. *Opns. Res.* **17**, 915–916.

Erlang, A. K. (1909). The theory of probabilities and telephone conversations. *Nyt. Tiddskrift Matematik B* **20**, 33–39.

Glynn, P. W., and Whitt, W. (1986a). A central-limit-theorem version of $L = \lambda W$. *Queueing Systems* **1**, 191–215.

Glynn, P. W., and Whitt, W. (1986b). Sufficient conditions for functional limit-theorem versions of $L = \lambda W$. *Queueing Systems* **1**, 279–287.

Glynn, P. W., and Whitt, W. (1989). Extensions of the queueing relations $L = \lambda W$ and $H = \lambda G$. *Opns. Res.* **37**, 634–644.

Heyman, D. P., and Stidham, S., Jr. (1980). The relation between customer average and time average in queues. *Opns. Res.* **28**, 983–994.

Jewell, W. S. (1967). A simple proof of $L = \lambda W$. *Opns. Res.* **17**, 1109–1116.

Karr, A. F. (1978). Markov chains and processes with a prescribed invariant measure. *Stoch. Proc. Appl.* **7**, 277–290.

Karr, A. F., and Pittenger, A. O. (1978). The inverse balayage problem for Markov chains. *Stoch. Proc. Appl.* **7**, 165–178.

Keller, J. B. (1976). The inverse problem. *Am. Math. Monthly* **83**, 107–118.

Kendall, D. G. (1951). Some problems in the theory of queues. *J.R.S.S. B* **13**, 151–185.

Kendall, D. G. (1953). Stochastic processes occurring in the theory of queues and their analysis by the method of imbedded Markov chains. *Ann. Math. Statist.* **24**, 338–354.

Kendall, D. G. (1964). Some recent work and further problems in the theory of queues. *Theo. Prob. Appl.* **9**, 1–15.

Krawkowski, M. (1973). Conservation methods in queueing theory. *Rev. Fran. d'Auto Inf. Rech. Oper.* 7e Année, **V-1**, 3–20.

Little, J. D. C. (1961). A proof of the queueing formula $L = \lambda W$. *Opns. Res.* **9**, 383–387.

Louchard, G. and Latouche, G. (Eds.) (1983). *Probability Theory and Computer Science. Part II Stochastic Modeling: Queueing Models.* Academic Press, New York.

Morse, P. M. (1958). *Queues, Inventories and Maintenance.* Wiley, New York.

Newell, G. F. (1971). *Applications of Queueing Theory*, (2nd ed., 1982), Chapman & Hall, London.

Newell, G. F. (1984). Approximations for superposition arrival processes in queues. *Mgmt. Sci.* **30**, 623–632.

Niu, Shun-Chen. (1984). Inequalities between arrival averages and time averages in stochastic processes arising from queueing theory. *Opns. Res.* **32**, 785–795.

Prabhu, N. U. (1986). Editorial introduction. *Queueing Systems* **1**, 1–4.

Prabhu, N. U. (1987). A bibliography of books and survey papers on queueing systems: theory and applications. *Queueing Systems* **2**, 393–398.

Rolski, T., and Stidham, S., Jr (1983). Continuous versions of the queueing formulas $L = \lambda W$ and $H = \lambda G$. *Opns. Res. Letters* **2**, 211–215.

Romalhoto, M. F., Amaral, J. A., and Cochito, M. T. (1983). A survey of J. Little's formula. *Inter. Stat. Review* **51**, 255–278.

Saaty, T. L. (1957). A. K. Erlang. *Opns. Res.* **5**, 293–294.

Stidham, S., Jr. (1974). A last word on $L = \lambda W$. *Opns. Res.* **22**, 417–421.

Strauch, R. E. (1970). When a queue looks the same to an arriving customer as to an observer. *Mgmt. Sci.* **17**, 140–141.

Whitt, W. (1982). Approximating a point process by a renewal process I: Two basic methods. *Opns. Res.* **30**, 125–147.

Whitt, W. (1983). Untold horrors of the waiting room: What the equilibrium distribution will never tell about the queuelength process. *Mgmt. Sci.* **29**, 395–408.

Wolff, R. W. (1982). Poisson arrivals see time averages. *Opns. Res.* **30**, 223–231.

3 Birth-and-Death Queueing Systems: Exponential Models

3.1. Introduction

Many simple but interesting queueing systems can be studied through birth–death processes, which have been discussed in Section 1.4. In such a process transitions take place from one state only to a neighboring state. With an arrival there is a transition from the state $k(\geq 0)$ to the state $(k + 1)$ and with a service completion there is a transition from the state j to the state $(j - 1)(j > 0)$, the state denoting the number in the system. Before discussing queueing models in terms of birth–death equations, we take up the simplest queueing system $M/M/1$ through a simple alternative approach. This approach is based on the *rate-equality principle*, which holds for systems in steady state. The principle is stated as follows:

Rate-Equality Principle. *The rate at which a process enters a state $n(\geq 0)$ equals the rate at which the process leaves that state n. In other words, the rate of entering and the rate of leaving a particular state are the same for every state, that is, rate in = rate out, for every state.*

We have already discussed this principle under Remark (3) following Eq. (3.23) in Chapter 1.

3.2. The Simple *M*/*M*/1 Queue

In such a queueing system the arrivals occur from an infinite source in accordance with a Poisson process with parameter (say, λ) that is, the interarrival times are independent exponential with mean $1/\lambda$; the service times are independently and exponentially distributed, with parameter (say, μ); and there is only one server. The queue discipline is FCFS; the traffic intensity is $\rho = \lambda/\mu$, λ and μ being the arrival and service rates, respectively.

3.2.1. Steady-State Solution of *M*/*M*/1

Assume that steady state exists and let

$$p_n = \lim_{t\to\infty} Pr\{N(t) = n\}, \qquad n = 0, 1, 2, \ldots,$$

$N(t)$ being the number in the system (in the service channel and in queue, if any) at instant t; p_n is also the proportion of time the process is in state n.

We proceed to derive the equations involving p_n by using the rate-equality principle; then proceed to solve the equations to find p_n.

Consider state 0. The process can leave or go out from state 0 only when there is an arrival. (It then goes from state 0 to 1.) The long-run proportion of time the process is in state 0 is p_0 and, since λ is the rate of arrival, the rate at which the process leaves state 0 (to go to state 1) is λp_0. Again, the process can enter or go to state 0 only from state 1 through a departure or service completion. The proportion of time the process is in state 1 is p_1 and μ is the rate of leaving state 1 though service completion; the rate at which the process can leave state 1 to go to state 0 is μp_1. Thus, the rate at which the process enters state 0 is μp_1. Using the rate-equality principle, we get

$$\lambda p_0 = \mu p_1. \tag{2.1}$$

Consider state 1. The process can leave or go out from state 1 in two ways, either through an arrival or through a departure. The proportion of time the process is in state 1 is p_1 and the total rate at which arrival or departure occurs is $\lambda p_1 + \mu p_1$, that is, the rate at which the process leaves state 1 is $\lambda p_1 + \mu p_1$. The process can enter or go to state 1 in two ways, either through an arrival from state 0 or through a departure from state 2. Thus, the rate at which the process enters state 1 is $\lambda p_0 + \mu p_2$. The same argument then gives

$$\lambda p_1 + \mu p_1 = \lambda p_0 + \mu p_2.$$

Proceeding in the same way, we find, for $n > 1$,

$$\lambda p_n + \mu p_n = \lambda p_{n-1} + \mu p_{n+1}. \tag{2.2}$$

Equations (2.1) and (2.2) are called *balance equations*, as two rates are balanced; these are also called *conservation-flow relationships.*

Equations (2.1) and (2.2) together with the relation $\sum_{n=0}^{\infty} p_n = 1$ give the complete solution p_n. There are various methods of solving the equations. Equation (2.2) is a homogeneous difference equation of order 2 and can be solved by using the theory of solution of difference equation.

A simple straightforward solution is given next. From (2.2) we get

$$\begin{aligned}\lambda p_n - \mu p_{n+1} &= \lambda p_{n-1} - \mu p_n \\ &= \lambda p_{n-2} - \mu p_{n-1}, \quad \text{putting} \quad n-1 \quad \text{for} \quad n, \\ &\cdots \qquad \cdots \\ &= \lambda p_0 - \mu p_1 = 0, \quad \text{by} \quad (2.1)\end{aligned}$$

Thus,

$$\begin{aligned}p_{n+1} &= \frac{\lambda}{\mu} p_n = \frac{\lambda}{\mu}\left(\frac{\lambda}{\mu} p_{n-1}\right) = \left(\frac{\lambda}{\mu}\right)^2 p_{n-1} \\ &= \left(\frac{\lambda}{\mu}\right)^3 p_{n-2} = \cdots = \left(\frac{\lambda}{\mu}\right)^{n+1} p_0, \qquad n = 0, 1, 2, \ldots .\end{aligned}$$

Using $\sum_{n=0}^{\infty} p_n = 1$, we get

$$p_0\left[1 + \sum_{n=1}^{\infty}\left(\frac{\lambda}{\mu}\right)^n\right] = 1.$$

If $\lambda/\mu = \rho < 1$, then we get

$$p_0\left[\frac{1}{1-\rho}\right] = 1, \tag{2.3}$$

that is, $p_0 = 1 - \rho$, so that $p_n = (1-\rho)\rho^n$, $n = 1, 2, \ldots$. The distribution is geometric and is memoryless.

3.2.1.1. Some Performance Measures. Let N be the number and W the waiting time in the system in steady state. We have

$$\begin{aligned}E\{N\} &= \sum_{n=0}^{\infty} np_n = \sum_{n=1}^{\infty} n(1-\rho)\rho^n \\ &= \rho(1-\rho)\sum_{n=1}^{\infty} n\rho^{n-1} = \frac{\rho(1-\rho)}{(1-\rho)^2} = \frac{\rho}{1-\rho}\end{aligned} \tag{2.4}$$

and

$$\begin{aligned}
E\{N^2\} &= \sum_{n=0}^{\infty} n^2 p_n = \sum_{n=1}^{\infty} n^2(1-\rho)\rho^n \\
&= (1-\rho)\sum \{n^2 - n + n\}\rho^n \\
&= (1-\rho)\{\sum n(n-1)\rho^n + \sum n\rho^n\} \\
&= (1-\rho)[\rho^2 \sum n(n-1)\rho^{n-2} + \rho \sum n\rho^{n-1}] \\
&= (1-\rho)\left[\rho^2 \frac{d^2}{d\rho^2}\{\sum \rho^n\} + \rho \frac{d}{d\rho}\{\sum \rho^n\}\right] \\
&= (1-\rho)\frac{2\rho^2}{(1-\rho)^3} + \frac{(1-\rho)\rho}{(1-\rho)^2} = \frac{2\rho^2}{(1-\rho)^2} + \frac{\rho}{1-\rho}
\end{aligned}$$

so that

$$\begin{aligned}
\operatorname{var}\{N\} &= E\{N^2\} - [E\{N\}]^2 \\
&= \frac{\rho}{(1-\rho)^2}. \qquad (2.5)
\end{aligned}$$

Using Little's formula $L = \lambda W$, we get that the expected waiting time in the system, $E\{W\}$, equals

$$E\{W\} = \frac{E\{N\}}{\lambda} = \frac{1}{\lambda}\frac{\rho}{(1-\rho)} = \frac{1}{\mu(1-\rho)}. \qquad (2.6)$$

Notes. 1. The condition $\lambda/\mu = \rho < 1$ is necessary to get a solution. If $\lambda/\mu > 1$, then $\sum_n (\lambda/\mu)^n$ increases without limit and $p_n = 0$ for $n = 0, 1, 2, \ldots$. Thus $\rho < 1$ is a necessary condition for existence of steady states.

2. When $\rho \to 1$ from below, $E\{N\}$ and $\operatorname{var}\{N\}$ tend to ∞; also $p_0 \to 0$, $p_n \to 0$, $n \geq 1$, so that the probability that the system is empty or contains a certain fixed number is very, very small. Large variance of the number in the system implies that a randomly observed system size is likely to be very different from the expected system size.

3.2.2. *Waiting-Time Distributions*

We may consider two types of waiting times: (1) waiting time W_q in the queue or *queueing time* and (2) waiting time W in the system, which includes queueing time plus service time, that is, total time spent in the system by the test unit (also called *sojourn time* or *response time*). To find these distributions, queue discipline has to be taken into account. Assume that it is FCFS.

We consider at first the distribution of the waiting time or queueing time W_q of a test unit. We have $W_q = 0$ when there is no such unit in the system just before arrival of the test unit, the probability of this event is $p_0 = 1 - \rho$. Thus, $Pr\{W_q > 0\} = 1 - p_0 = \rho$, i.e., the probability that an arrival has to wait in queue is ρ; this probability is large for large ρ. If the test unit finds, on arrival, $n(>1)$ units in the system then $W_q = S_n$, where

$$S_n = v_1' + v_2 + \cdots + v_n,$$

v_1' being the residual service time of the customer being served, at the epoch of his arrival and $v_2, \ldots, v_n$ being the service times of the $(n-1)$ units in the queue. The residual service time v_1' of the exponential service-time distribution with mean $1/\mu$ has an exponential distribution with the same mean $(1/\mu)$. The RVs $v_2, \ldots, v_n$ are independent exponential with mean $1/\mu$. Thus, S_n is the sum of n identically and independently distributed exponential RVs, each with mean $(1/\mu)$. The RV S_n therefore has a gamma distribution having PDF

$$\frac{\mu^n x^{n-1} e^{-\mu x}}{\Gamma(n)}, \qquad x > 0. \tag{2.7}$$

Denote

$$w_q(x)dx = P\{x \le W_q < x + dx\}.$$

Conditioning on the number of units that the test unit finds on arrival, we get

$$\begin{aligned} w_q(x)dx &= \sum_{n=1}^{\infty} Pr\{x \le W_q < x + dx \mid \text{the test unit finds } n \text{ in the system}\} \\ &\quad \times Pr\{\text{the test unit finds } n \text{ in the system}\} \\ &= \sum_{n=1}^{\infty} \frac{\mu^n x^{n-1} e^{-\mu x}}{\Gamma(n)} dx(a_n). \end{aligned}$$

Now since $a_n = p_n$ for the Poisson arrival process, we get

$$\begin{aligned} w_q(x) &= \mu e^{-\mu x}(1-\rho)\rho \sum_{n=1}^{\infty} \frac{(\mu\rho x)^{n-1}}{(n-1)!} \\ &= \mu\rho(1-\rho)e^{-\mu(1-\rho)x}, \qquad x > 0. \end{aligned}$$

Thus, the PDF of W_q is given by

$$\begin{aligned} w_q(x) &= p_0 = 1 - \rho, \qquad x = 0 \\ &= \mu\rho(1-\rho)e^{-\mu(1-\rho)x}, \qquad x > 0, \end{aligned} \tag{2.8}$$

which is a modified exponential distribution.

The LST of W_q can be obtained from (2.8).

We can use the same argument to obtain directly the LST of waiting-time distribution. Let $w_q^*(s)$ be the LST of the waiting time W_q in the queue and $w_q^*(s|n)$ be the conditional LST of the waiting time in the queue given that the test unit finds n in the system on his arrival. We have

$$w_q^*(s) = \sum_{n=0}^{\infty} a_n w_q^*(s|n),$$

where a_n is the probability that the arrival finds n in the system. For Poisson arrivals, however, $a_n = p_n = (1-\rho)\rho^n$. If he finds $n = 0$ customers, he does not wait, if he finds $n(\geq 1)$, he waits until the services of n are completed. The service time of n being equal to the sum of n IID exponential variables with mean $1/\mu$, we get

$$w_q^*(s|n) = \left(\frac{\mu}{s+\mu}\right)^n, \qquad n \geq 1.$$

Thus,

$$\begin{aligned} w_q^*(s) &= (1-\rho) + (1-\rho)\sum_{n=1}^{\infty} \rho^n \left(\frac{\mu}{s+\mu}\right)^n & (2.9) \\ &= (1-\rho) + (1-\rho)\frac{\rho\mu/(s+\mu)}{1 - \dfrac{\rho\mu}{s+\mu}} \\ &= (1-\rho) + \frac{\mu\rho(1-\rho)}{s+\mu(1-\rho)}. & (2.9a) \end{aligned}$$

The LST of the waiting time in the system W (or sojourn time) is given by

$$\begin{aligned} w^*(s) &= w_q^*(s)\frac{\mu}{s+\mu} \\ &= \frac{\mu(1-\rho)}{s+\mu}\frac{s+\mu}{s+\mu(1-\rho)} \\ &= \frac{\mu(1-\rho)}{s+\mu(1-\rho)} \end{aligned} \tag{2.10}$$

so that the PDF of the waiting time (in the system) is obtained as

$$w(x) = \mu(1-\rho)e^{-\mu(1-\rho)x}, \qquad x \geq 0; \tag{2.11}$$

the distribution is exponential with parameter $\mu(1-\rho)$.

Notes.

(1) From (2.9), we get

$$W_q(x) = Pr\{W_q \leq x\} = (1-\rho)\delta(x) + (1-\rho)\sum_{n=1}^{\infty} \rho^n \{B(x)\}^{n*} \tag{2.12}$$

where $\{B(x)\}^{n*}$ is the n-fold convolution of the exponential service time DF $B(x)$.

The form (2.9) implies that $w_q^*(s)$ is the geometric compounding of the exponential service time; so also (2.12) does imply the same about $W_q(x)$.

(2) Moments of W_q can be easily obtained as follows: we have

$$\begin{aligned} E[W_q] &= -\frac{d}{ds} w_q^*(s)\bigg|_{s=0} \\ &= \frac{(1-\rho)\lambda}{(s-\lambda+\mu)^2}\bigg|_{s=0} = \frac{\lambda}{\mu(\mu-\lambda)} \\ &= \frac{\rho}{\mu(1-\rho)} \quad \text{and} \end{aligned} \tag{2.13}$$

$$\begin{aligned} E[W_q^2] &= \frac{d^2}{ds^2} w_q^*(s)\bigg|_{s=0} \\ &= \frac{2(1-\rho)\lambda}{(s-\lambda+\mu)^3}\bigg|_{s=0} \\ &= \frac{2(1-\rho)\lambda}{(\mu-\lambda)^3} = \frac{2\lambda}{\mu(\mu-\lambda)^2}. \end{aligned} \tag{2.14}$$

Hence,

$$\operatorname{var}[W_q] = \frac{2\lambda}{\mu(\mu-\lambda)^2} - \left[\frac{\lambda}{\mu(\mu-\lambda)}\right]^2 = \frac{\lambda(2\mu-\lambda)}{\mu^2(\mu-\lambda)^2} = \frac{\rho(2-\rho)}{(\mu-\lambda)^2}. \tag{2.15}$$

(3) In the following table, values of $E(N)$, var(N), $E(W)$, and var(W) are given for certain values of $\mu = 1$, and $\lambda = \rho$.

$\rho(=\lambda)$	0.4	0.6	0.8	0.9	0.98	0.99	0.995
$E(N)$	0.67	1.5	4.0	9.0	49.0	99.0	497.5
var(N)	1.11	3.75	20.0	90.0	2450.0	$10^2 \times 99.0$	$10^3 \times 199.0$
$E(W)$	1.67	2.50	5.0	10.0	50.0	100.0	200.0
var(W)	2.78	6.25	25.0	100.0	2500.0	$10^4 \times 1.0$	$10^4 \times 4.0$

(4) The preceding gives the unconditional distribution of the queueing time of a test unit. The conditional distribution of the queueing time given that the test unit has to wait, has the PDF $f_c(x)$ given by

$$\begin{aligned} f_c(x)dx &= P\{x \le W_q < x + dx \mid \text{test unit has to wait}\} \\ &= \frac{\mu\rho(1-\rho)e^{-\mu(1-\rho)x}}{\rho}\,dx \\ &= \mu(1-\rho)e^{-\mu(1-\rho)x}\,dx, \qquad x > 0. \end{aligned} \tag{2.16}$$

The conditional distribution is exponential with mean $1/\mu(1-\rho)$.

3.2.3. The Output Process

The problem of output (efflux, departures) from a queueing system was first considered by Morse (1955). He observed that the output of a Poisson input queue with a single channel having exponential service time and in steady state must be Poisson with the same rate as the input. A formal proof was first given by Burke (1956). The interesting result is considered next.

Theorem 3.1. *In an M/M/1 queueing system in steady state, the interdeparture times are independently and identically distributed exponential random variables with mean $1/\lambda$, where λ is the parameter of the input (Poisson) process. In other words, the output process is Poisson with the same parameter as the input process.*

Proof. Let $N(t)$ be the number in the system at time t and $t'_1, t'_2, \ldots, t'_n, t'_{n+1} \ldots$ denote the successive departure instants so that $L = t'_{n+1} - t'_n$ is the nth interdeparture interval or period. Let

$$F_k(t) = P\{N(t'_n + t) = k, \quad t'_{n+1} - t'_n > t\}, \quad t > 0, k = 0, 1, \ldots, \tag{2.17}$$

be the joint probability distribution of $N(t)$ and $t'_{n+1} - t'_n$. Since the probability that a departing customer leaves k in the system is equal to the probability that the number in the system is k for Poisson input (that is, $d_k = p_k$), we have

$$F_k(0) = p_k = (1 - \rho)\rho^k, \qquad k = 0, 1, 2, \ldots. \tag{2.18}$$

For an infinitesimal interval of length dt

$$F_0(t + dt) = F_0(t)(1 - \lambda\, dt) + o(dt) \tag{2.19}$$

and

$$F_k(t + dt) = F_k(t)[1 - \lambda\, dt - \mu\, dt] + F_{k-1}(t)\lambda\, dt + o(dt).$$

These equations reduce to

$$\begin{aligned} F_0'(t) &= -\lambda F_0(t) \quad \text{and} \\ F_k'(t) &= \lambda F_{k-1}(t) - (\lambda + \mu)F_k(t), \qquad k = 1, 2, 3, \ldots. \end{aligned} \tag{2.20}$$

Subject to the initial condition $F_k(0) = p_k$, the unique solution of (2.20) is given by

$$\begin{aligned} F_k(t) &= p_k e^{-\lambda t} \\ &= (1 - \rho)\rho^k e^{-\lambda t}. \end{aligned} \tag{2.21}$$

Again

$$\begin{aligned} Pr\{N(t'_{n+1} + 0) = k, \qquad & t \le t'_{n+1} - t'_n < t + dt\} \\ &= F_{k+1}(t) \cdot Pr\{\text{one service completion in } (t, t + dt)\} \\ &= F_{k+1}(t)[\mu\, dt + o(dt)] \\ &= (1 - \rho)\rho^{k+1} e^{-\lambda t}\mu\, dt + o(dt) \\ &= (1 - \rho)\rho^k \lambda e^{-\lambda t}\, dt + o(dt), \end{aligned} \tag{2.22}$$

which proves the independence of $N(t'_{n+1} + 0)$ and $(t'_{n+1} - t'_n)$. It follows that $L = t'_{n+1} - t'_n$ has density $\lambda e^{-\lambda t}$.

Now we look at the independence of the interdeparture intervals. Let Λ represent the set of lengths of an arbitrary number of interdeparture intervals subsequent to the interval of length L. Let $P(.)$ represent the probability function of the variable within ( ). The Markov property implies that

$$P[\Lambda \mid N(L)] = P[\Lambda \mid N(L), L]. \tag{2.23}$$

Since $N(L)$ and L are independent,

$$P[N(L), L] = P[N(L)]P(L). \tag{2.24}$$

The joint probability function of the initial interval length, the state at the end of the interval, and the set of subsequent interval lengths may be expressed (using (2.23) and (2.24)) as

$$\begin{aligned} P[L, N(L), \Lambda] &= P[\Lambda \mid N(L), L]P[N(L), L] \\ &= P[\Lambda \mid N(L)]P[N(L)]P(L) \\ &= P[\Lambda, N(L)]P(L). \end{aligned}$$

Thus,

$$P(L, \Lambda) = \sum_{N(L)=0}^{\infty} P[\Lambda, N(L)]P(L) = P(L)P(\Lambda)$$

which implies the mutual independence of all intervals. This completes the proof.

Notes.

(1) For the existence of steady state, the condition $\lambda < \mu$ is essential. The output process is, for large t, Poisson, even for $\lambda \geq \mu$, when steady state does not hold. For $M/M/1$ queue the output process, for large t, is Poisson with rate $\min(\lambda, \mu)$, (Goodman and Massey, 1984).

(2) That the expected inter-output (interdeparture) rate for a single server Poisson input queue (with arrival rate λ) is λ can also be shown as follows.

Conditioning on the state in which a departure leaves the system, (the state $n > 0$ or $n = 0$), we have

$$\begin{aligned} E(L) &= E\{L \mid N = n(>0)\}[Pr\{N = n(>0)\}] \\ &\quad + E(L \mid N = 0)\ Pr\{N = 0\} \\ &= (1/\mu)(1 - p_0) + (1/\lambda + 1/\mu)p_0 \\ &= (1/\mu) + (1/\lambda)(1 - \rho) \\ &= 1/\lambda. \end{aligned}$$

(3) Robustness of the $M/M/1$ Queue

Albin (1984) examines the robustness or insensitivity of the $M/M/1$ queueing model to specific perturbations of the input process. Perturbations considered are deviations from the exponential distribution of the interarrival times and from the assumption of independence between successive interarrival times. Let θ be a parameter of the system and let $L(\theta)$ be the expected number in the system in steady state in the standard $M/M/1$ queue, and let

$L(\theta + \varepsilon)$ be the expected number in the perturbed system. Then expanding by Taylor's series, we get

$$L(\theta + \varepsilon) \simeq L(\theta) + \varepsilon L'(\theta).$$

$L'(\theta)$ is called the perturbation rate, and the system is called sensitive when $|L'(\theta)|$ is large compared with $L(\theta)$. For example, for an $M/M/1$ system with ρ as parameter,

$$L(\rho) = \frac{\rho}{(1 - \rho)}$$

$$L'(\rho) = \frac{1}{(1 - \rho)^2} = \left[\frac{1}{\rho^2}\right][L(\rho)]^2,$$

which shows that the expected number in the $M/M/1$ system is sensitive to small changes in ρ, in particular, for ρ near to 1. Her results in the specific cases examined indicate that the expected number in the system is robust or insensitive to specific perturbations in the arrival process, such as perturbations by insertion of a few short interarrival times, of an occasional batch arrival, or of small dependencies between successive interarrival times. For details, refer to Albin (1984). See also Zolotarev (1977).

3.2.4. *Semi-Markov Process Analysis*

We shall now consider the approach through semi-Markov process (discussed in Section 1.9.) for this system. Here we consider that a transition occurs with the arrival or departure of a unit, i.e., t_n $(n = 0, 1, 2, \ldots)$ is the epoch at which the nth transition (through an arrival or service completion) occurs. With the notation of Section 1.9., $\{Y(t), t \geq 0\}$ is a semi-Markov process having $\{X_n, n \geq 0\}$ [where $X_n = N(t_n + 0)$, $Y(t) = X_n$, $t_n \leq t < t_{n+1}$] as its embedded Markov chain. If u and v denote the interarrival and service times, respectively, then $Pr\{u < v\} = \lambda/(\lambda + \mu)$, $Pr\{u > v\} = \mu/(\lambda + \mu)$, and $\{\min(u, v)\}$ is exponential with parameter $(\lambda + \mu)$. (See Problem 1.17.)

We have

$$\begin{aligned} Q_{0,1}(t) &= Pr\{\text{transition occurs from state 0 to state 1 by time } t\} \\ &= Pr\{\text{an arrival occurs by time } t\} \\ &= 1 - e^{-\lambda t} \end{aligned}$$

$$
\begin{aligned}
Q_{j,j+1} &= Pr\{\text{transition occurs from state } j \text{ to state } j+1 \text{ by time } t\} \\
&= Pr\{\text{one transition (arrival or service completion) occurs by time } t\}. \\
&\quad\ Pr\{\text{transition is through an arrival}\} \\
&= Pr\{\min(u, v) \le t\} Pr\{u < v\} \\
&= [1 - e^{-(\lambda + \mu)t}] \frac{\lambda}{\lambda + \mu}, \qquad j > 0.
\end{aligned}
$$

Similarly,

$$Q_{j,j-1}(t) = [1 - e^{-(\lambda+\mu)t}]\left[\frac{\mu}{\lambda + \mu}\right], \qquad j > 0.$$

Thus,

$$p_{i,j} = \lim_{t \to \infty} Q_{i,j}(t) \quad \text{gives}$$

$$p_{0,1} = 1, \qquad p_{j,j+1} = \frac{\lambda}{\lambda + \mu}, \qquad p_{j,j-1} = \frac{\mu}{\lambda + \mu}, \qquad j \ge 1$$

and

$$p_{i,j} = 0 \text{ in all other cases.}$$

Thus, $v_k = \lim_{n \to \infty} p_{i,j}^{(n)}$ are given as the unique solution of $\mathbf{V} = \mathbf{V}P$, i.e.,

$$v_j = \sum_k v_k p_{kj}.$$

We get

$$
\begin{aligned}
v_0 &= v_1 p_{1,0} = \left(\frac{\mu}{\lambda + \mu}\right) v_1 \\
v_1 &= v_0 p_{0,1} + v_2 p_{2,1} \\
&= v_0 + v_2\left(\frac{\mu}{\lambda + \mu}\right)
\end{aligned}
$$

and for $j > 1$

$$
\begin{aligned}
v_j &= v_{j-1} p_{j-1,j} + v_{j+1} p_{j+1,j} \\
&= v_{j-1}\left(\frac{\lambda}{\lambda + \mu}\right) + v_{j+1}\left(\frac{\mu}{\lambda + \mu}\right).
\end{aligned}
$$

Solving the above difference equation, we get

$$v_j = \mathrm{A} + \mathrm{B}(\lambda/\mu)^j, \qquad j = 1, 2, \ldots,$$

where A, B are constants. Evaluating the constants with the help of expressions for v_0, v_1, we get

$$v_j = \left(\frac{\lambda + \mu}{\lambda}\right)\left(\frac{\lambda}{\mu}\right)^j v_0, \qquad j = 1, 2, \ldots.$$

From (9.4) of Chapter 1, we get

$$\begin{aligned} p_k &= \lim_{t \to \infty} Pr\{Y(t) = k\} \\ &= \frac{v_k \mu_k}{\sum v_j \mu_j}, \qquad k = 0, 1, \ldots. \end{aligned}$$

where μ_k = expected sojourn time in state k. We get

$$\begin{aligned} \mu_0 &= E\{u\} = 1/\lambda \\ \mu_k &= \text{expected time for a transition} \\ &= E\{\min(u, v)\} = \frac{1}{\lambda + \mu}, \qquad j \geq 1 \end{aligned}$$

and

$$v_j \mu_j = \frac{1}{\lambda}\left(\frac{\lambda}{\mu}\right)^j v_0$$

so that

$$\begin{aligned} \sum_{j=0}^{\infty} v_j \mu_j &= (v_0/\lambda)\frac{1}{1 - \lambda/\mu} \\ &= \frac{\mu}{\lambda}\frac{v_0}{(\mu - \lambda)} \end{aligned}$$

Thus,

$$\begin{aligned} p_k &= \frac{(1/\lambda)(\lambda/\mu)^k v_0}{(\mu/\lambda)\{1/(\mu - \lambda)\} v_0} \\ &= (1 - \lambda/\mu)(\lambda/\mu)^k \\ &= (1 - \rho)\rho^k, \qquad k = 0, 1, 2, \ldots. \end{aligned}$$

Notes.

(1) Here the system size $N(t)$ is semi-Markovian (as well as Markovian).

(2) $v_0 < p_0$ and $v_k > p_k$, $k = 1, 2, \ldots$.

3.3. System with Limited Waiting Space: The *M/M/1/K* Model

For the simple queue $M/M/1$, the assumption is that the system can accommodate any number of units. In this model, we assume that the system can accommodate a finite number of units, say K, including the one being served, if any. Customers arrive in accordance with a Poisson process with rate, say, λ; a customer will join the system whenever he finds less than K in the system and a customer who arrives when there are K in the system leaves the system and is lost to the system. Service time is exponential with rate μ.

3.3.1. Steady-State Solution

Here the system will behave as an ordinary $M/M/1$ queue so long as the number in the system is less than K. Then when it reaches the state K or the system contains K customers, no more arrival is allowed to the system, and the number in the system cannot exceed K. From state K, only departure is possible. Using the rate-equality principle, the balance equations can be written as follows:

$$\begin{aligned} \lambda p_0 &= \mu p_1 \\ \lambda p_n + \mu p_n &= \lambda p_{n-1} + \mu p_{n+1}, \qquad n = 1, 2, \ldots, K-1 \\ \lambda p_{K-1} &= \mu p_K. \end{aligned} \tag{3.1}$$

Solving the first two equations recursively, we get

$$p_n = p_0 \rho^n, \qquad n = 0, 1, 2, \ldots, K-1.$$

From the last one, we get

$$p_K = \rho p_{K-1} = \rho(p_0 \rho^{K-1}) = p_0 \rho^K.$$

Thus,

$$p_n = p_0 \rho^n, \qquad n = 0, 1, 2, \ldots, K. \tag{3.2}$$

Using the normalizing condition

$$\sum_{n=0}^{K} p_n = 1, \quad \text{we get} \quad p_0 \sum_{n=0}^{K} \rho^n = 1$$

so that

$$p_0 = \left[\sum_{n=0}^{K} \rho^n\right]^{-1} = \frac{1-\rho}{1-\rho^{K+1}}, \qquad \rho \neq 1,$$

$$= \frac{1}{K+1}, \qquad \rho = 1.$$

Thus, we have, for $n = 0, 1, \ldots, K$,

$$p_n = p_0 \rho^n = \frac{(1-\rho)\rho^n}{1-\rho^{K+1}}, \qquad \rho \neq 1,$$

$$= \frac{1}{K+1}, \qquad \rho = 1. \tag{3.3}$$

Its PGF is given by

$$G(s) = \sum_{n=0}^{K} p_n s^n = \frac{1-\rho}{1-\rho^{K+1}} \left[\frac{1-(\rho s)^{K+1}}{1-\rho s}\right] \quad \text{for} \quad \rho \neq 1. \tag{3.4}$$

The distribution of the number in the system is uniform for $\rho = 1$ and truncated geometric for $\rho \neq 1$.

3.3.2. Expected Number in the System L_K

We have for $\rho = 1$,

$$L_K = \sum_{n=0}^{K} n p_n$$

$$= \sum_{n=0}^{K} \frac{n}{K+1} = \frac{K}{2};$$

for $\rho \neq 1$,

$$L_K = \frac{(1-\rho)\rho}{1-\rho^{K+1}} \sum_{n=0}^{K} n\rho^{n-1}$$

$$= \frac{(1-\rho)\rho}{1-\rho^{K+1}} \sum_{n=0}^{K} \frac{d}{d\rho}(\rho^n)$$

$$= \frac{(1-\rho)\rho}{1-\rho^{K+1}} \frac{d}{d\rho}\left(\sum_{n=0}^{K} \rho^n\right)$$

$$= \frac{(1-\rho)\rho}{1-\rho^{K+1}} \frac{1-(K+1)\rho^K + K\rho^{K+1}}{(1-\rho)^2}$$

$$= \frac{\rho}{1-\rho} - \frac{(K+1)\rho^{K+1}}{1-\rho^{K+1}}. \tag{3.5}$$

Remarks.

(1) It may be noted that the finite series $\sum_{n=0}^{K} \rho^n$ has a sum for all values of ρ; thus, the steady-state solution exists for all values of ρ.
(2) Assuming $\rho < 1$ and taking limit as $K \to \infty$, we get the corresponding results for an $M/M/1$ model.
(3) The system with $K = 1$ (with no waiting room at all, as it happens with a telephone line) is known as a system with *blocked calls cleared* with a single server.
(4) The effective input rate λ' to the system $M/M/1/K$ is the expected number of customers joining the system in unit time and equals

$$\lambda' = \lambda(1 - p_K) = \frac{\lambda(1 - \rho^K)}{1 - \rho^{K+1}}, \qquad \rho \neq 1$$

$$= \frac{\lambda K}{K + 1}, \qquad \rho = 1.$$

(5) The average number of customers diverted from the system or lost to the system in unit time equals

$$\lambda - \lambda(1 - p_K) = \lambda p_K = \frac{\lambda(1 - \rho)\rho^K}{1 - \rho^{K+1}}, \qquad \rho \neq 1,$$

$$= \frac{\lambda}{K + 1}, \qquad \rho = 1.$$

(6) The utilization factor of the service station is not equal to ρ as in usual models, but equals

$$b = \sum_{n=1}^{K} p_n = 1 - p_0 = \frac{\rho(1 - \rho^K)}{1 - \rho^{K+1}}.$$

(7) The expected number of customers leaving the service station (after being served) in unit time equals

$$\mu b = \mu(1 - p_0) = \mu\left[1 - \frac{1 - \rho}{1 - \rho^{K+1}}\right] = \lambda'.$$

(8) Naor (1969) has used the model $M/M/1/K$ to study the regulation of queue size by levying tolls. (See Problems and Complements 3.4.) Rue and Rosenshine (1981) have extended Naor's arguments to obtain a policy for "individual optimum" in case of M classes of customers. Bounds on this policy and a numerical example with three classes of customers are also considered.

(9) Though the *arrival* process is Poisson, the *input* for this truncated system is not truly Poisson. PASTA does not hold in this case. The probability a_n that an effective arrival finds n in the system [in an infinitesimal interval $(t, t+h)$] can be obtained by applying Baye's theorem. We have

$$\begin{aligned} a_n &= Pr\{\text{an arrival finds } n \text{ in system}\} \\ &= Pr\{n \text{ in system} \mid \text{an arrival is about to occur}\} \\ &= \frac{Pr\{\text{an arrival is about to occur} \mid n \text{ in system}\} \times p_n}{\sum_{k=0}^{K-1} Pr\{\text{an arrival is about to occur} \mid k \text{ in system}\}} \\ &= \lim_{h \to 0} \frac{\{\lambda h + o(h)\} p_n}{\sum_{k=0}^{K-1} \{\lambda h + o(h)\} p_k} \\ &= \frac{\lambda p_n}{\sum_{k=0}^{K-1} \lambda p_k} \qquad (3.6) \\ &= \frac{p_n}{1 - p_K}, \quad n = 0, 1, \ldots, K-1. \qquad (3.7) \end{aligned}$$

3.3.3. *Equivalence of an M/M/1/K Model with a Two-Stage Cyclic Model*

Consider a cyclic model with a fixed number K of jobs circulating endlessly between two servers I and II. (See Fig. 3.1.) The two servers have independent exponential service-time distributions with rates μ and λ, respectively, the

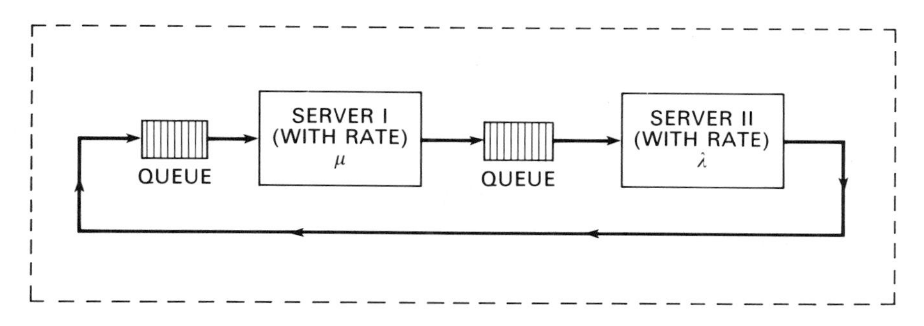

FIGURE 3.1. Diagram of a two-stage cyclic queue.

order of service at each of the service counters being FCFS. As long as the number of customers at server I is less than K, customers will arrive there from server II, the interarrival times being distributed as the service time of server II, that is, exponential with rate λ. When all the customers are at server I, no further arrival can take place. Thus, the model is the same as the $M/M/1/K$ model; the event that there are n customers either in queue or in service with server I and $K - n$ customers with server II in the cyclic model is the same as that of n customers in a $M/M/1/K$ model, and so, for $n = 0, 1, 2, \ldots, K$,

$$\begin{aligned} p(n, K - n) &= Pr(n \text{ customers with server I and } K - n \text{ with server II}) \\ &= p_n, \text{ as given by (3.3).} \end{aligned}$$

The utilization factors of servers I and II are given by

$$\rho_1 = 1 - p_0 \quad \text{and} \quad \rho_2 = 1 - p_K, \qquad \frac{\rho_1}{\rho_2} = \rho.$$

The expected numbers N_1 and N_2 with the subsystems (servers I and II) are given by

$$E(N_1) = \sum_{n=0}^{K} np_n$$

and

$$E(N_2) = K - E(N_1).$$

The average time taken for a customer to go through *both* the servers to make a complete cycle is

$$\frac{K}{(\rho_1 \mu)} = \frac{K}{(\rho_2 \lambda)}.$$

The preceding model is useful for computer systems where the CPU can be taken as server I and I/O unit as server II. (See Problems and Complements 5.6 for the model with general service time.)

3.4. Birth-and-Death Processes: Exponential Models

Consider a queueing system where the arrivals occur from an infinite source in accordance with a Poisson process and the service times are independently and exponentially distributed. The rates of arrival and service are state-dependent: when there are n in the system, the rate of arrival is λ_n and the rate of service is μ_n. The queue discipline is FCFS and there is no restriction as to the number of servers.

Let $N(t)$ be the number in the system as time t. Assume that steady state exists and let

$$p_n = \lim_{t \to \infty} P\{N(t) = n\}, \qquad n = 0, 1, 2, \ldots.$$

We shall obtain the balance equations by using the rate-equality principle.

Consider state 0. The process can leave or go out from state 0 only when there is an arrival when the change occurs from state 0 to state 1. It can enter the state 0 from state 1 only with a departure. Using the rate-equality principle we get

$$\lambda_0 p_0 = \mu_1 p_1. \tag{4.1}$$

Consider an arbitrary state n ($n = 1, 2, \ldots$). The process can leave state n in two ways, either through an arrival (when the state changes from n to $n + 1$) or through a departure (when the state changes from n to $(n - 1)$). The rate at which the process leaves state n is thus $(\lambda_n + \mu_n)$; the proportion of time it is in state n is p_n and the total rate at which the process leaves state n is $(\lambda_n + \mu_n)p_n$. On the other hand, the process can enter the state either from state $(n - 1)$ through an arrival or from state $(n + 1)$ through a departure. The proportion of time the process is in state $(n - 1)$ is p_{n-1} and the rate at which it can enter state n through an arrival from state $(n - 1)$ is λ_{n-1}. Thus, the rate of entering state n from state $(n - 1)$ through an arrival is $\lambda_{n-1}p_{n-1}$; similarly the rate of entering state n from state $(n + 1)$ through a departure is $\mu_{n+1}p_{n+1}$, so that the total rate at which the process enters state n is $\lambda_{n-1}p_{n-1} + \mu_{n+1}p_{n+1}$. Thus, we have

$$(\lambda_n + \mu_n)p_n = \lambda_{n-1}p_{n-1} + \mu_{n+1}p_{n+1}, \qquad n = 1, 2, \ldots. \tag{4.2}$$

Equations (4.1) and (4.2) are the balance equations (conservation-flow equations) for the model. Solving the equations and using $\sum p_n = 1$, we get the complete solutions. We now proceed to solve them. From (4.2), we get

$$\begin{aligned}
\lambda_n p_n - \mu_{n+1}p_{n+1} &= \lambda_{n-1}p_{n-1} - \mu_n p_n \\
&= \lambda_{n-2}p_{n-2} - \mu_{n-1}p_{n-1}, \quad \text{putting} \quad n - 1 \text{ for } n \\
&\quad \ldots \qquad \ldots \\
&= \lambda_0 p_0 - \mu_1 p_1 = 0 \quad \text{by} \quad (4.1).
\end{aligned}$$

Thus,

$$\begin{aligned} p_{n+1} &= \frac{\lambda_n}{\mu_{n+1}} p_n \\ &= \frac{\lambda_n}{\mu_{n+1}} \frac{\lambda_{n-1}}{\mu_n} p_{n-1} \\ &= \frac{\lambda_n}{\mu_{n+1}} \frac{\lambda_{n-1}}{\mu_n} \cdots \frac{\lambda_1}{\mu_2} \frac{\lambda_0}{\mu_1} p_0 \\ &= \prod_{k=0}^{n} \frac{\lambda_k}{\mu_{k+1}} p_0, \qquad n = 0, 1, 2, \ldots, \quad \text{or} \end{aligned}$$

$$p_n = \prod_{k=0}^{n-1} \frac{\lambda_k}{\mu_{k+1}} p_0, \qquad n = 1, 2, \ldots. \tag{4.3}$$

Using $\sum_{n=0}^{\infty} p_n = 1$, we get

$$p_0 = \frac{1}{1 + \sum\limits_{n=1}^{\infty} \prod\limits_{k=0}^{n-1} \frac{\lambda_k}{\mu_{k+1}}}. \tag{4.4}$$

The necessary and sufficient condition for the existence of a steady state is the convergence of the infinite series $\sum_{n=1}^{\infty} \prod_{k=0}^{n-1} \lambda_k/\mu_{k+1}$ which occurs in the denominator of (4.4). When the series converges, p_0 can be obtained by using (4.4).

The preceding birth–death process model can be used to study various queueing systems. For example, the $M/M/1$ model considered in Sec. 3.2 is the model with constant arrival and service rates

$$\begin{aligned} \lambda_n &= \lambda, \qquad n = 0, 1, 2, \ldots, \quad \text{and} \\ \mu_n &= \mu, \qquad n = 1, 2, \ldots \end{aligned}$$

Using (4.4) and (4.3) with these values of λ_n, μ_n, we at once get the steady-state results. The $M/M/1/K$ queueing model can be obtained by putting

$$\begin{aligned} \lambda_n &= \lambda,\ n = 0, 1, \ldots, K-1 \\ &= 0,\ n = K, \\ \mu_n &= \mu,\ n = 1, 2, \ldots, K. \end{aligned}$$

3.5. The $M/M/\infty$ Model: Exponential Model with an Infinite Number of Servers

We consider here the exponential model with an infinite number of servers. An example of where such a model can be appropriate is a service facility with provision of self-service. It is a birth–death model with

$$\lambda_n = \lambda, \qquad n = 0, 1, 2, \ldots, \quad \text{and}$$

$$\mu_n = n\mu, \qquad n = 1, 2, \ldots .$$

The solution is given by

$$p_n = \prod_{k=0}^{n-1} \frac{\lambda_k}{\mu_{k+1}} = p_0 \prod_{k=0}^{n-1} \frac{\lambda}{(k+1)\mu}$$

$$= p_0 \frac{\lambda^n}{\mu(2\mu)\cdots(n\mu)} = p_0 \frac{(\lambda/\mu)^n}{n!}, \qquad n = 0, 1, 2, \ldots .$$

To find p_0, we use

$$1 = \sum_{n=0}^{\infty} p_n = \left[\sum_{n=0}^{\infty} \frac{\left(\dfrac{\lambda}{\mu}\right)^n}{n!} \right] p_0$$

$$= e^{\lambda/\mu} p_0$$

so that

$$p_0 = e^{-\lambda/\mu}$$

and, thus,

$$p_n = \frac{e^{-\lambda/\mu}(\lambda/\mu)^n}{n!}, \qquad n = 0, 1, 2, \ldots . \tag{5.1}$$

The distribution is Poisson with mean λ/μ. The expected number of customers in the system is λ/μ and the expected response time is $1/\mu = [(\lambda/\mu)/\lambda]$, the average time required for service (as is obvious). The result holds irrespective of the magnitude of λ/μ.

Notes.

(1) The result holds irrespective of the form of the service time distribution as well (as will be shown in Section 6.8.).

(2) Newell (1984) discusses in his book several interesting aspects of this model.

3.6. The Model *M/M/c*

3.6.1. Steady-State Distribution

We consider a queue with Poisson input (having rate, say, λ) and with c $(1 \leq c \leq \infty)$ parallel service channels having IID exponential service time distribution, each with rate, say, μ. If there are n units in the system, and n is less than c, then, in all, n channels are busy and the interval between two consecutive service completions is again exponential with rate $n\mu$ (see Problem 1.17(b)). If there are n $(\geq c)$ in the system then all the c channels are busy and the interval between two consecutive service completions is exponential with rate $c\mu$. Thus, we have a birth–death model having constant arrival (birth) rate λ and state dependent service (death) rate

$$\begin{aligned} \mu_n &= n\mu, \qquad n = 0, 1, 2, \ldots, c \\ &= c\mu, \qquad n = c + 1, c + 2, \ldots. \end{aligned}$$

Denote $\rho = \lambda/c\mu$. Assume that steady state exists, and that the system is in steady state. Putting the values of λ_n and μ_n in (4.3) and (4.4), we get, for $n = 1, 2, \ldots, c$,

$$\begin{aligned} p_n &= \frac{\lambda\lambda\cdots\lambda}{(\mu)(2\mu)\cdots(n\mu)}\, p_0 = \frac{(\lambda/\mu)^n}{n!}\, p_0, \\ &= \frac{\lambda}{n\mu}\, p_{n-1} \end{aligned} \tag{6.1}$$

and for $n = c, c + 1, c + 2, \ldots,$

$$\begin{aligned} p_n &= \frac{(\lambda)(\lambda)\cdots(\text{to } n \text{ factors})}{[(\mu)(2\mu)\cdots(c\mu)][(c\mu)(c\mu)\cdots(\text{to } (n-c) \text{ factors})]}\, p_0 \\ &= \frac{\lambda^n}{c!\,\mu^c c^{n-c} \mu^{n-c}}\, p_0 = \frac{(\lambda/\mu)^n}{c!\,c^{n-c}}\, p_0 \\ &= \frac{\lambda}{c\mu}\, p_{n-1} = \rho^{n-c} p_c. \end{aligned} \tag{6.2}$$

The normalizing condition $\sum_{n=0}^{\infty} p_n = 1$ yields

$$p_0^{-1} = 1 + \sum_{n=1}^{c-1} \frac{\left(\frac{\lambda}{\mu}\right)^n}{n!} + \sum_{n=c}^{\infty} \frac{\left(\frac{\lambda}{\mu}\right)^n}{c!c^{n-c}} = \sum_{n=0}^{c-1} \frac{\left(\frac{\lambda}{\mu}\right)^n}{n!} + \frac{1}{c!c^{-c}} \sum_{n=c}^{\infty} \left(\frac{\lambda}{c\mu}\right)^n. \tag{6.3}$$

In order that steady-state solutions exist, the series $\sum_{n=c}^{\infty} (\lambda/c\mu)^n$ must be convergent; this, in turn, implies that ρ is less than 1. Thus, when $\rho < 1$

$$p_0 = \left[\sum_{n=0}^{c-1} \frac{(\lambda/\mu)^n}{n!} + \frac{(\lambda/\mu)^c}{c!(1 - \lambda/c\mu)}\right]^{-1} \tag{6.3a}$$

and the steady-state distribution is given by (6.1) and (6.2) with p_0 given by (6.3a).

Notes.

(1) The steady-state probabilities p_n satisfy the recurrence relations:

$$\begin{aligned} p_n &= \frac{\lambda}{n\mu} p_{n-1} = \frac{c}{n} \rho p_{n-1}, \qquad n = 1, 2, \ldots, c-1 \\ &= \frac{\lambda}{c\mu} p_{n-1} = \rho p_{n-1}, \qquad n = c, c+1 \ldots. \end{aligned}$$

Thus, for $n < c$, $p_n/p_{n-1} > 1$ if $(n/c) < \rho < 1$; in this case p_n is monotone increasing in n until n exceeds $c\rho$ and then is monotone decreasing until $n = c$.

For $n > c$, p_n is monotone decreasing in n. The mode of the distribution is n for which $n < c\rho \le n + 1$, and also $n + 1$, when $c\rho$ is an integer.

(2) For finite n, $\{p_n\}$ behaves like a Poisson distribution for $n \le c$ and like a geometric distribution for $n > c$.

(3) The probability that an arriving unit has to wait on arrival is given by

$$\begin{aligned} C \equiv C\left(c, \frac{\lambda}{\mu}\right) &= Pr(N \ge c) = \sum_{n=c}^{\infty} p_n \\ &= \frac{\left(\frac{\lambda}{\mu}\right)^c}{c!(1-\rho)} p_0 = \frac{p_c}{1-\rho}. \end{aligned} \tag{6.4}$$

This is known as Erlang's C formula (second formula). Tables for different values of c and (λ/μ) are available (Descloux, 1962).

(4) The $M/M/\infty$ results are useful as an approximation for an $M/M/c$ queue when $\rho < 1$.

3.6.2. Expected Number of Busy and Idle Servers

The expected number of busy servers $E(B)$ is given by

$$
\begin{aligned}
E(B) &= \sum_{n=0}^{c-1} np_n + \sum_{n=c}^{\infty} cp_n \\
&= \left[\sum_{n=0}^{c-1} \frac{n\left(\frac{\lambda}{\mu}\right)^n}{n!} + \frac{c\left(\frac{\lambda}{\mu}\right)^c}{c!(1-\rho)}\right] p_0 \\
&= \frac{\lambda}{\mu}\left[\sum_{n=1}^{c-1} \frac{\left(\frac{\lambda}{\mu}\right)^{n-1}}{(n-1)!} + \frac{\left(\frac{\lambda}{\mu}\right)^{c-1}}{(c-1)!(1-\rho)}\right] p_0 \\
&= \frac{\lambda}{\mu}\left[\sum_{m=0}^{c-2} \frac{\left(\frac{\lambda}{\mu}\right)^{m}}{m!} + \frac{\{(1-\rho)+\rho\}\left(\frac{\lambda}{\mu}\right)^{c-1}}{(c-1)!(1-\rho)}\right] p_0 \\
&= \frac{\lambda}{\mu}\left[\sum_{m=0}^{c-2} \frac{\left(\frac{\lambda}{\mu}\right)^{m}}{m!} + \frac{\left(\frac{\lambda}{\mu}\right)^{c-1}}{(c-1)!} + \frac{\rho\left(\frac{\lambda}{\mu}\right)^{c-1}}{(c-1)!(1-\rho)}\right] p_0 \\
&= \frac{\lambda}{\mu}\left[\sum_{m=0}^{c-1} \frac{\left(\frac{\lambda}{\mu}\right)^{m}}{m!} + \frac{\left(\frac{\lambda}{\mu}\right)^{c}}{c!(1-\rho)}\right] p_0 \\
&= \frac{\lambda}{\mu} p_0^{-1} p_0 = \frac{\lambda}{\mu} = c\rho. \qquad (6.5)
\end{aligned}
$$

Hence, the expected number of idle servers $E(I)$ is given by

$$
\begin{aligned}
E(I) &= E(c-B) = E(c) - E(B) \\
&= c - c\rho = c(1-\rho). \qquad (6.6)
\end{aligned}
$$

3.6.2.1. Expected Number in the System $E(N)$

$$E(N) = E(B) + E(Q)$$

where $E(Q)$ is the expected number in queue. We have

$$
\begin{aligned}
E(Q) &= \sum_{n=c}^{\infty} (n-c)p_n \\
&= \sum_{n=c}^{\infty} (n-c) \frac{\left(\frac{\lambda}{\mu}\right)^n}{c!c^{n-c}} p_0 \\
&= \frac{\left(\frac{\lambda}{\mu}\right)^c}{c!} \sum_{n=c}^{\infty} (n-c)\left(\frac{\lambda}{c\mu}\right)^{n-c} p_0 \\
&= \frac{\left(\frac{\lambda}{\mu}\right)^c}{c!} \frac{\lambda}{c\mu} \sum_{m=0}^{\infty} m\left(\frac{\lambda}{c\mu}\right)^{m-1} p_0 \qquad (6.7) \\
&= \frac{\left(\frac{\lambda}{\mu}\right)^c}{c!} \frac{\lambda}{c\mu} p_0 \frac{1}{\left(1-\frac{\lambda}{c\mu}\right)^2} \\
&= \frac{\left(\frac{\lambda}{\mu}\right)^c}{c!} p_0 \frac{\rho}{(1-\rho)^2} \\
&= \frac{\rho p_c}{(1-\rho)^2} = \frac{\rho}{(1-\rho)} Pr(N \geq c).
\end{aligned}
$$

Thus,

$$
\begin{aligned}
E(N) &= E(B) + E(Q) \\
&= c\rho + \rho \frac{p_c}{(1-\rho)^2} \qquad (6.8) \\
&= c\rho + \frac{\rho C}{1-\rho}, \quad \text{where} \quad C = Pr\{N \geq c\}.
\end{aligned}
$$

Using Little's formula, we can find $E(W_Q)$, expected waiting time in the queue, and $E(W)$, expected waiting time in the system (sojourn or response time). We have

$$E(W_Q) = \frac{E(Q)}{\lambda} = \frac{p_c}{c\mu(1-\rho)^2} = \frac{1}{c\mu(1-\rho)} Pr(N \geq c) \quad \text{and} \tag{6.9}$$

$$E(W) = \frac{E(N)}{\lambda} = \frac{1}{\mu} + \frac{p_c}{c\mu(1-\rho)^2}. \tag{6.10}$$

Note. The relations (6.5) and (6.6) are fairly general and hold for any $G/G/c$ queue with $\rho = \lambda/c\mu < 1$. It can be seen simply as follows. We have

$$L = \lambda W \quad \text{and}$$

$$L_Q = \lambda W_Q.$$

Subtracting, we get $L - L_Q = \lambda(W - W_Q)$. The left-hand side gives the average number in the service channels or the average number of busy channels $E(B)$; $(W - W_Q)$ gives the average time spent in service and so equals $1/\mu$. Thus,

$$E(B) = \frac{\lambda}{\mu} = c\rho.$$

(For a direct proof, see Harris, 1974.)

3.6.3. *Waiting-Time Distributions*

We can find the waiting-time distributions for the $M/M/c$ queue using the same arguments as were used for the $M/M/1$ system.

Let $w_q(x)$ and $w(x)$ be the PDF of the waiting time W_q and W_s in the queue and in the system, respectively, of the test unit, and let $w_q^*(s)$ and $w^*(s)$ be their LTs. Further, let $w_q^*(s|n)$ and $w^*(s|n)$ be the LTs of the PDF of the conditional distributions of the respective waiting times given that the test unit finds, on arrival, n in the system. We obtain $w^*(s)$ by conditioning on the number of units the test unit finds on arrival. If the test unit finds, on arrival, $n < c$ units, he does not have to wait and his waiting time in the system equals his service time, that is,

$$w^*(s|n) = \frac{\mu}{s + \mu}, \qquad \text{for } n < c. \tag{6.11}$$

If he finds $n \geq c$ units in the system, he has to wait in the queue until the completion of service of $(n - c + 1)$ units; all the c service channels being busy then, the rate of service is $c\mu$. Taking into consideration his own service time, he thus has to wait in the system for completion of $(n - c + 1)$ services at the rate $c\mu$ and his own service at the rate μ, that is,

$$w^*(s|n) = \left(\frac{c\mu}{s + c\mu}\right)^{n-c+1}\left(\frac{\mu}{s+\mu}\right), \quad n \geq c. \tag{6.12}$$

Further, $a_n = p_n$ for this Poisson arrival process. Thus,

$$\begin{aligned} w^*(s) &= \sum_{n=0}^{c-1} w^*(s|n)p_n + \sum_{n=c}^{\infty} w^*(s|n)p_n \\ &= \frac{\mu}{s+\mu}\left[\sum_{n=0}^{c-1} p_n + \sum_{n=c}^{\infty} p_n\left(\frac{c\mu}{s+c\mu}\right)^{n-c+1}\right]. \end{aligned} \tag{6.13}$$

Putting in the values of p_n, we get

$$\begin{aligned} w^*(s) &= \frac{\mu}{s+\mu}\left[\sum_{n=0}^{c-1}\frac{\left(\frac{\lambda}{\mu}\right)^n}{n!} + \sum_{n=c}^{\infty}\frac{\left(\frac{\lambda}{\mu}\right)^n}{c!c^{n-c}}\left(\frac{c\mu}{s+c\mu}\right)^{n-c+1}\right]p_0 \\ &= \frac{\mu}{s+\mu}\left[\sum_{n=0}^{c-1}\frac{\left(\frac{\lambda}{\mu}\right)^n}{n!} + \frac{\left(\frac{\lambda}{\mu}\right)^c}{c!}\left(\frac{c\mu}{s+c\mu}\right)\sum_{r=0}^{\infty}\left(\frac{\lambda}{c\mu}\right)^r\left(\frac{c\mu}{s+c\mu}\right)^r\right]p_0 \\ &= \frac{\mu}{s+\mu}\left[\sum_{n=0}^{c-1}\frac{\left(\frac{\lambda}{\mu}\right)^n}{n!} + \frac{\left(\frac{\lambda}{\mu}\right)^c}{c!}\frac{c\mu}{(s+c\mu)}\frac{(s+c\mu)}{(s+c\mu-\lambda)}\right]p_0 \\ &= \frac{\mu}{s+\mu}\left[\sum_{n=0}^{c-1}\frac{\left(\frac{\lambda}{\mu}\right)^n}{n!} + \frac{\left(\frac{\lambda}{\mu}\right)^c}{c!}\frac{c\mu}{s+c\mu-\lambda}\right]p_0. \end{aligned} \tag{6.14}$$

We get $w(x)$ by inverting $w^*(s)$. From (6.14) we get

$$\begin{aligned} w^*(s) &= \sum_{n=0}^{c-1}\frac{\left(\frac{\lambda}{\mu}\right)^n}{n!}p_0\frac{\mu}{s+\mu} + \frac{\left(\frac{\lambda}{\mu}\right)^c}{c!}p_0\left(\frac{\mu}{s+\mu}\right)\frac{c\mu}{s+c\mu-\lambda} \\ &= \sum_{n=0}^{c-1}\frac{\left(\frac{\lambda}{\mu}\right)^n}{n!}p_0\frac{\mu}{s+\mu} + \frac{\left(\frac{\lambda}{\mu}\right)^c}{c!}p_0\frac{c\mu^2}{(c-1)\mu-\lambda}\left[\frac{1}{s+\mu} - \frac{1}{s+c\mu-\lambda}\right]. \end{aligned}$$

Inverting the transform, we get

$$w(x) = \sum_{n=0}^{c-1} \frac{\left(\frac{\lambda}{\mu}\right)^n}{n!} p_0 \mu e^{-\mu x} + \frac{\left(\frac{\lambda}{\mu}\right)^c}{c!} p_0 \frac{c\mu^2}{(c-1)\mu - \lambda} \times [e^{-\mu x} - e^{-(1-\rho)c\mu x}].$$

For $c = 1$,

$$w(x) = \mu p_0 e^{-\mu x} + \frac{\left(\frac{\lambda}{\mu}\right) p_0 \mu^2}{-\lambda} [e^{-\mu x} - e^{-(1-\rho)\mu x}]$$
$$= \mu(1-\rho)e^{-(1-\rho)\mu x}.$$

Since $w^*(s) = w_q^*(s)[\mu/(s+\mu)]$, we get from (6.13)

$$w_q^*(s) = \sum_{n=0}^{c-1} p_n + \sum_{n=c}^{\infty} p_n \left(\frac{c\mu}{s + c\mu}\right)^{n-c+1}$$
$$= \left[\sum_{n=0}^{c-1} \frac{\left(\frac{\lambda}{\mu}\right)^n}{n!} + \frac{\left(\frac{\lambda}{\mu}\right)^c}{c!} \frac{c\mu}{s + c\mu - \lambda}\right] p_0. \tag{6.15}$$

Inverting (6.15) we get the PDF

$$w_q(x) = \left(\sum_{n=0}^{c-1} p_n\right)\delta(x) + \sum_{n=c}^{\infty} p_n \frac{c\mu(c\mu x)^{n-c} e^{-c\mu x}}{(n-c)!}, \qquad x \geq 0 \tag{6.16}$$

where δ is Dirac delta (or unit impulse) function. Putting in the expressions for p_n and simplifying, we get

$$w_q(x) = \left(1 - \frac{p_c}{1-\rho}\right)\delta(x) + c\mu p_c e^{-c\mu(1-\rho)x}. \tag{6.17}$$

3.6.3.1. Complementary Distribution Function. We have, for $t \geq 0$,

$$Pr\{W_q > t\} = \int_t^{\infty} w_q(x)dx$$
$$= \frac{\left(\frac{\lambda}{\mu}\right)^c}{(c-1)!} \mu \frac{e^{-c(\mu - \lambda)t}}{c(\mu - \lambda)} p_0 \tag{6.18}$$
$$= C\left(c, \frac{\lambda}{\mu}\right) e^{-(1-\rho)c\mu t}$$

where $C(c, \lambda/\mu) = Pr(W_q > 0)$ is the blocking probability, given by (6.4). Further,

$$Pr\{W_q > t | W_q > 0\} = \frac{Pr\{W_q > t\}}{Pr\{W_q > 0\}} = e^{-(1-\rho)c\mu t}. \tag{6.19}$$

The distribution of the conditional waiting time in the queue, given that the test unit has to wait, is exponential with mean

$$E\{W_q | W_q > 0\} = \frac{1}{(1-\rho)c\mu}. \tag{6.20}$$

Values of $E\{W_q\}$ and $E\{W_q | W_q > 0\}$ for different values of c, λ, and μ are tabulated in Descloux (1962).

3.6.3.2. *Expected Waiting Time.* We have

$$\begin{aligned}
E(W_s) &= -\frac{d}{ds} w^*(s)\Big|_{s=0} \\
&= \frac{1}{\mu}\left[\sum_{n=0}^{c-1} \frac{\left(\frac{\lambda}{\mu}\right)^n}{n!} p_0 + \frac{\left(\frac{\lambda}{\mu}\right)^c}{c!} \frac{c\mu}{c\mu - \lambda} p_0\right] + \frac{\left(\frac{\lambda}{\mu}\right)^c}{c!} \frac{c\mu}{(c\mu - \lambda)^2} p_0 \\
&= \frac{1}{\mu}\left[\sum_{n=0}^{c-1} \frac{\left(\frac{\lambda}{\mu}\right)^n}{n!} + \frac{\left(\frac{\lambda}{\mu}\right)^c}{c!} \frac{1}{1-\rho}\right] p_0 + \frac{\left(\frac{\lambda}{\mu}\right)^c}{c!(c\mu)} \frac{1}{(1-\rho)^2} p_0.
\end{aligned}$$

Thus,

$$E(W_s) = \frac{1}{\mu} + \frac{\left(\frac{\lambda}{\mu}\right)^c}{c!c\mu} \frac{1}{(1-\rho)^2} p_0 = \frac{1}{\mu} + \frac{p_c}{c\mu(1-\rho)^2}. \tag{6.21a}$$

$$= \frac{1}{\mu} + \frac{C}{c\mu(1-\rho)}. \tag{6.21b}$$

It follows that

$$E(W_q) = \frac{\left(\frac{\lambda}{\mu}\right)^c}{c!c\mu} \frac{1}{(1-\rho)^2} p_0 = \frac{p_c}{c\mu} \frac{1}{(1-\rho)^2}. \tag{6.21c}$$

Similarly, we have

$$
\begin{aligned}
E(W_s^2) &= \frac{d^2}{ds^2} w^*(s)\Big|_{s=0} \\
&= \frac{2}{\mu^2}\left[\left\{\sum_{n=0}^{c-1} \frac{\left(\frac{\lambda}{\mu}\right)^n}{n!} + \frac{\left(\frac{\lambda}{\mu}\right)^c}{c!} \frac{c\mu}{c\mu - \lambda}\right\} p_0\right] \\
&\quad + \frac{\left(\frac{\lambda}{\mu}\right)^c}{c!} \frac{2c\mu}{(c\mu - \lambda)^3} p_0 \\
&= \frac{2}{\mu^2} + \left(\frac{2\left(\frac{\lambda}{\mu}\right)^c}{c!(c\mu)^2} \frac{p_0}{(1-\rho)^3}\right) \\
&= \frac{2}{\mu^2} + \frac{2p_c}{(c\mu)^2(1-\rho)^3}.
\end{aligned}
\tag{6.22}
$$

Putting $c = 1$, we get the corresponding results for an $M/M/1$ queue. In particular, we have, for the $M/M/1$ queue,

$$
E(W_s^2) = \frac{2}{\mu^2} + \frac{2\left(\frac{\lambda}{\mu}\right)}{\mu^2(1-\rho)^2}. \tag{6.23}
$$

Note. *Convexity properties* of performance measures (which have been found useful in the analysis of optimization problems (see Section 8.4)) have been receiving increasing attention. Grassman (1983) shows that $E(N)$ is convex for fixed μ and $c \geq 2$; Harel and Zipkin (1987) show that $f_c(\rho) = 1/E(W_s)$ is strictly concave in ρ for fixed μ and $c \geq 2$, while $f_1(\rho) = \mu(1-\rho)$ is linear in ρ (for $c = 1$).

3.6.4. The Output Process

Here we consider the output process of an $M/M/c$ queueing system. The result is the same as that of an $M/M/1$ system and the proof is similar.

Theorem 3.2. *In an $M/M/c$ queueing system in steady state with rates of arrival and service λ and μ, respectively, the interdeparture times are in-*

dependently and identically distributed as an exponential random variable with mean $1/\lambda$, i.e., the output process is Poisson with parameter λ.

Proof. Let $N(t)$ be the number in the system at time t and let $t'_1, t'_2, \ldots$ denote the successive departure instants, so that $L = t'_{n+1} - t'_n$ is the nth interdeparture interval. Let

$$F_k(t) = Pr\{N(t'_n + t) = k, \quad L > t\}, \quad t > 0, \quad k = 0, 1, 2, \ldots. \tag{6.24}$$

We have, because of Poisson input,

$$\begin{aligned} F_k(0) = p_k &= \frac{\left(\frac{\lambda}{\mu}\right)^k}{k!} p_0, \qquad 0 \le k \le c, \\ &= \frac{\left(\frac{\lambda}{\mu}\right)^k}{c!c^{k-c}} p_0, \qquad k \ge c. \end{aligned} \tag{6.25}$$

For an infinitesimal interval of length dt

$$\begin{aligned} F_0(t + dt) &= F_0(t)[1 - \lambda\, dt] + o(dt) \quad \text{and} \\ F_k(t + dt) &= F_k(t)[1 - \lambda\, dt - j\mu\, dt] + F_{k-1}(t)\lambda\, dt + o(dt) \end{aligned} \tag{6.26}$$

where $j = k$ for $k < c$ and $j = c$ for $k \ge c$; the equations reduce to

$$\begin{aligned} F'_0(t) &= -\lambda F_0(t) \quad \text{and} \\ F'_k(t) &= \lambda F_{k-1}(t) - (\lambda + j\mu)F_k(t), \qquad k = 1, 2, 3, \ldots. \end{aligned} \tag{6.27}$$

Subject to the initial condition $F_k(0) = p_k$, the unique solution of (6.27) is given by

$$F_k(t) = p_k e^{-\lambda t}. \tag{6.28}$$

Again,

$$\begin{aligned} &Pr\{N(t'_{n+1} + 0) = k, \qquad t \le t'_{n+1} - t'_n < t + dt\} \\ &\qquad = F_{k+1}(t)\, Pr\{\text{one service completion in } (t, t + dt)\}. \end{aligned} \tag{6.29}$$

For $k + 1 \le c$, $(k + 1)$ $(\le c)$ channels are busy and the rate of service of $(k + 1)\mu$, while for $k + 1 > c$, all the c channels are busy and the rate of service is $c\mu$. Thus, the RHS of (6.29) reduces to

$$F_{k+1}(t)(k + 1)\mu\, dt + o(dt), \qquad k + 1 \le c, \quad \text{and}$$

$$F_{k+1}(t)c\mu\, dt + o(dt), \qquad k + 1 > c.$$

Now for $k + 1 \le c$.

$$
\begin{aligned}
F_{k+1}(t)[(k+1)\mu\, dt] + o(dt) &= p_{k+1} e^{-\lambda t}[(k+1)\mu\, dt] + o(dt) \\
&= \frac{\left(\dfrac{\lambda}{\mu}\right)^{k+1}}{(k+1)!} p_0 e^{-\lambda t}(k+1)\mu\, dt + o(dt) \\
&= \frac{\left(\dfrac{\lambda}{\mu}\right)^{k}}{k!} p_0 \lambda e^{-\lambda t} + o(dt) \\
&= p_k \lambda e^{-\lambda t} + o(dt),
\end{aligned}
$$

and for $k + 1 > c$

$$
\begin{aligned}
F_{k+1}(t) c\mu\, dt + o(dt) &= p_{k+1} e^{-\lambda t}[c\mu\, dt] + o(dt) \\
&= \frac{\left(\dfrac{\lambda}{\mu}\right)^{k+1}}{c!c^{k+1-c}} p_0 e^{-\lambda t} c\mu\, dt + o(dt) \\
&= \frac{\left(\dfrac{\lambda}{\mu}\right)^{k}}{c!c^{k-c}} p_0 \lambda e^{-\lambda t} + o(dt) \\
&= p_k \lambda e^{-\lambda t} + o(dt).
\end{aligned}
$$

Thus,

$$
\begin{aligned}
Pr\{N(t'_{n+1} + 0) = k, \quad & t \le t'_{n+1} - t'_n < t + dt\} \\
&= p_k \lambda e^{-\lambda t}\, dt + o(dt),
\end{aligned}
\tag{6.30}
$$

which proves the independence of $N(t'_{n+1} + 0)$ and $t'_{n+1} - t'_n$, and it follows that $L = t'_{n+1} - t'_n$ has the density $\lambda e^{-\lambda t}$.

The proof of the independence of the interdeparture intervals is the same as that given for the $M/M/1$ queueing output process (given in Section 3.2.3).

Notes.

(1) The theorem is known as Burke's theorem. Burke also shows that the output process is independent of all other processes associated with the system and that of all the systems with FCFS service discipline, $M/M/c$ is the only system to possess this property.

(2) For an alternative proof of Burke's theorem and for generalizations, see Reich (1957, 1965).

3.7. The $M/M/c/c$ Model: Loss System

Consider a c-server model with Poisson input and exponential service time such that when all the c-channels are busy an arrival leaves the system without waiting for service. This is called a (c-channel) *loss system* and was first investigated by Erlang.

This is a birth and death queueing model with

$$\begin{aligned} \lambda_n &= \lambda, & \mu_n &= n\mu, & n &= 0, 1, 2, \ldots, c-1 \\ \lambda_n &= 0, & \mu_n &= c\mu, & n &\geq c. \end{aligned} \tag{7.1}$$

Using (4.3) and (4.4) we get

$$\begin{aligned} p_n &= \frac{\left(\frac{\lambda}{\mu}\right)^n}{n!} p_0, \qquad n = 1, \ldots, c \\ &= 0, \qquad n > c \end{aligned}$$

and

$$p_0 = \left[\sum_{k=0}^{c} \frac{\left(\frac{\lambda}{\mu}\right)^k}{k!}\right]^{-1}.$$

Thus,

$$p_n = \frac{\left(\frac{\lambda}{\mu}\right)^n \Big/ n!}{\sum_{k=0}^{c} \left(\frac{\lambda}{\mu}\right)^k \Big/ k!}, \qquad n = 0, 1, 2, \ldots, c. \tag{7.2}$$

The above formula is known as Erlang's first (or delay) formula. An arriving unit is lost to the system, when he finds, on arrival, that all the channels are busy: the probability of this event is

$$p_c = \frac{\left(\frac{\lambda}{\mu}\right)^c \Big/ c!}{\sum_{k=0}^{c} \left(\frac{\lambda}{\mu}\right)^k \Big/ k!}. \tag{7.3}$$

The above formula is known as *Erlang's loss* (or blocking, or overflow) *formula* or B-formula and is denoted by $B(c, \lambda/\mu)$.

Notes.

(1) The above formulas hold irrespective of the form of service time distribution (see Section 6.8.3). Some properties of Erlang's loss function are obtained by Jagerman (1974).
(2) Newell (1984) discusses asymptotic approximations for Erlang's loss formula when the number of servers as well as the offered load are large.
(3) Messerli (1972) shows that Erlang's blocking probability [as given by B-formula (7.3)] is decreasing convex in c, the number of servers. This implies that the overflow or loss can be reduced by adding extra servers; however, the marginal decrease is itself decreasing. Yao (1986) considers a more general loss system $G/M/m/m$ (with general interarrival time, ordered-entry, and heterogeneous servers), and examines convexity properties of the overflow.

Example 3.1. Expected number of busy channels. Let B be the RV denoting the number of busy channels. We have

$$
\begin{aligned}
E\{B\} &= \sum_{n=1}^{c} np_n = \sum_{n=1}^{c} \frac{n\left(\dfrac{\lambda}{\mu}\right)^n}{n!} p_0 \\
&= \left(\frac{\lambda}{\mu}\right) p_0 \sum_{n=1}^{c} \frac{\left(\dfrac{\lambda}{\mu}\right)^{n-1}}{(n-1)!} = \left(\frac{\lambda}{\mu}\right) p_0 \left[\sum_{n=0}^{c} \frac{\left(\dfrac{\lambda}{\mu}\right)^n}{n!} - \frac{\left(\dfrac{\lambda}{\mu}\right)^c}{c!}\right] \\
&= \frac{\lambda}{\mu}[1 - p_c] = \frac{\lambda}{\mu}\left[1 - B\left(c, \frac{\lambda}{\mu}\right)\right].
\end{aligned}
\tag{7.4}
$$

If I is the RV denoting the number of idle channels, then

$$
\begin{aligned}
E\{I\} &= E\{c - B\} = c - E(B) \\
&= c - \frac{\lambda}{\mu}(1 - p_c) \qquad (7.5) \\
&= c - \frac{\lambda}{\mu}\left[1 - B\left(c, \frac{\lambda}{\mu}\right)\right].
\end{aligned}
$$

Example 3.2. Busy probability (Kaufman, 1979). Let X_i be the indicator variable for the ith randomly chosen channel: $X_i = 1$ or 0 according to whether the ith channel is busy or free. Let $P_c\{A\}$ denote the probability of an event A in an equilibrium $M/M/c/c$ loss system. Then

$$\text{(i)}\quad P_c\{X_1 = 1, \ldots, X_k = 1\} = \frac{B\left(c, \dfrac{\lambda}{\mu}\right)}{B\left(c-k, \dfrac{\lambda}{\mu}\right)}, \qquad 1 \le k \le c$$

$$\text{(ii)}\quad P_c\{X_1 = 1\} = \frac{\left(\dfrac{\lambda}{\mu}\right)\left[1 - B\left(c, \dfrac{\lambda}{\mu}\right)\right]}{c}, \quad \text{and}$$

$$\text{(iii)}\quad P_c\{X_{k+1} = 1 \mid X_1 = 1, X_2 = 1, \ldots, X_k = 1\} = P_{c-k}\{X_1 = 1\}.$$

Proof (i). Conditioning on the number of busy channels, we get

$$\begin{aligned}
P_c\{X_1 = 1, \ldots, X_k = 1\} &= \sum_{j=k}^{c} Pr\{X_1 = 1, \ldots, X_k = 1 \mid n = j\} p_j \\
&= \sum_{j=k}^{c} \frac{\binom{j}{k}}{\binom{c}{k}} p_0 \frac{\left(\dfrac{\lambda}{\mu}\right)^j}{j!} = \frac{(c-k)!}{c!} p_0 \sum_{j=k}^{c} \frac{\left(\dfrac{\lambda}{\mu}\right)^j}{(j-k)!} \\
&= \frac{(c-k)!}{c!} \left(\frac{\lambda}{\mu}\right)^{k-c} \sum_{j=k}^{c} \left[\left(\frac{\lambda}{\mu}\right)^c p_0\right] \frac{\left(\dfrac{\lambda}{\mu}\right)^{j-k}}{(j-k)!} \qquad (7.6) \\
&= \left[\frac{\left(\dfrac{\lambda}{\mu}\right)^c}{c!} p_0\right] \left[\frac{(c-k)!}{\left(\dfrac{\lambda}{\mu}\right)^{c-k}} \sum_{r=0}^{c-k} \frac{\left(\dfrac{\lambda}{\mu}\right)^r}{r!}\right] \\
&= \frac{Pr\{c \text{ channels busy in } M/M/c/c\}}{Pr\{c-k \text{ channels busy in } M/M/c-k/c-k\}} \\
&= \frac{B\left(c, \dfrac{\lambda}{\mu}\right)}{B\left(c-k, \dfrac{\lambda}{\mu}\right)}.
\end{aligned}$$

Proof (ii).

$$P_c\{X_1 = 1\} = \sum_{j=1}^{c} P_c\{X_1 = 1 | B = j\} Pr\{B = j\}$$

($B \equiv$ number of busy channels)

$$= \sum_{j=1}^{c} \frac{j}{c} p_j = \frac{1}{c} \sum_{j=1}^{c} j p_j$$

$$= \frac{1}{c} E(B)$$

$$= \frac{\lambda/\mu}{c}\left[1 - B\left(c, \frac{\lambda}{\mu}\right)\right] \quad \text{by Eq. (7.4).} \tag{7.7}$$

Proof (iii). Again,

$$P_c\{X_{k+1} = 1 | X_1 = 1, X_2 = 1, \ldots, X_k = 1\}$$

$$= \frac{P_c\{X_1 = 1, X_2 = 1, \ldots, X_k = 1, X_{k+1} = 1\}}{P_c\{X_1 = 1, X_2 = 1, \ldots, X_k = 1\}}$$

$$= \frac{B\left(c, \frac{\lambda}{\mu}\right)}{B\left(c-k-1, \frac{\lambda}{\mu}\right)} \frac{B\left(c-k, \frac{\lambda}{\mu}\right)}{B\left(c, \frac{\lambda}{\mu}\right)} = \frac{B\left(c-k, \frac{\lambda}{\mu}\right)}{\mathrm{B}\left(\mathrm{c}-k-1, \frac{\lambda}{\mu}\right)}$$

$$= \frac{\left[\left(\frac{\lambda}{\mu}\right)^{c-k} (c-k)!\right]\left[\sum_{r=0}^{c-k}\left(\frac{\lambda}{\mu}\right)^r \Big/ r!\right]^{-1}}{\left[\left(\frac{\lambda}{\mu}\right)^{c-k-1} \Big/ (c-k-1)!\right]\left[\sum_{r=0}^{c-k-1}\left(\frac{\lambda}{\mu}\right)^r \Big/ r!\right]^{-1}} \tag{7.8}$$

$$= \frac{\frac{\lambda}{\mu}}{(c-k)}\left[1 - \frac{\left(\frac{\lambda}{\mu}\right)^{c-k} \Big/ (c-k)!}{\sum_{r=0}^{c-k}\left(\frac{\lambda}{\mu}\right)^r \Big/ r!}\right]$$

$$= \frac{\frac{\lambda}{\mu}}{c-k}\left[1 - B\left(c-k, \frac{\lambda}{\mu}\right)\right]$$

$$= P_{c-k}\{X_1 = 1\}.$$

Now (iii) implies that if $(k+1)$ channels in an $M/M/c/c$ system in equilibrium are randomly chosen without replacement and the first k channels are busy, then the conditional probability that the $(k+1)$st channel is busy equals the *a priori* probability that one randomly chosen channel in an $M/M/c-k/(c-k)$ system is busy.

Note. Kaufman shows that the results of Examples 3.1 and 3.2 also hold for the loss systems with Poisson input (with rate λ) and general service-time distribution (with mean rate μ).

3.8. Model with Finite Input Source

3.8.1. Steady-State Distribution

Our assumption so far has been that the source of population from which the arrivals occur is infinite. Now we shall examine the case of arrivals from a source with finite population, say, of size m. A unit may be in the system or outside the system. The system consists of a fixed number, say, of c parallel servers, where $c \le m$; a unit entering the system starts receiving service from one of the parallel servers, if there is any free server available, or joins the queue, if there is none. The service time distribution of each of the c servers is IID exponential with parameter μ; the total service rate when n servers are busy is $n\mu$, when $n \le c$ and $c\mu$, when $n \ge c$. If at any instant there are n in the system, (either receiving service or some in the queue while c are receiving service) then there are $(m-n)$ outside the system from which arrivals to the system occur, the average arrival rate being $\lambda(m-n)$; the distribution of interarrival time is IID exponential with parameter λ. We have state dependent arrival and service rates. The system may be denoted by $M/M/c/\ /m$ (Kleinrock, 1975). To fix our ideas, we can consider the following example. Suppose that there are m machines and c repair-persons (servers) to repair them, when required. A machine is in the system when it is in a failed (or non-working) state requiring repair facility of one of the c servers; when all the repair-persons are busy, the machine joins the queue. The time required to repair a machine is IID exponential with parameter μ. A machine in working order is outside the system, the time for breakdown (or life time) of a machine being exponential with parameter λ. The above situation can be appropriately described by a birth–death model with rates

$$\begin{aligned}\lambda_n &= (m-n)\lambda, \qquad n = 0, 1, \ldots, m-1\\ &= 0, \qquad n \geq m\end{aligned}$$

and

$$\begin{aligned}\mu_n &= n\mu, \qquad n = 1, 2, \ldots, c-1\\ &= c\mu, \qquad n \geq c.\end{aligned} \tag{8.1}$$

Assume that the system (the whole system) is in steady state. Denote

$$p_n = Pr\{\text{the number in the system is } n\}.$$

Clearly, $p_n = 0$ for $n > m$.

Using (4.3), we have, for $n = 0, 1, \ldots, c-1$,

$$\begin{aligned}p_n &= \prod_{i=0}^{n-1} \frac{\lambda_i}{\mu_{i+1}} p_0 = \prod_{i=0}^{n-1} \frac{(m-i)\lambda}{(i+1)\mu} p_0\\ &= \frac{(m\lambda)((m-1)\lambda)((m-2)\lambda)\cdots((m-n+1)\lambda)}{(\mu)(2\mu)\cdots(n\mu)} p_0\\ &= \frac{m(m-1)(m-2)\cdots(m-n+1)}{n!}\left(\frac{\lambda}{\mu}\right)^n p_0\\ &= \frac{m!}{n!(m-n)!}\left(\frac{\lambda}{\mu}\right)^n p_0\\ &= \binom{m}{n}\left(\frac{\lambda}{\mu}\right)^n p_0;\end{aligned} \tag{8.2}$$

and for $n = c, c+1, \ldots, m$,

$$\begin{aligned}p_n &= \prod_{i=0}^{n-1} \frac{(m-i)\lambda}{(i+1)\mu} p_0\\ &= \frac{m!}{(m-n)!} \cdot \frac{1}{c!c^{n-c}} \cdot \left(\frac{\lambda}{\mu}\right)^n p_0.\end{aligned} \tag{8.3}$$

Using the normalizing condition

$$\sum_{n=0}^{m} p_n = 1,$$

we get

$$p_0 = \left[\sum_{n=0}^{c-1} \binom{m}{n} \left(\frac{\lambda}{\mu}\right)^n + \sum_{n=c}^{m} \frac{m!}{(m-n)!} \cdot \frac{1}{c!c^{n-c}} \left(\frac{\lambda}{\mu}\right)^n \right]^{-1}. \tag{8.4}$$

Notes.

(1) In the context of the machine repairperson problem,

$$p_n = Pr\{\text{number of machine under or awaiting repair is } n\}$$

$$q_n = Pr\{\text{number of machines in working order is } n\} = p_{m-n}.$$

(2) When $\rho < 1$, then taking limit as $m \to \infty$, we get the results of the $M/M/c$ queue.

(3) For a system with a finite population of size, m, and infinite number of servers, we shall get

$$p_n = \frac{\binom{m}{n}\left(\frac{\lambda}{\mu}\right)^n}{\left[1 + \left(\frac{\lambda}{\mu}\right)\right]^m}, \qquad n = 0, 1, \ldots, m.$$

(4) Bunday and Scarton (1980) show that the results [(8.2), (8.3), (8.4)] hold for any system having finite input source and having exponential service time (with parameter μ) and general independent identical interarrival times or lifetimes (with parameter λ); i.e., for a $G/M/c$ model with finite input source of size m.

Example 3.3. An application of the M/M/N//N model in electronics (Koenigsberg, 1980). Semiconductor noise as a queueing problem was first formulated by Bell. Koenigsberg discusses cyclic queue models of semiconductor noise. One model considered by him is a finite-input-source model with N customers (electrons) and N servers (impurity levels), where N is large. Each server can serve one customer at a time and each customer in the queue can be served by one vacant or free server. The situation is similar to having N machines subject to breakdown and repairs by N repairpeople. Let

x = rate of electron withdrawal from the conduction band (rate of breakdown of working machines), which depends on the number of vacant impurity levels,

y = rate of electron excitation to the conduction band (rate of repair of broken-down machines), which depends on the number of busy impurity levels, and

$p_n = Pr\{n$ electrons are in the conduction band$\}$.

Then putting $m = c = N$, $\lambda = x$, $\mu = \lambda$ in (8.2) and (8.4) we have

$$p_n = \binom{N}{n}\left(\frac{x}{y}\right)^n p_0, \qquad n = 0, 1, \ldots, N-1, N,$$

$$p_0 = \left[\sum_{n=0}^{N}\binom{N}{n}\left(\frac{x}{y}\right)^n\right]^{-1} = \left(1 + \frac{x}{y}\right)^{-N}, \quad \text{and}$$

$$L = \sum_{n=0}^{N} np_n = \left(\frac{Nx}{y}\right)\left(1 + \frac{x}{y}\right)^{-1} = \frac{Nx}{x+y}.$$

Note. Some other models of electron excitation are considered by Koenigsberg. The second model is equivalent to that of the machine-interference problem, in which machines subject to breakdown can be repaired in the "transition zone" by one of the N repairpeople; however, at any time $(N - m)$ of the repairpeople are engaged in other duties and are not available for repairs. The other duties arise in accordance with a Poisson process of rate say, z, and completions take place in accordance with an independent Poisson process of rate, say, s.

The third model is equivalent to one in which the repairperson moves the machine from the point of breakdown to a repair facility where it is placed in service (excitation). Then the repairperson returns to the base, perhaps carrying out other duties en route.

3.8.2. Waiting-Time Distribution for an M/M/c//m Model (Wong, 1979)

Consider the exponential model described in the preceding section with c servers and input from a finite source of m units $(m > c)$. Wong (1979) has obtained the waiting time distribution for such a model.

Let a_n be the probability that an arrival finds n units in the system; in case of a machine-interference problem, a_n is the probability that n machines are *not* in working order. Here the input is quasi-random, being from a finite source. We can find a_n from (3.6) on replacing λ by $(m-n)\lambda$; then

$$a_n = \frac{(m-n)p_n}{\sum_{k=0}^{m}(m-k)p_k} = \frac{(m-n)p_n}{m-L}, \tag{8.5}$$

where $L = \sum kp_k$ is the average number of machines in the system that are not in working order and $m - L$ is the average number of machines in working order.

Define $w^*(s)$ to be the LST of the waiting time in the system (response time). The LST $w^*(s)$ can be obtained by conditioning on the number of units n that an arrival finds. Define $w^*(s|n)$ to be the LST of the conditional distribution of the waiting time in the system (response time) given that the test unit finds n in the system. Let $w_q^*(s)$ and $w_q^*(s|n)$ be the corresponding quantities for waiting time in the queue. We have

$$w^*(s) = \sum_{n=0}^{m-1} w^*(s|n)a_n. \tag{8.6}$$

Now

$$\begin{aligned} w^*(s|n) &= \frac{\mu}{s+\mu} \quad \text{for} \quad n < c \\ &= \left(\frac{c\mu}{s+c\mu}\right)^{n-c+1}\left(\frac{\mu}{s+\mu}\right) \quad \text{for} \quad c \le n \le m-1 \end{aligned} \tag{8.7}$$

so that from (8.6) we get

$$w^*(s) = \frac{\mu}{s+\mu}\left[\sum_{n=0}^{c-1} a_n + \sum_{n=c}^{m-1} a_n\left(\frac{c\mu}{s+c\mu}\right)^{n-c+1}\right] \tag{8.8}$$

where a_n is given by (8.5). It follows that

$$w_q^*(s) = \sum_{n=0}^{c-1} a_n + \sum_{n=c}^{m-1} a_n\left(\frac{c\mu}{s+c\mu}\right)^{n-c+1}. \tag{8.9}$$

Inverting the transform we get

$$w_q(x) = \left(\sum_{n=0}^{c-1} a_n\right)\delta(x) + \sum_{n=c}^{m-1} a_n \frac{(c\mu)(c\mu x)^{n-c}e^{-c\mu x}}{(n-c)!} \tag{8.10}$$

Inversion of $w^*(s)$ is done next. Before taking this up, we find the moments of W_s and W_q. We have

$$
\begin{aligned}
E\{W_s\} &= -\frac{d}{ds} w^*(s)\bigg|_{s=0} \\
&= \frac{1}{\mu} + \sum_{n=c}^{m-1} a_n \frac{n-c+1}{c\mu} \\
&= \frac{1}{\mu} + \sum_{n=c}^{m-1} \frac{(m-n)(n-c+1)}{c\mu(m-L)} p_n, \quad \text{and}
\end{aligned}
\tag{8.11}
$$

$$E\{W_q\} = E\{W_s\} - \frac{1}{\mu}.$$

3.8.2.1. Inversion of w*(s). The inversion can be carried in two ways: (i) as the inversion of the colvolution and (ii) as the inversion of terms obtained by resolution of the expression into partial fractions. The method given by Jordan (1950, Section 13c, p. 38) may be used for resolution into partial fractions. We consider the first method next.

Consider the inversion of

$$\frac{\mu}{s+\mu}\left(\frac{c\mu}{s+c\mu}\right)^{n-c+1}.$$

The inverse transform of the first factor

$$f_1^*(s) = \frac{\mu}{\mu + s}$$

is

$$f_1(x) = \mu e^{-\mu x} \tag{8.12}$$

and that of the second factor

$$f_2^*(s) = \left(\frac{c\mu}{s+c\mu}\right)^{n-c+1}$$

is

$$f_2(x) = (c\mu)^{n-c+1} \frac{x^{n-c} e^{-c\mu x}}{(n-c)!}. \tag{8.13}$$

Using these and noting that the inverse of $f_1^*(s)f_2^*(s)$ is the convolution of $f_1(x)$ and $f_2(x)$, we find that the inverse LT $L^{-1}[f_1^*(s)f_2^*(s))]$ of $f_1^*(s)f_2^*(s)$ is given by

$$L^{-1}[f_1^*(s)f_2^*(s)] = \mu(c\mu)^{n-c+1}\left[\int_0^x e^{-\mu(x-t)}\frac{t^{n-c}e^{-c\mu t}}{(n-c)!}\,dt\right]$$

$$= \frac{\mu(c\mu)^{n-c+1}}{(n-c)!}e^{-\mu x}\int_0^x t^{n-c}e^{-\mu(c-1)t}\,dt.$$

Using the representation of the incomplete gamma function

$$\int_0^x \frac{\alpha^k t^{k-1}e^{-\alpha t}}{\Gamma(k)} = 1 - \sum_{r=0}^{k-1} e^{-\alpha x}\frac{(\alpha x)^r}{r!} \tag{8.14}$$

we get

$$L^{-1}\left[\frac{\mu}{s+\mu}\left(\frac{c\mu}{s+c\mu}\right)^{n-c+1}\right]$$

$$= \frac{\mu(c\mu)^{n-c+1}e^{-\mu x}}{(n-c)![\mu(c-1)]^{n-c+1}}\left[1 - \sum_{r=0}^{n-c} e^{-\mu(c-1)x}\frac{[\mu(c-1)x]^r}{r!}\right]$$

$$= \mu\left(\frac{c}{c-1}\right)^{n-c+1} e^{-\mu x}\left[1 - e^{\mu x}\sum_{r=0}^{n-c} e^{-c\mu x}\frac{[\mu(c-1)x]^r}{r!}\right]. \tag{8.15}$$

Thus, we find from (8.8), (8.12), and (8.15) that the inverse Laplace transform $w(x)$ of $w^*(s)$ given by (8.8) can be written as

$$w(x) = \mu e^{-\mu x}\sum_{n=0}^{c-1} a_n + \sum_{n=c}^{m-1} a_n\left[\left(\frac{c}{c-1}\right)^{n-c+1}\mu e^{-\mu x}\right.$$

$$\left. -\mu\left(\frac{c}{c-1}\right)^{n-c+1}\sum_{r=0}^{n-c} e^{-\mu x}\frac{[\mu(c-1)x]^r}{r!}\right].$$

The expression for the distribution function $F(x) = \int_0^x w(t)dt$ can be written down.

Note. The model has been used to describe various important situations arising out of *machine-interference problems,* and to find solutions of design problems such as determination of the optimal number of repairpeople for a given set of machines, the maximum length of time during which all m machines are working and so on.

3.9. Transient Behavior

3.9.1. Introduction

Most of the known results in queueing theory pertain to steady state. This is because the equations involved become considerably simplified in the limit when time t from the initialization becomes very large. Analytically tractable results can then be obtained. Nevertheless, steady-state conditions do not hold well in many applied situations as the time horizon of operation terminates and remains finite. For example, the repairperson at a service facility leaves after a fixed duration of time; so also is the case of a bankteller. Steady-state analysis and performance measures obtained thereof do not then make much sense. Transient behavior is more meaningful under such circumstances: an analysis that deals with a system's operating behavior for a fixed, finite amount of time and takes into account the initial conditions is more relevant.

Transient results are, however, more difficult to obtain and not many such results are found in the literature. Morse (1955) studied the $M/M/1$ queue and obtained the transient state probabilities of the number in the system at time t. The problem was studied and complete solution obtained in the 1950s by several researchers using different methods (as indicated in the next section). Transient state behavior of $M/M/s$ queue was studied by Saaty (1960) and Jackson and Henderson (1966).

Note. In the 1980s, there has been a fresh interest in the transient behavior of $M/M/1$ queue as indicated by such papers as Abate and Whitt (1987, 1988), Hubbard *et al.* (1986), Parthasarathy (1987), Syski (1988), Pegden and Rosenshine (1982), Sharma and Gupta (1982), and Mohanty and Panny (1990) (who consider discrete time approach).

Similar interest has been evinced in case of $M/M/c$ queue also, see, for example, Van Doorn (1981, Chapter 6), Kelton and Law (1985) and Parthasarathy and Sharafali (1989).

While most authors consider continuous-time framework, Kelton and Law (1985) carry out their analysis of transient behavior in discrete time through indexing of customer number. (See Problems and Complements 3.19.).

We examine below transient state behavior through continuous-time framework.

3.9.1.1. Transient State Distribution for the M/M/1 Model. This single server model envisages Poisson input and exponential service time with FCFS queue discipline. The arrivals occur in accordance with a Poisson process with parameter (or intensity), say, λ, i.e., the probability of one arrival in an infinitesimal interval of length h is $\lambda h + o(h)$, while that of more than one arrival is $o(h)$. The distribution of service time is exponential with parameter, say, μ, i.e., the probability of one service completion in an interval of infinitesimal length h is $\mu h + o(h)$ and that of more than one service completion is $o(h)$. The model corresponds to that of a birth–death process with rates $\lambda_n = \lambda$, $n = 0, 1, 2, \ldots$ and $\mu_n = \mu$, $n = 1, 2, \ldots$. Denote

$$p_n(t) = Pr\{N(t) = n\}, \qquad n \geq 0$$

where $N(t)$ is the number in the system at time t. Using (4.5) and (4.6) of Chapter 1, it can be easily seen that $p_n(t)$ satisfies the differential–difference equations given below:

$$p_0'(t) = -\lambda p_0(t) + \mu p_1(t) \tag{9.1}$$

$$p_n'(t) = -(\lambda + \mu)p_n(t) + \lambda p_{n-1}(t) + \mu p_{n+1}(t), \qquad n \geq 1. \tag{9.2}$$

These equations in transient state have been solved by using a number of techniques: by the spectral method [Ledermann and Reuter (1954)], by the generating function method [Bailey (1954)], by the combinatorial method [Champernowne (1956)], by the difference equation solution technique (Conolly (1958), [Feller (1966)], and so on. We consider here two methods: the difference– equation technique and the method of generating function.

3.9.2. Difference–Equation Technique

Knowledge of the initial state is required for obtaining transient state distribution. For simplicity, suppose that the number of units present at $t = 0$ is 0, i.e., at time $t = 0$ the system is empty. This implies that $p_0(0) = 1$ and $p_n(0) = 0$, $n \neq 0$.

Let $\bar{p}_n(s)$ be the LT of $p_n(t)$. Then taking LT of (9.1) and (9.2) and using

$$L\{p_n'(t)\} = sL\{p_n(t)\} - p_n(0),$$

we get

$$(s + \lambda)\bar{p}_0(s) = 1 + \mu\bar{p}_1(s) \quad \text{and} \tag{9.3}$$

$$(s + \lambda + \mu)\bar{p}_n(s) = \lambda\bar{p}_{n-1}(s) + \mu\bar{p}_{n+1}(s), \qquad n \geq 1. \tag{9.4}$$

The equation (9.4) is a difference equation of order two having for its characteristic equation

$$\mu z^2 - (s + \lambda + \mu)z + \lambda = 0. \tag{9.5}$$

Let z_1, z_2 be the roots of (9.5):

$$z_i \equiv z_i(s) = \frac{(s + \lambda + \mu) \pm \sqrt{\{(s + \lambda + \mu)^2 - 4\lambda\mu\}}}{2\mu}$$

with $i = 1$ (with +ve sign before the radical), $i = 2$ (with −ve sign before the radical). We have

$$z_1 + z_2 = (s + \lambda + \mu)/\mu, \qquad z_1 z_2 = \lambda/\mu, \qquad |z_1| > |z_2|.$$

Further $|z_1| > 1$ and $|z_2| < 1$ as can be seen by applying:

Rouché's Theorem. *If $f(z)$ and $g(z)$ are functions analytic inside and on a closed contour C, and if $|g(z)| < |f(z)|$ on C, then $f(z)$ and $f(z) + g(z)$ have the same number of zeros inside C.*

Here we take C as the unit circle $|z| = 1$, $f(z) = (s + \lambda + \mu)z$, $g(z) = \lambda + \mu z^2$, $\text{Re}(s) > 0$.

It follows that

$$\begin{aligned} |f(z)| &= |(s + \lambda + \mu)z| = |s + \lambda + \mu| \\ &\geq |\lambda + \mu| = \lambda + \mu \\ &\geq |\lambda + \mu z^2| = |g(z)|. \end{aligned}$$

As $f(z)$ has only one zero inside $|z| = 1$, the equation

$$f(z) + g(z) = \mu z^2 - (s + \lambda + \mu)z + \lambda = 0$$

will also have only one root inside C.

Now the root z_2 being of smaller modulus must be such that $|z_2| < 1$. The solution of the difference equation (9.4) can be written as

$$\bar{p}_n(s) = A z_1^n + B z_2^n, \qquad n \geq 1. \tag{9.6}$$

Taking LT of

$$\sum_{n=0}^{\infty} p_n(t) = 1,$$

we get

$$\sum \bar{p}_n(s) = 1/s.$$

Thus,

$\sum \bar{p}_n(s) = \sum (Az_1^n + Bz_2^n)$ converges and since $|z_1| > 1$, A must be identically equal to 0. Thus,

$$\bar{p}_n(s) = Bz_2^n, \qquad n \geq 1.$$

We can choose B to yield the correct value of $\bar{p}_0(s)$ such that (9.4) is satisfied for $n = 1$; this gives $\bar{p}_0(s) = B$. Thus we have

$$\bar{p}_n(s) = \bar{p}_0(s) z_2^n, \qquad n \geq 0. \tag{9.7}$$

Using $\sum \bar{p}_n(s) = 1/s$, we get

$$\bar{p}_0(s) = (1 - z_2)/s$$

and

$$\bar{p}_n(s) = (1 - z_2) z_2^n / s, \qquad n \geq 0. \tag{9.7a}$$

3.9.2.1. Steady-State Distribution. When it exists, the steady-state probability p_n can be obtained by taking the limit of $p_n(t)$ as $t \to \infty$. We have

$$\lim_{s \to 0} z_2 = \lim_{s \to 0} \frac{(s + \lambda + \mu) - \sqrt{\{(s + \lambda + \mu)^2 - 4\lambda\mu\}}}{2\mu}$$

$$= \begin{cases} \dfrac{(\lambda + \mu) - (\mu - \lambda)}{2\mu} = \rho, & \text{when} \quad \lambda < \mu \\ \dfrac{(\lambda + \mu) - (\lambda - \mu)}{2\mu} = 1, & \text{when} \quad \lambda \geq \mu \end{cases}. \tag{9.8}$$

By applying the initial value theorem of Laplace transforms, we get

$$\begin{aligned} p_n &= \lim_{t \to \infty} p_n(t) = \lim_{s \to 0} s\bar{p}_n(s) \\ &= \lim_{s \to 0} (1 - z_2) z_2^n \\ &= \begin{cases} (1 - \rho)\rho^n & \rho < 1 \\ 0, & \rho \geq 1 \end{cases} \qquad n = 0, 1, 2, \ldots . \end{aligned} \tag{9.9}$$

The interpretation of the result $p_n = 0$, $\rho \geq 1$ is that when the traffic intensity is greater than or equal to 1, then the probability that the system contains a finite number of units n is zero, as should be intuitively clear (because of the increasing queue length). Note that when $\rho = 1$, the states are persistent null with infinite recurrence time.

3.9.2.2. Transient-State Distribution. Assume that $p_0(0) = 1$. We have from (9.7a)

$$
\begin{aligned}
\bar{p}_n(s) &= \frac{(1 - z_2)z_2^n}{s} \\
&= \frac{(1 - z_2)\, z_2^n}{\{\mu(z_1 - 1)(1 - z_2)\}}, \quad \text{since } -s = \mu(1 - z_1)(1 - z_2) \\
&= \left(\frac{z_2^{n+1}}{\lambda}\right)\left[1 + \sum_{r=0}^{\infty}\left(\frac{1}{z_1}\right)^{r+1}\right], \quad \text{since } |1/z_1| < 1 \\
&= \left(\frac{1}{\lambda}\right)\left[z_2^{n+1} + \sum_{r=0}^{\infty} z_2^{n+1}\left(\frac{\mu}{\lambda}\right)^{r+1} z_2^{r+1}\right] \\
&= \left(\frac{1}{\lambda}\right)\left[z_2^{n+1} + \left(\frac{\lambda}{\mu}\right)^{n+1} \sum_{k=n+2}^{\infty}\left(\frac{\mu}{\lambda}\right)^k z_2^k\right].
\end{aligned}
\tag{9.10}
$$

We need the inverse Laplace transform of z_2^n to find $p_n(t)$. We have

$$
\begin{aligned}
z_2 &= \frac{(s + \lambda + \mu) - \sqrt{\{(s + \lambda + \mu)^2 - 4\lambda\mu\}}}{(2\mu)} \\
&= \frac{4\lambda\mu}{2\mu\left[(s + \lambda + \mu) + \sqrt{\{(s + \lambda + \mu)^2 - 4\lambda\mu\}}\right]}.
\end{aligned}
$$

Writing

$$\frac{4\lambda\mu}{2\mu} = 2\sqrt{\lambda\mu}\,\sqrt{\frac{\lambda}{\mu}}$$

we get

$$z_2^n = \left(\frac{\lambda}{\mu}\right)^{n/2}\left[\frac{2\sqrt{\lambda\mu}}{(s + \lambda + \mu) + \sqrt{\{(s + \lambda + \mu)^2 - 4\lambda\mu\}}}\right]^n.$$

From the Table of Laplace transform, we note that

$$\text{LT of } \left\{\frac{n}{t}\, I_n(at)\right\} \quad \text{is} \quad \left\{\frac{a}{s + \sqrt{(s^2 - a^2)}}\right\}^n.$$

Using the translation property of LT we get that LT of

$$\left[\left(\frac{\lambda}{\mu}\right)^{n/2} e^{-(\lambda+\mu)t}\left\{\frac{n}{t}\, I_n(2t\sqrt{\lambda\mu})\right\}\right] \quad \text{is} \quad z_2^n. \tag{9.11}$$

Hence from (9.10) and (9.11) we get

$$p_n(t) = \frac{e^{-(\lambda+\mu)t}}{\lambda}\left[\left(\frac{\lambda}{\mu}\right)^{(n+1)/2}\left(\frac{n+1}{t}\right)I_{n+1}(at) + \left(\frac{\lambda}{\mu}\right)^{n+1}\sum_{k=n+2}^{\infty}\left(\frac{\mu}{\lambda}\right)^{k+2}\left(\frac{k}{t}\right)I_k(at)\right], \tag{9.12}$$

where $2\sqrt{\lambda\mu} = a$.

Using the property

$$\frac{2m}{z} I_m(z) = I_{m-1}(z) - I_{m+1}(z),$$

we get

$$\frac{2k}{at} I_k(at) = I_{k-1}(at) - I_{k+1}(at).$$

Substituting in (9.12), we get

$$\begin{aligned}
p_n(t) &= e^{-(\lambda+\mu)t}\frac{\sqrt{\lambda\mu}}{\lambda}\left[\left(\frac{\lambda}{\mu}\right)^{(n+1)/2}\{I_n(at) - I_{n+2}(at)\}\right. \\
&\quad \left. + \left(\frac{\lambda}{\mu}\right)^{n}\left\{\sum_{k=n+2}^{\infty}\left(\frac{\mu}{\lambda}\right)^{(k-2)/2}(I_{k-1}(at) - I_{k+1}(at))\right\}\right] \\
&= e^{-(\lambda+\mu)t}\left[\left(\frac{\lambda}{\mu}\right)^{n/2} I_n(at) - \left(\frac{\lambda}{\mu}\right)^{n/2} I_{n+2}(at)\right. \\
&\quad + \left(\frac{\lambda}{\mu}\right)^{n}\left\{\left(\frac{\mu}{\lambda}\right)^{(n+1)/2}\{I_{n+1}(at) - I_{n+3}(at)\}\right. \\
&\quad \left.\left. + \left(\frac{\mu}{\lambda}\right)^{(n+2)/2}\{I_{n+2}(at) - I_{n+4}(at)\}\right\} + \cdots\right] \\
&= e^{-(\lambda+\mu)t}\left[\left(\frac{\mu}{\lambda}\right)^{-n/2} I_n(at) + \left(\frac{\mu}{\lambda}\right)^{(-n+1)/2} I_{n+1}(at)\right. \\
&\quad + \left(\frac{\lambda}{\mu}\right)^{n}\left(1 - \frac{\lambda}{\mu}\right)\left\{\left(\frac{\mu}{\lambda}\right)^{(n+2)/2} I_{n+2}(at)\right. \\
&\quad \left.\left. + \left(\frac{\mu}{\lambda}\right)^{(n+3)/2} I_{n+3}(at) + \cdots\right\}\right].
\end{aligned}$$

Finally,

$$p_n(t) = e^{-(\lambda+\mu)t}\Bigg[\rho^{n/2} I_n(at) + \rho^{(n-1)/2} I_{n+1}(at)$$
$$+ \{(1-\rho)\rho^n\}\Bigg\{\sum_{k=n+2}^{\infty} \rho^{-k/2} I_k(at)\Bigg\}\Bigg], \qquad n \geq 0. \tag{9.13}$$

So far we assumed that $p_0(0) = 1$. The general case $p_i(0) = 1$, $p_j(0) = 0, j \neq i$ can be treated similarly; when $p_i(0) = 1$, we shall get, for all $n \geq i$, $n < i$

$$p_n(t) = e^{-(\lambda+\mu)t}\Bigg[\rho^{(n-i)/2} I_{n-i}(at) + \rho^{(n-i-1)/2} I_{n+i+1}(at)$$
$$+ (1-\rho)\rho^n \sum_{k=n+i+2}^{\infty} \rho^{-k/2} I_k(at)\Bigg] \qquad n \geq 0. \tag{9.14}$$

3.9.2.3. Derivation of Steady-State Distribution. Assume that $\rho < 1$. Using the limiting property as $t \to \infty$,

$$\lim I_r(at) \to \frac{\exp(at)}{\sqrt{(2\pi at)}} \text{ (independent of } r),$$

it can be shown that, as $t \to \infty$, the first two terms of both the expression on the RHS of (9.13) and (9.14) tend to 0 and the third term tends to unity. Thus, when $\rho < 1$,

$$p_n = \lim_{t\to\infty} p_n(t) = (1-\rho)\rho^n, \qquad n \geq 0.$$

3.9.3. Method of Generating Function

We now consider the popular method introduced by Bailey.

Denote

$$P(z, t) = \sum_{n=0}^{\infty} p_n(t) z^n, \qquad |z| < 1$$

$$\bar{P}(z, s) = \text{LT of } P(z, t) = \sum_{n=0}^{\infty} \bar{p}_n(s) z_n.$$

Multiplying (9.2) by z^n, $n = 1, 2, \ldots$ and adding with (9.1) we get

$$\sum_{n=0}^{\infty} p_n'(t) z^n = -(\lambda+\mu) \sum_{n=0}^{\infty} p_n(t) z^n + \mu p_0(t)$$
$$+ \lambda \sum_{n=1}^{\infty} p_{n-1}(t) z^n + \mu \sum_{n=0}^{\infty} p_{n+1}(t) z^n$$

or

$$\frac{\partial}{\partial t} P(z, t) = -(\lambda + \mu)P(z, t) + \mu p_0(t) + \lambda z P(z, t)$$
$$+ \left(\frac{\mu}{z}\right)\{P(z, t) - p_0(t)\}$$

or

$$z \frac{\partial}{\partial t} P(z, t) = \{\lambda z^2 - (\lambda + \mu)z + \mu\}P(z, t) - \mu(1 - z)p_0(t). \tag{9.15}$$

Taking LT and noting that

$$L\left\{\frac{\partial}{\partial t} P(z, t)\right\} = s\bar{P}(z, s) - P(z, 0)$$

we get

$$z[s\bar{P}(z, s) - P(z, 0)] = \{\lambda z^2 - (\lambda + \mu)z + \mu\}\bar{P}(z, s) - \mu(1 - z)\bar{p}_0(s)$$

whence

$$\bar{P}(z, s) = \frac{\mu(1 - z)\bar{p}_0(s) - zP(z, 0)}{\lambda z^2 - (s + \lambda + \mu)z + \mu}. \tag{9.16}$$

The relation involves $P(z, 0)$ (which can be found from the initial condition) as well as $\bar{p}_0(s)$. To find $\bar{p}_0(s)$ we note that the denominator of the RHS of (9.16) has two roots, say, ξ_1 and ξ_2,

$$\xi_i = \frac{(s + \lambda + \mu) \mp \sqrt{\{(s + \lambda + \mu)^2 - 4\lambda\mu\}}}{2\lambda}, \qquad i = 1, 2.$$

Now the roots of $\lambda z^2 - (s + \lambda + \mu)z + \mu = 0$ are the reciprocals of the roots of the equation $\mu z^2 - (s + \lambda + \mu)z + \lambda = 0$. Thus $\xi_1 = 1/z_1$, $\xi_2 = 1/z_2$, $|\xi_1| < 1$, $|\xi_2| > 1$.

As $\bar{P}(z, s)$ converges in the region $|z| \le 1$, the zero (in the unit disc) of the numerator and denominator of $\bar{P}(z, s)$ must coincide. Thus, $z = \xi_1$ must also be a root of the numerator, so that

$$\mu(1 - \xi_1)\bar{p}_0(s) = \xi_1 P(\xi_1, 0)$$

or

$$\bar{p}_0(s) = \frac{\xi_1 P(\xi_1, 0)}{\mu(1 - \xi_1)} \tag{9.17}$$

and finally, putting the value of $\bar{p}_0(s)$ in (9.16), we get

$$\bar{P}(z, s) = \frac{(1-z)\xi_1 P(\xi_1, 0) - z(1-\xi_1)P(z, 0)}{\lambda(z-\xi_1)(z-\xi_2)(1-\xi_1)}. \tag{9.18}$$

$P(z, 0)$ can be found from the initial condition. Suppose that $p_i(0) = 1$, then $P(z, 0) = z^i$ and

$$\bar{P}(z, s) = \frac{(1-z)\xi_1^{i+1} - z^{i+1}(1-\xi_1)}{\lambda(z-\xi_1)(z-\xi_2)(1-\xi_1)}. \tag{9.19}$$

3.9.3.1. Particular Case. When $i = 0$, $p_0(0) = 1$, $p_n(0) = 0$, $n \neq 0$, and $P(z, 0) = 1$, so also $P(\xi_1, 0) = 1$. Then

$$\begin{aligned}\bar{P}(z, s) &= \frac{(1-z)\xi_1 - z(1-\xi_1)}{\lambda(z-\xi_1)(z-\xi_2)(1-\xi_1)} \\ &= \frac{z_1 z_2}{\lambda(z_1 - 1)} \sum_{n=0}^{\infty} z_2^n z^n \\ &= \frac{(1-z_2)}{s} \sum_{n=0}^{\infty} z_2^n z^n, \quad \text{since} \quad -s = \mu(z_1 - 1)(z_2 - 1).\end{aligned}$$

Hence,

$$\begin{aligned}\bar{p}_n(s) &= \text{coefficient of } z^n \text{ in } P(z, s) \\ &= \frac{(1-z_2)z_2^n}{s}, \qquad n = 0, 1, 2, \ldots \text{ (as in 9.7a).}\end{aligned}$$

Remark. (1) Abate and Whitt (1988) show how the transform analysis can be continued to obtain a better description of the transient behavior.
(2) The approach through application of Rouché's theorem, which can be used for some other models as well, could be quite complicated in practice. Neuts (1979) formulates an alternative approach for queues solvable without Rouché's theorem.

3.9.4. Busy Period Analysis

We define a busy period as the interval of time from the instant a unit arrives at an empty system and its service begins, to the instant when the server becomes *free* for the *first time*. A busy period is a RV, being the first passage time from state 1 to state 0. Denote

T = length of the busy period

$b(t)$ = PDF of T

$N^*(t)$ = number present at time t during a busy period

$\{N^*(t), t \geq 0\}$ is a zero-avoiding state process

$q_n(t) = Pr\{N^*(t) = n\}, \qquad n = 0, 1, 2, \ldots$

$\bar{q}_n(s)$ = LT of $q_n(t)$.

We have $q_1(0) = 1$, $q_n(0) = 0$, $n \neq 1$, For $n \geq 2$, $q_n(t)$ will satisfy the same differential equations as $p_n(t)$, i.e., the equations (9.2) will hold good also for $q_n(t)$, $n = 2, 3, \ldots$. Thus,

$$q_n'(t) = -(\lambda + \mu)q_n(t) + \lambda q_{n-1}(t) + \mu q_{n+1}(t), \qquad n \geq 2. \tag{9.20}$$

As the term $q_0(t)$ will not occur, the equation corresponding to $n = 1$ will be

$$q_1'(t) = -(\lambda + \mu)q_1(t) + \mu q_2(t). \tag{9.21}$$

Taking LT of (9.21) we get

$$\mu\bar{q}_{n+1}(s) - (s + \lambda + \mu)\bar{q}_n(s) + \lambda\bar{q}_{n-1}(s) = 0, \qquad n \geq 2. \tag{9.22}$$

This is a difference equation of order 2 having the same character-isitic equation (9.5). Thus,

$$\bar{q}_n(s) = Az_1^n + Bz_2^n, \qquad n \geq 2.$$

Since $\sum \bar{q}_n(s)$ converges and $|z_1| > 1$, $A \equiv 0$. We can choose the constant B such that (9.22) is satisfied for $n = 2$. Thus, $\bar{q}_1(s) = Bz_2$ and

$$\bar{q}_n(s) = q_1(s)z_2^{n-1}, \qquad n \geq 1. \tag{9.23}$$

The LT of (9.21) yields

$$s\bar{q}_1(s) = 1 - (\lambda + \mu)\bar{q}_1(s) + \mu z_2\bar{q}_1(s)$$

or

$$[(s + \lambda + \mu) - \mu z_2]\bar{q}_1(s) = 1.$$

Thus

$$\bar{q}_1(s) = \frac{1}{(s + \lambda + \mu) - \mu z_2}$$

$$= \frac{z_2}{\lambda}, \qquad \text{(since } z_2 \text{ is a root of (9.5))}$$

so that

$$\bar{q}_n(s) = \bar{q}_1(s) z_2^{n-1} = \frac{z_2^n}{\lambda}.$$

Inversion of the LT yields

$$q_n(t) = \left(\frac{\lambda}{\mu}\right)^{n/2} \frac{n}{\lambda t} e^{-(\lambda+\mu)t} I_n(2t\sqrt{\lambda\mu}), \qquad n = 1, 2, \ldots. \tag{9.24}$$

Conditioning on the number of units present at instant t, all of which complete their service in $(t, t + dt)$, we have

$$\begin{aligned} b(t)dt &= Pr\{t \le T < t + dt\} \\ &= \sum_{j=1}^{\infty} Pr\{t \le T < t + dt | N^*(t) = j\} \times Pr\{N^*(t) = j\} \\ &= Pr\{t \le T < t + dt | N^*(t) = 1\} Pr\{N^*(t) = 1\} \\ &\quad + \sum_{j=2}^{\infty} Pr\{t \le T < t + dt | N^*(t) = j\} Pr\{N^*(t) = j\}. \end{aligned}$$

The first term implies that there is only one unit (at the instant t) whose service is completed between $(t, t + dt)$, the probability of this event being $\mu dt + o(dt)$. The second term implies service completion of two or more units in $(t, t + dt)$ and the probability of this event is $o(dt)$. Thus, taking limit as $dt \to 0$,

$$\begin{aligned} b(t) &= [\mu q_1(t)] \\ &= \frac{1}{t} \rho^{-1/2} e^{-(\lambda+\mu)t} I_1(2t\sqrt{\lambda\mu}). \end{aligned} \tag{9.25}$$

The LST of T is given by

$$\begin{aligned} b^*(s) &= L\{b(t)\} = \mu \bar{q}_1(s) \\ &= \mu \frac{z_2}{\lambda} = \frac{z_2}{\rho}. \end{aligned} \tag{9.26}$$

Now

$$\begin{aligned} \lim_{s \to 0} z_2 &= \rho \quad \text{if} \quad \rho < 1 \\ &= 1 \quad \text{if} \quad \rho \ge 1 \end{aligned}$$

so that

$$b^*(0) = 1 \quad \text{if} \quad \rho < 1$$
$$= \frac{1}{\rho} \quad \text{if} \quad \rho \geq 1.$$

This shows that there is non-zero probability that the busy period, when $\rho > 1$, is infinitely large, as should be intuitively clear.

3.9.4.1. Moments of the Busy Period. Writing

$$K = \sqrt{\{(s + \lambda + \mu)^2 - 4\lambda\mu\}},$$

we have

$$\frac{d}{ds} z_2 = [1 - 2(s + \lambda + \mu)/2K]/(2\mu) = -z_2/K$$

so that

$$E(T) = -\frac{d}{ds} b^*(s)\bigg|_{s=0}$$
$$= \frac{1}{\mu - \lambda} = \frac{1}{\mu(1 - \rho)}.$$

Again

$$\frac{d^2}{ds^2} z_2 = \frac{z_2}{K^2} + \frac{z_2}{K^3}(s + \lambda + \mu)$$
$$= \frac{2\lambda}{K^3}$$

so that

$$E(T^2) = \frac{d^2}{ds^2} b^*(s)\bigg|_{s=0} = \frac{2\lambda}{\rho} \frac{1}{(\mu - \lambda)^3}$$
$$= \frac{2}{\mu^2(1 - \rho)^3}.$$

We have

$$\text{var}(T) = \frac{1 + \rho}{2(1 - \rho)^3}.$$

Remarks.

(1) It was the celebrated French mathematician Emile Borel (1871–1956) who introduced the concept of busy period. He obtained the joint distribution of the busy period T and the number N served during the busy period for an $M/D/1$ model.

(2) The LST of the busy period can also be obtained from a certain functional equation that it satisfies (see Section 6.4.2).

(3) *Idle period* is the interval I from the instant the server becomes free to the instant of the next arrival (when the server resumes service). This interval is the residual interarrival time. Since the interarrival time is exponential, the idle period I is also exponential with the same parameter λ as the interarrival time. Thus, $E(I) = 1/\lambda$ for a model with Poisson input.

(4) The expected duration of the busy period can also be obtained from a result of renewal theory. (See Section 1.8, Chapter I, relation (8.1)). The busy period T and the idle period I form an alternating renewal process. Thus,

$$\frac{E(T)}{E(I)} = \frac{1 - p_0}{p_0}.$$

Since $p_0 = 1 - \rho$ for a single server system, we get

$$E(T) = \frac{\rho}{1 - \rho}\left(\frac{1}{\lambda}\right) = \frac{1}{\mu(1 - \rho)}.$$

(5) We have examined the busy period initiated by a single unit. The busy period initiated by r units (with $r \geq 1$ units in the systsem at the commencement of the busy period) is given by the sum of r IID random variables T_i ($\equiv T$, the busy period initiated by a single unit). Unless otherwise stated, we shall take $r = 1$.

(6) Let $E(N)$ denote the average number of customers served during a busy period T. For every $E(N)$ arrivals during a busy period exactly one arrival (the first customer during a busy period) will find the system empty. Hence the probability a_0 that an arrival finds the system empty is given by

$$a_0 = \frac{1}{E(N)}.$$

As PASTA holds, $a_0 = p_0 = 1 - \rho$ so that $E(N) = 1/(1 - \rho)$.

We can get the result intuitively as follows: since the server remains continuously busy serving during a busy period, we have

$$\begin{aligned} E(N) &= E(T) \times \{\text{rate of service}\} \\ &= \frac{1}{\mu(1-\rho)} \times \mu \\ &= \frac{1}{1-\rho}. \end{aligned} \tag{9.27}$$

3.9.4.2. Number Served during a Busy Period of an M/M/1 Queue. Let N be the number served during a busy period that starts with one customer. We find the distribution of N.

Consider the system-size process observed at each arrival and departure epoch as a one-dimensional random walk with a reflecting barrier at the origin. Let X_n be the system size at the nth arrival/departure epoch. The transition probabilities are given by

$$\begin{aligned} p_{i,i+1} &= Pr\{X_n = i+1 \mid X_n = i\} \\ &= Pr\{\text{an arrival occurs before a departure occurs}\} \\ &= \frac{\lambda}{\lambda+\mu} = \frac{\rho}{1+\rho} = p \text{ (say), and} \end{aligned} \tag{9.28}$$

$$\begin{aligned} p_{i,i-1} &= Pr\{X_n = i-1 \mid X_n = i\} \\ &= Pr\{\text{a departure occurs before an arrival occurs}\} \\ &= \frac{\mu}{\lambda+\mu} = \frac{1}{1+\rho} = 1-p. \end{aligned} \tag{9.29}$$

The event that $\{N = n\}$ is equivalent to the event that the *first* return to the origin through the positive axis of the random walk occurs at epoch $2n$. Now $Pr\{N = n\} = P\{\text{the first return to origin occurs at epoch } 2n\}$ is given by

$$\varphi_{1,2n-1} = \frac{1}{2n-1}\binom{2n-1}{\frac{2n-1+1}{2}} p^n (1-p)^{n-1} \tag{9.30}$$

(Feller, 1968, vol. I, Th. 4, p. 90). That is,

$$Pr\{N = n\} = \frac{1}{2n-1}\binom{2n-1}{n}\frac{\rho^{n-1}}{(1+\rho)^{2n-1}}$$

$$= \frac{1}{n}\binom{2n-2}{n-1}\frac{\rho^{n-1}}{(1+\rho)^{2n-1}}, \qquad n = 1, 2, 3, \ldots. \tag{9.31}$$

The PGF of N is given by

$$P(s) = \sum_{n=1}^{\infty} Pr(N = n)s^n = \frac{2s}{1 + \rho + [(1+\rho)^2 - 4\rho s]}. \tag{9.32}$$

We have

$$E(N) = \frac{1}{(1-\rho)} \quad \text{and} \tag{9.33}$$

$$\text{var}(N) = \frac{\rho(1+\rho)}{(1-\rho)^3}. \tag{9.34}$$

When the busy period starts with m customers, then

$$Pr\{N = n\} = \frac{m}{n}\binom{2n-m-1}{n-1}\frac{\rho^{n-m}}{(1+\rho)^{2n-m}}, \qquad n = m, m+1, \ldots. \tag{9.35}$$

Its PGF is

$$\begin{aligned} P(s) &= \sum_{n=m}^{\infty} Pr(N = n)s^n \\ &= \frac{(2s)^m}{[1 + \rho + \{(1+\rho)^2 - 4\rho s\}]^m} \end{aligned} \tag{9.36}$$

with

$$E(N) = \frac{m}{(1-\rho)} \quad \text{and} \tag{9.37}$$

$$\text{var}(N) = \frac{m\rho(1+\rho)}{(1-\rho)^3}. \tag{9.38}$$

Haight (1961) has described the distribution of N which is "analogous to the Borel–Tanner distribution" relating to the number served during the busy period of an $M/D/1$ queue.

Example 3.3. ***First passage time distribution.*** For an $M/M/1$ queue in transient state let τ be the random variable denoting the time taken for the system to *fall* from the initial state a to another state $b(<a)$ for the first time. Let $\{X(t)\}$ be the process denoting such states of the system, given $X(0) = a$,

$$q_n(t) = P\{X(t) = n \mid X(0) = a\}. \tag{9.39}$$

Using similar arguments it easily can be shown that the state probabilities satisfy the following equations:

$$\begin{aligned} q_n'(t) &= -(\lambda + \mu)q_n(t) + \lambda q_{n-1}(t) + \mu q_{n+1}(t), \qquad n \geq (b+2) \\ q_{b+1}'(t) &= -(\lambda + \mu)q_{b+1}(t) + \mu q_{b+2}(t) \\ q_b'(t) &= \mu q_{b+1}(t). \end{aligned} \tag{9.40}$$

The PDF $f(t)$ of the first passage time distribution from state a to state b is given by $f(t) = \mu q_{b+1}(t) = q_b'(t)$. Putting $a = 1$ and $b = 0$, we can get back to the initial busy-period distribution.

3.9.5. Waiting-Time Process

Let $W(t)$ be the time required to serve all the units present in the system at the instant t, given that $W(0) = 0$. Then, if $N(t)$ is the number present at instant t, $(N(0) = 0)$, then

$$W(t) = \begin{cases} 0 & \text{if } N(t) = 0 \\ v_1' + v_2 + \cdots + v_{N(t)} & \text{if } N(t) > 0, \end{cases}$$

where v_1' is the residual service time of the unit being served at the instant t, and $v_2, \ldots, v_{N(t)}$ are the service times of the units waiting at the instant t. $\{W(t), t > 0\}$, which is known as *virtual waiting time*, is a Markov process (with continuous state space). Given $W(0) = 0$, then proceeding as in the case of waiting time in the system in steady state, it can be seen that its probability element $f(x, t)dx = P\{x \leq W(t) < x + dx\}$, $0 < x < \infty$, $0 < t < \infty$, is

$$f(x, t)dx = \sum_{n=0}^{\infty} \mu \frac{(\mu x)^n}{\Gamma(n+1)} e^{-\mu x}\, dx\; p_n(t).$$

Its LT can be put in a closed form. (See Prabhu, 1965).

Example 3.4. ***Transient solution of the M/M/1/1 model.*** Here $\lambda_0 = \lambda$, $\mu_0 = 0$, and $\lambda_1 = 0$, $\mu_1 = \mu$; if $N(t)$ denotes the number in the system at time t, then $Pr\{N(t) = n\} = p_n(t) = 0$ for all $n > 1$, i.e., we are concerned with only

$p_0(t)$ and $p_1(t)$ such that $p_0(t) + p_1(t) = 1$. The differential-difference equations of the model then become

$$p_0'(t) = -\lambda p_0(t) + \mu p_1(t)$$

$$p_1'(t) = -\mu p_1(t) + \lambda p_0(t).$$

Writing $p_1(t) = 1 - p_0(t)$ in the first equation, we get $p_0'(t) + (\lambda + \mu)p_0(t) = \mu$. The solution of this first-order linear differential equation with constant coefficients is given by

$$p_0(t) = \mathrm{C}e^{-(\lambda+\mu)t} + \frac{\mu}{(\lambda + \mu)},$$

where C is constant. Given the initial distribution $p_i(0) = Pr\{N(0) = i\}$, we get

$$p_0(t) = p_0(0)e^{-(\lambda+\mu)t} + \frac{\mu}{\lambda + \mu}\{1 - e^{-(\lambda+\mu)t}\}.$$

Similarly,

$$p_1(t) = p_1(0)e^{-(\lambda+\mu)t} + \frac{\lambda}{\lambda + \mu}\{1 - e^{-(\lambda+\mu)t}\}.$$

The steady-state solutions are

$$p_0 = \lim_{t\to\infty} p_0(t) = \frac{\mu}{(\lambda + \mu)}$$

$$p_1 = \lim_{t\to\infty} p_1(t) = \frac{\lambda}{(\lambda + \mu)}$$

irrespective of whether the value of $\rho = \lambda/\mu < 1$ or not.

Assume that the initial distribution is identical with the steady-state distribution so that

$$p_0(0) = p_0 = \frac{\mu}{(\lambda + \mu)} \quad \text{and} \quad p_1(0) = p_1 = \frac{\lambda}{(\lambda + \mu)}.$$

Then we find that for all $t > 0$,

$$p_0(t) = \frac{\mu}{(\lambda + \mu)} = p_0 \quad \text{and} \quad p_1(t) = \frac{\lambda}{(\lambda + \mu)} = p_1.$$

That is, if the process is in equilibrium (steady state) initially, then it will be always (for all $t > 0$) in steady state. This is true for any other ergodic system.

Note. In case of the $M/M/c/c$ model, if the system is in equilibrium initially, i.e., $Pr\{N(0) = n\} = p_n(0) = p_n$, $0 < n < c$, where p_n are steady-state probabilities (given by relation (7.2)), then $\{N(t), t > 0\}$ becomes a stationary process for which $N(t)$ has the same distribution for all $t > 0$, i.e.,

$$p_n(t) = Pr\{N(t) = n\} = p_n$$

(given by (7.2)) for all $t > 0$. (See Takács, 1969.)

3.10. Transient-State Distribution of the $M/M/c$ Model

3.10.1. Solution of the Differential-Difference Equations

Consider a c-server queueing system with Poisson input (with parameter λ) and exponential service time (with parameter μ) for each of the c-servers. The model corresponds to that of a birth–death process with rates

$$\begin{aligned} \lambda_n &= \lambda, & n &= 0, 1, 2, \ldots, \\ \mu_n &= n\mu, & n &= 1, 2, \ldots, c-1, \\ &= c\mu, & n &= c, c+1, c+2, \ldots. \end{aligned}$$

Let $N(t)$ be the number in the system at time t and let

$$p_n(t) = Pr\{N(t) = n\}.$$

Then p_n's satisfy the differential-difference equations

$$p'_0(t) = -\lambda p_0(t) + \mu p_1(t) \tag{10.1}$$

$$p'_n(t) = -(\lambda + n\mu)p_n(t) + \lambda p_{n-1}(t) + (n+1)\mu p_{n+1}(t), \quad 1 \le n \le c-1, \tag{10.2}$$

$$p'_n(t) = -(\lambda + c\mu)p_n(t) + \lambda p_{n-1}(t) + c\mu p_{n+1}(t), \qquad n \ge c. \tag{10.3}$$

These equations in transient state have been solved by Saaty (1960) and Jackson and Henderson (1966). We consider here the difference-equation technique of Jackson and Henderson. Denote the initial condition by

$$p_n(0) = \delta_{in} \text{ (i being the number at time 0).}$$

Consider that $i = c$ at time 0; time is reckoned from the instant when all the servers become busy with none in the queue. Let $p_n^*(s)$ denote the LT of $p_n(t)$.

Taking the LT of (10.1) through (10.3), we get

$$(\lambda + s)p_0^*(s) = \mu p_1^*(s), \tag{10.4}$$

$$(\lambda + s + n\mu)p_n^*(s) = \lambda p_{n-1}^*(s) + (n+1)\mu p_{n+1}^*(s), \qquad 1 \le n \le c-1, \tag{10.5}$$

$$(\lambda + s + c\mu)p_n^*(s) - \delta_{cn} = \lambda p_{n-1}^*(s) + c\mu p_{n+1}^*(s), \qquad n \ge c. \tag{10.6}$$

The second difference equation is one with variable coefficients and as such it has to be solved by using special techniques as follows. Assume that solutions exist, i.e., the equations are consistent. The solutions $p_n^*(s)$, $0 \le n \le c-1$ of the first two equations can be obtained independently of $p_n^*(s)$, $n \ge c$. We shall use some interesting properties of generating functions. Let us write $p_n^*(s) = f(n)$, then (10.5) can be written as

$$(n+1)\mu f(n+1) = (\lambda + s + n\mu)f(n) - \lambda f(n-1). \tag{10.7}$$

Let

$$G[f(n)] = \sum_{n=0}^{\infty} f(n)t^n = F(t) \tag{10.8}$$

be the generating function of $\{f(n)\}$. Multiplying both sides of (10.7) by t^n for $n = 1, 2, \ldots$ and adding, we get

$$\mu \sum_{n=1}^{\infty} (n+1)f(n+1)t^n = (\lambda + s)\sum_{n=1}^{\infty} f(n)t^n + \mu \sum_{n=1}^{\infty} nf(n)t^n - \lambda t \sum_{n=1}^{\infty} f(n-1)t^{n-1}. \tag{10.9}$$

We have

$$\sum_{n=1}^{\infty} nf(n)t^n = t\sum_{n=1}^{\infty} nf(n)t^{n-1} = tF'(t)$$

$$\sum_{n=1}^{\infty} (n+1)f(n+1)t^n = \sum_{m=1}^{\infty} mf(m)t^{m-1} - f(1) \quad (\text{putting } n+1=m)$$

$$= F'(t) - f(1).$$

Thus, from (10.9), we have

$$\mu[F'(t) - f(1)] = (\lambda + s)[F(t) - f(0)] + \mu t F'(T) - tF(t) \quad \text{or}$$

$$\mu(1 - t)F'(t) = \{(\lambda + s) - \lambda t\}F(t) + \{\mu f(1) - (\lambda + s)f(0)\}$$

$$= \{(\lambda + s) - \lambda t\}F(t), \quad \text{because of (10.4),}$$

or

$$\frac{F'(t)}{F(t)} = \frac{(\lambda + s) - \lambda t}{\mu(1 - t)}$$

$$= \frac{\lambda}{\mu} + \frac{s}{\mu(1 - t)}.$$

Integrating we get

$$\log F(t) = \log A + \frac{\lambda}{\mu} t - \frac{s}{\mu} \log (1 - t) \quad \text{or}$$

$$F(t) = \frac{A \exp\left(\dfrac{\lambda t}{\mu}\right)}{(1 - t)^{s/\mu}}.$$

Putting $t = 0$, $F(0) = A$, but $F(0) = f(0) = p_0^*(s)$, so that

$$F(t) = p_0^*(s) e^{\lambda t/\mu} (1 - t)^{-s/\mu}; \tag{10.10}$$

expanding $F(t)$ and comparing the coefficients of t^n, we get

$$p_n^*(s) = p_0^*(s) \sum_{j=0}^{n} \frac{\left(\dfrac{\lambda}{\mu}\right)^{n-j}}{(n-j)!} \frac{\left(\dfrac{s}{\mu}\right)\left(\dfrac{s}{\mu} + 1\right) \cdots \left(\dfrac{s}{\mu} + j - 1\right)}{j!}$$

$$= p_0^*(s) \sum_{j=0}^{n} \frac{\left(\dfrac{\lambda}{\mu}\right)^{n-j} \Gamma\left(\dfrac{s}{\mu} + j\right)}{(n - j)!\, j!\, \Gamma\left(\dfrac{s}{\mu}\right)}, \qquad 0 \le n \le c - 1,\ s \ne 0. \tag{10.11}$$

Now, we are to solve (10.6). Putting $n = c$ and $n = c + r$, $r = 1, 2, \ldots$, we get

$$(\lambda + s + c\mu)p_c^*(s) - 1 = \lambda p_{c-1}^*(s) + c\mu p_{c+1}^*(s) \quad \text{and} \tag{10.12}$$

$$(\lambda + s + c\mu)p_{c+r}^*(s) = \lambda p_{c+r-1}^*(s) + c\mu p_{c+r+1}^*(s). \tag{10.13}$$

Denote $\sum_{r=0}^{\infty} p_{c+r}^* t^r = V(t)$. Multiplying (10.13) by t^r and adding for $r = 1, 2, 3, \ldots$, we get

$$(\lambda + s + c\mu) \sum_{r=1}^{\infty} p_{c+r}^*(s)t^r = \lambda t \sum_{r=1}^{\infty} p_{c+r-1}^*(s)t^{r-1} + \frac{c\mu}{t} \sum_{r=1}^{\infty} p_{c+r+1}^* t^{r+1} \quad \text{or}$$

$$(\lambda + s + c\mu)[V(t) - p_c^*(s)] = \lambda t V(t) + \frac{c\mu}{t}[V(t) - tp_{c+1}^*(s) - p_c^*(s)]$$

$$= \lambda t V(t) + \frac{c\mu}{t}[V(t) - p_c^*(s)]$$

$$+ \lambda p_{c-1}^*(s) - \{(\lambda + s + c\mu)p_c^*(s) - 1\} = 0$$

(using (10.12)); or

$$[c\mu - (\lambda + s + c\mu)t + \lambda t^2]V(t) = p_c^*(s)(c\mu) + t\{-1 - \lambda p_{c-1}^*(s)\}.$$

Thus,

$$V(t) = \frac{c\mu p_c^*(s) - t\lambda p_{c-1}^*(s) - t}{c\mu - (\lambda + s + c\mu)t + \lambda t^2}. \tag{10.14}$$

To get $p_{c+r}^*(s)$ we have to expand the RHS of (10.14) in powers of t. Writing

$$[c\mu - (\lambda + s + c\mu)t + \lambda t^2] = \lambda(t - \alpha_1)(t - \alpha_2)$$

where

$$\begin{matrix}\alpha_1 \\ \alpha_2\end{matrix} = \frac{[(\lambda + s + c\mu) \pm \sqrt{\{(\lambda + s + c\mu)^2 - 4\lambda c\mu\}}]}{2\lambda},$$

($\alpha_1(\alpha_2)$ corresponding to the positive (negative) sign before the radical sign), we get

$$\frac{1}{c\mu - (\lambda + s + \mu)t + \lambda t^2} = \frac{1}{\lambda(t - \alpha_1)(t - \alpha_2)}$$

$$= \frac{1}{\lambda(\alpha_1 - \alpha_2)}\left[\frac{1}{t - \alpha_1} - \frac{1}{t - \alpha_2}\right]$$

$$= \frac{1}{\lambda(\alpha_1 - \alpha_2)}\left[\frac{-1}{\alpha_1}\left(1 - \frac{t}{\alpha_1}\right)^{-1} + \frac{1}{\alpha_2}\left(1 - \frac{t}{\alpha_2}\right)^{-1}\right]$$

$$= \frac{1}{\lambda(\alpha_1 - \alpha_2)}\left[\frac{1}{\alpha_2}\sum_{k=0}^{\infty}\left(\frac{t}{\alpha_2}\right)^k - \frac{1}{\alpha_1}\sum_{k=0}^{\infty}\left(\frac{t}{\alpha_1}\right)^k\right].$$

Thus,

$$
\begin{aligned}
p_{c+r}^*(s) &\equiv \text{coeff of } t^r \text{ in } V(t) \\
&\equiv \text{coeff of } t^r \text{ on the RHS of (10.14)} \\
&= \frac{1}{\lambda(\alpha_1 - \alpha_2)} \left[c\mu p_c^*(s) \left\{ \frac{1}{\alpha_2^{r+1}} - \frac{1}{\alpha_1^{r+1}} \right\} \right. \\
&\quad - \lambda p_{c-1}^*(s) \left\{ \frac{1}{\alpha_2^r} - \frac{1}{\alpha_1^r} \right\} \\
&\quad \left. - \left(\frac{1}{\alpha_2^r} - \frac{1}{\alpha_1^r} \right) \right], \qquad r = 0, 1, 2, \ldots. \qquad (10.15)
\end{aligned}
$$

Thus, we get all the coefficients $p_n^*(s)$, $n > c$ in terms of $p_c^*(s)$ and $p_{c-1}^*(s)$. Again using (10.5) we get $p_c^*(s)$ in terms of $p_{c-1}^*(s)$ and $p_{c-2}^*(s)$ and all the $p_n^*(s)$, $n \geq 1$ in terms of $p_0^*(s)$. The term $p_0^*(s)$ can be obtained by using the relation

$$\sum_{n=0}^{c-1} p_n^*(s) + \sum_{n=c}^{\infty} p_n^*(s) = \frac{1}{s},$$

or alternatively as follows. Write (10.5) as

$$
\begin{aligned}
&(\lambda + s + c\mu)p_n^*(s) + (n - c)\mu p_n^*(s) \\
&\quad = \lambda p_{n-1}^*(s) + c\mu p_{n+1}^*(s) - (c - n - 1)\mu p_{n+1}^*(s), \quad 1 \leq n \leq c - 1. \qquad (10.16)
\end{aligned}
$$

Multiplying (10.16) by z^n for $1 \leq n \leq c - 1$ and (10.6) by z^n for $n \geq c$ and adding for $n = 1, 2, \ldots$ to (10.4) and writing

$$P(z, s) = \sum_{n=0}^{\infty} p_n^*(s) z^s, \qquad (10.17)$$

we get on simplification

$$P(z, s) = \frac{\mu(1 - z) \sum_{n=0}^{c-1} (c - n) z^n p_n^*(s) - z^{c+1}}{\lambda z^2 - (\lambda + s + \mu) z + c\mu}. \qquad (10.18)$$

Noting that $\lambda z^2 - (\lambda + s + \mu)z + c\mu = \lambda(z - \alpha_1)(z - \alpha_2)$, and considering that $P(z, s)$ exists in the unit circle, we find that the numerator of $F(z, s)$ must also vanish for $z = \alpha_2$. Thus, we have

$$\sum_{n=0}^{c-1} (c - n) \alpha_2^n p_n^*(s) = \frac{\alpha_2^{c+1}}{\mu(1 - \alpha_2)}. \qquad (10.19)$$

Now using the expression of $p_n^*(s), 0 \le n \le c-1$ as given in (10.11), we get

$$p_0^*(s) = \frac{\alpha_2^{c+1}}{\mu(1-\alpha_2)} \left[\sum_{n=0}^{c-1} (c-n)\alpha_2^n \left\{ \sum_{j=0}^{n} \frac{\left(\frac{\lambda}{\mu}\right)^{n-j} \Gamma\left(\frac{s}{\mu}+j\right)}{(n-j)!j!\Gamma\left(\frac{s}{\mu}\right)} \right\} \right]^{-1}. \quad (10.20)$$

Thus, all $p_n^*(s)$ are known from (10.11), (10.15), and (10.20).

Note. Here we found the solution of the difference equations (10.4)–(10.6). For $n \ge c$, (10.6) is a difference equation of order 2 with constant coefficients and can be solved in the usual manner to get $p_n^*(s)$, $n > c$. Equation (10.5) is a difference equation in $p_n^*(s)$ with coefficients as functions of n; a special technique is needed (as used here) to find $p_n^*(s)$, $n \le c-1$. Then $p_0^*(s)$ can be obtained by either of the methods indicated. The special technique adopted here for the solution of (10.5) with (10.4) will be used subsequently in discussing a more complicated model $M/M(1, b)/c$ (Section 4.6).

3.10.1.1. *Steady-State Distribution of the M/M/c Model.*

Assume that $\rho = \lambda/c\mu < 1$. Then we can easily find

$$p_n = \lim_{t\to 0} p_n(t) = \lim_{s\to 0} sp_n^*(s)$$

from (10.10), (10.15), and (10.20). From (10.8) and (10.10) we get

$$\begin{aligned}
\lim_{s\to 0} sF(t) &= \lim_{s\to 0} \sum_{n=0}^{\infty} sp_n^*(s)t^n \\
&= \sum p_n t^n \\
&= p_0 e^{(\lambda/\mu)t} \\
&= \sum p_0 \left\{ \frac{\left(\frac{\lambda}{\mu}\right)^n}{n!} \right\} t^n.
\end{aligned}$$

Thus, for $0 \le n \le c-1$,

$$p_n = p_0 \frac{\left(\frac{\lambda}{\mu}\right)^n}{n!}. \quad (10.21)$$

From (10.5) for $n = c - 1$

$$[\lambda + s + (c - 1)\mu]p^*_{c-1}(s) = \lambda p^*_{c-2}(s) + c\mu p^*_c(s);$$

we get, multiplying both sides by s and taking limits as $s \to 0$,

$$[\lambda + (c - 1)\mu]p_{c-1} = \lambda p_{c-2} + c\mu p_c,$$

whence using (10.21), we get

$$p_c = \frac{1}{c\mu}\left\{[\lambda + (c - 1)\mu]\left[\frac{\left(\frac{\lambda}{\mu}\right)^{c-1}}{(c - 1)!}\right] - \lambda\frac{\left(\frac{\lambda}{\mu}\right)^{c-2}}{(c - 2)!}\right\}p_0$$

$$= \frac{1}{c\mu}\left[\frac{\lambda\left(\frac{\lambda}{\mu}\right)^{c-1}}{(c - 1)!}\right]p_0$$

$$= \frac{\left(\frac{\lambda}{\mu}\right)^{c}}{c!}\,p_0,$$

so that $p_n = p_0(\lambda/\mu)^n/n!$ holds for $n = 0, 1, \ldots, c$. Now as $s \to 0$,

$$\alpha_1 \to \frac{[(\lambda + c\mu) + (\lambda - c\mu)]}{2\lambda} = 1 \quad \text{and}$$

$$\alpha_2 \to \frac{[(\lambda + c\mu) - (\lambda - c\mu)]}{2\lambda} = \frac{1}{\rho}.$$

Thus, from (10.15), we get

$$p_{c+r} = \lim_{s \to 0} sp^*_{c+r}(s)$$

$$= \frac{1}{\lambda\left(1 - \frac{1}{\rho}\right)}[c\mu p_c(\rho^{r+1} - 1)] - \lambda p_{c-1}(\rho^r - 1)]$$

$$= \rho^r p_c = \frac{\left(\frac{\lambda}{\mu}\right)^{c+r}}{c!c^r}\,p_0, \qquad r = 0, 1, 2, \ldots, \quad \text{or}$$

$$p_n = \frac{\left(\frac{\lambda}{\mu}\right)^{n}}{c!c^{n-c}}\,p_0, \qquad n = c, c + 1, c + 2, \ldots. \tag{10.22}$$

From (10.21) and (10.22), using

$$\sum_{n=0}^{\infty} p_n = 1,$$

we get

$$p_0 = \left[\sum_{n=0}^{c-1} \frac{\left(\frac{\lambda}{\mu}\right)^n}{n!} + \sum_{n=c}^{\infty} \frac{\left(\frac{\lambda}{\mu}\right)^n}{c!c^{n-c}}\right]^{-1}$$

$$= \left[\sum_{n=0}^{c-1} \frac{\left(\frac{\lambda}{\mu}\right)^n}{n!} + \frac{\left(\frac{\lambda}{\mu}\right)^c}{c!(1-\rho)}\right]^{-1} \tag{10.23}$$

Thus, we get all p_n, $n = 0, 1, 2, 3, \ldots$ (as have been obtained in Section 3.6.1).

3.10.2. Busy Period of an M/M/c Queue

We consider here the approach by Chaudhry and Templeton (1973). A busy period of a multiserver queue may be defined as the interval of time commencing from the instant of arrival of a unit that makes a fixed number $k(\leq c)$ of channels busy to the first subsequent instant when the number of busy channels drops down to $(k-1)$. Let $\{N^*(t), t > 0\}$ be the stochastic process denoting the number of units present at the instant during the busy period T. This process avoids the states $0, 1, \ldots, k-1$, i.e., the states are k, $k+1, \ldots$. Now $Pr\{N^*(0) = k\} = 1$ and the duration of the busy period is the interval $t(>0)$ for which $N^*(t)$ becomes $(k-1)$ for the first time. Let

$$q_n(t) = Pr\{N^*(t) = n | N^*(0) = k\}, \qquad n \geq k$$

and $b(t)$ be the PDF of the busy period T. Then is can be easily seen that

$$b(t) \equiv q'_{k-1}(t) = k\mu q_k(t).$$

Now q_n's satisfy the same equations as $p_n(t)$ (10.2) restricted to $k \leq n \leq c-1$ and (10.3). That is,

$$q'_n(t) = -(\lambda + n\mu)q_n(t) + \lambda q_{n-1}(t) + (n+1)\mu q_{n+1}(t), \quad k \leq n \leq (c-1), \quad \text{and} \tag{10.24}$$

$$q'_n(t) = -(\lambda + c\mu)q_n(t) + \lambda q_{n-1}(t) + c\mu q_{n+1}(t), \qquad n \geq c, \tag{10.25}$$

where $\lambda q_{n-1}(t)$ is to be omitted from (10.24) when $n = k$ or from (10.25) when $n = k = c$, in which case (10.24) becomes redundant. Let $g_n(s)$ be the LT of

$q_n(t)$. Then taking the LT of (10.24) and (10.25) we get

$$(\lambda + s + n\mu)g_n(s) - 1 = \lambda g_{n-1}(s) + (n+1)\mu g_{n+1}(s), \quad k \le n \le c-1, \quad \text{and} \tag{10.26}$$

$$(\lambda + s + c\mu)g_n(s) = \lambda g_{n-1}(s) + c\mu g_{n+1}(s), \qquad n \ge c. \tag{10.27}$$

It is to be noted that (-1) in the LHS of (10.26) is to be used only when $n = k$. Further, when $n = k = c$, then (10.26) becomes redundant and (10.27) will have a term (-1) on the LHS of (10.27). Define

$$V(z, s) = \sum_{n=k}^{\infty} g_n(s)z^n.$$

Multiplying (10.26) and (10.27) by z^n and adding for appropriate values of n, we get

$$(\lambda + s + c\mu)V(z, s) - \sum_{n=k}^{c-1} \mu(c-n)g_n(s)z^n - z^k$$

$$= \lambda z V(z, s) + \frac{c\mu}{z} V(z, s) + \sum_{n=k}^{c-1} (n+1-c)\mu g_{n+1}(s)z^n$$

which can be written as

$$V(z, s) = \frac{z^{k+1} + \mu(z-1)\sum_{n=k}^{c-1}(c-n)g_n(s)z^n - k\mu g_k(s)z^k}{-\lambda z^2 + (\lambda + s + c\mu)z - c\mu}. \tag{10.28}$$

The denominator put to zero

$$\lambda z^2 - (\lambda + s + c\mu) + c\mu = 0$$

has two roots, α_1 and α_2,

$$\begin{matrix}\alpha_1 \\ \alpha_2\end{matrix} = \frac{\lambda + s + c\mu \pm \sqrt{(\lambda + s + c\mu)^2 - 4\lambda c\mu}}{2\lambda}$$

of which α_2 is of modulus less than 1. Considering that $V(z, s)$ exists in the unit circle, we see that the numerator of (10.28) must also vanish for $z = \alpha_2$. Putting $z = \alpha_2$ in the numerator of (10.28) equated to zero, we get

$$\alpha_2^{k+1} + \mu(\alpha_2 - 1)\sum_{n=k}^{c-1}(c-n)g_n(s)\alpha_2^n - k\mu g_k(s)\alpha_2^k = 0 \quad \text{or}$$

$$(1-\alpha_2)\sum_{n=k}^{c-1}(c-n)g_n(s)\alpha_2^n + kg_k(s)\alpha_2^k = \frac{(\alpha_2^{k+1})}{\mu}. \tag{10.29}$$

The equation involves $(c - k)$ unknowns $g_n(s)$, $(k \le n \le c - 1)$. These $(c - k)$ unknowns can be determined from (10.29) and (10.26). Thus, $g_k(s)$ can be found; its inversion gives $q_k(t)$. Note that (10.29) is sufficient to determine $g_k(s)$ in the cases $k = c$ or $k = (c - 1)$. When $k = c$ the term $\sum_k^{c-1}$ will not occur.

3.10.2.1. Particular Cases

I. $k = c$, i.e., the busy period starts from the instant that all the servers get busy to the instant when one of the servers becomes free for the first time.

Putting $k = c$ in (10.29) we get

$$c\mu g_c(s) = \alpha_2.$$

Inverting the LT we get

$$b(t) = c\mu q_c(t) = \left[\frac{e^{-(\lambda + c\mu)t}}{t}\right]\sqrt{\frac{c\mu}{\lambda}}\, I_1(2t\sqrt{c\lambda\mu})$$

$$= \left(\frac{1}{t}\right)\rho^{-1/2} e^{-(\lambda + c\mu)t} I_1(2t\sqrt{c\lambda\mu}). \tag{10.30}$$

We have

$$E(T) = -\frac{d}{ds}\, b^*(s)\bigg|_{s=0} = \frac{1}{c\mu - \lambda}.$$

Comparing the corresponding result for the single channel case (corresponding to $c = 1$) given in (9.25) we see the busy-period distribution for the c-channel case (when the busy period is defined as the period during which all the servers remain busy, i.e., $k = c$) can be obtained from the single-channel case by replacing μ by $c\mu$. This should be intuitively clear as the whole set of c-servers can then be considered as a compact set of "a single server" with rate $c\mu$.

II. $k = c - 1$.

Then from (10.29), we get

$$(1 - \alpha_2) g_{c-1}(s)\alpha_2^{c-1} + (c - 1) g_{c-1}(s)\alpha_2^{c-1} = \frac{\alpha_2^c}{\mu} \quad \text{or} \tag{10.31}$$

$$\mu g_{c-1}(s) = \frac{\alpha_2}{c - \alpha_2}$$

$$= \frac{c\mu}{\lambda}\frac{1}{\alpha_1}\left(1 - \frac{c\mu}{\lambda}\frac{1}{\alpha_1}\right)^{-1}$$

$$= \sum_{n=0}^{\infty}\left(\frac{\mu}{\lambda}\right)^{n+1}\frac{1}{\alpha_1^{n+1}},$$

since

$$\left|\frac{\mu}{\lambda}\frac{1}{\alpha_1}\right| = \left|\left(\frac{1}{c}\right)\alpha_2\right| < 1.$$

Inverting the LT, we get

$$\begin{aligned} b(t) &= (c-1)\mu g_{c-1}(t) \\ &= \frac{c-1}{t} e^{-(\lambda+c\mu)t} \sum_{n=0}^{\infty} (n+1)\rho^{(-1/2)(n+1)} I_{n+1}(2t\sqrt{c\lambda\mu}). \end{aligned} \tag{10.32}$$

In particular, when $c = 2$, $k = 1$, we get from (10.32)

$$b(t) = \frac{1}{t} e^{-(\lambda+2\mu)t} \sum_{n=0}^{\infty} (n+1)\rho^{(-1/2)(n+1)} I_{n+1}(2t\sqrt{2\lambda\mu}). \tag{10.33}$$

(where $\rho = \lambda/2\mu$).

3.10.3. Transient-State Distribution of the Output of an $M/M/c$ Queue

In Section 3.6.4 we saw that under steady state the output process of an $M/M/c$ queueing system is Poisson with the same rate as that of the input Poisson process. In Section 3.9.1 we discussed the transient-state properties of the $M/M/1$ queueing system. We also discussed the importance of the transient-state results. The transient-state properties of the output of an $M/M/c$ queue with c-servers have been investigated by Everitt and Downs (1984). We discuss their approach next.

Let λ and μ be the arrival and service rates, respectively, of an $M/M/c$ system. Let $N(t)$ denote the number of customers in the system at instant t, $A(t)$ denote the number of arrivals in $(0, t)$, and $D(t)$ denote the number of departures in $(0, t)$, so that

$$N(t) = N(0) + A(t) - D(t). \tag{10.34}$$

Now it can be easily seen that $\{N(t), t > 0\}$ is a discrete state space and continuous-time Markov process. Let

$$\begin{aligned} Pr\{D(t) = k, A(t) = n \mid N(0) = m\} \\ = P_t\{k, n \mid m\}, \end{aligned} \tag{10.35}$$

$$\begin{aligned} Pr\{D(t) = k \mid A(t) = n, N(0) = m\} \\ = P_t\{k \mid n, m\}, \text{ and} \end{aligned} \tag{10.36}$$

$$Pr\{A(t) = n \mid N(0) = m\} = P_t(n \mid m);$$

then

$$P_t\{k|n, m\} = \frac{P_t\{k, n|m\}}{P_t\{n|m\}} = \frac{P_t\{k, n|m\}}{\dfrac{e^{-\lambda t}(\lambda t)^n}{n!}} = \frac{e^{\lambda t}\lambda^{-n}P_t\{k, n|m\}}{\left(\dfrac{t^n}{n!}\right)}. \tag{10.37}$$

Denote

$$R_t(k|n, m) = \frac{t^n}{n!} P_t(k|n, m) = \lambda^{-n}e^{\lambda t}P_t(k, n|m). \tag{10.38}$$

Theorem 3.3. *The probabilities $P_t(k, n|m)$ and $R_t(k|n, m)$ satisfy the following differential-difference equations:*

$$\frac{d}{dt} P_t(k, n|m) = -\{\lambda + \mu \min(c, m + n - k)\}P_t(k, n|m) + P_t(k, n - 1|m) + \mu \min(c, m + n - k - 1)P_t(k - 1, n|m) \quad \text{and} \tag{10.39}$$

$$\frac{d}{dt} R_t(k|n, m) = -\mu \min(c, m + n - k)R_t(k|n, m) + R_t(k|n - 1, m) + \mu \min(c, m + n - k - 1)R_t(k - 1|n, m). \tag{10.40}$$

Proof. When the number of arrivals and departures are r and s, respectively, at any time, the number of servers at that time is $\min(c, m + r - s)$, and so the rate of service is $\mu \min(c, m + r - s)$. Using this fact, it can be seen that forward Chapman–Kolmogorov equations lead to the relation (10.39).

$$\text{Writing } P_t(k, n|m) = \lambda^n e^{-\lambda t}R_t(k|n, m)$$

and putting the expressions in terms of R_t in (10.39) and on simplifying, we at once get (10.40).

Note. The two relations (10.39) and (10.40) are equivalent. We get the behavior of the output process from both of them. But the relation (10.40) is simpler as it holds for all arrival rates λ (10.40 is independent of λ).

The general solution of (10.40) in time domain is difficult. A solution in terms of LST of R_t is given by Everitt and Downs (1984). (See Problems and Complements 3.15.)

3.11. Multichannel Queue with Ordered Entry

Two assumptions in usual multichannel queueing problems are:

(1) There is a single queue or waiting line for all the channels;

(2) Any unit who finds on entering the system that more than one service channel is free (or server idle) chooses a channel at random.

There is a class of problems, with practical applications, where neither of these two assumptions holds. Here each channel has its own queue or waiting line and the entering unit cannot choose the channel at random, but rather must use the first channel it comes to. We number the service channels $1, 2, \ldots, n$. The entering unit tries channel 1 first and if it is free he must go into it; if channel 1 is busy the arriving unit joins the queue in front of channel 1 unless the queue already there reaches the maximum queue length that can be accommodated there. In other words, if the maximum queue length is $(M - 1)$ in channel 1 so that the total capacity in channel 1 is M with the unit in service and if the entering unit finds fewer than M in the system in channel 1, he joins channel 1, while, if he finds M in the system in channel 1, he tries the next channel in order, that is, channel 2. In this way each arriving unit proceeds to test each service facility in order until he finds *one* channel with a number in the system for that channel less than the capacity of that channel. An arriving unit examines each channel from 1 to n in order; if he finds that each channel has reached the maximum capacity for that channel, then the unit leaves and is lost to the system.

This is a *multiqueue* problem with one separate queue before each channel and with *ordered entry*.

The queueing behavior in a supermarket is more or less similar to this problem. This model is useful in studying conveyor systems in industrial engineering. Disney (1962, 1963) first considered this model, We follow his approach here.

We make the following assumptions:

(1) Units arrive into the system from an infinite Poisson source with rate λ, i.e., arrivals are in accordance with a Poisson process with parameter λ.

(2) Service-time distribution in each channel is exponential with parameter μ.

(3) Each service facility contains one server and the queue discipline in each channel is FCFS.

We confine to steady-state analysis.

3.11.1. Two-Channel Model with Ordered Entry (with Finite Capacity)

Let M and N be the maximum capacity (for the queue and the unit under service) of channels 1 and 2, respectively. Assume that the system is in steady state. Let (i, j) be the state of the system where i and j denote the number in channel 1 and channel 2, respectively. Each value i, j includes the unit in service, if any, $0 \le i \le M$, $0 \le j \le N$, $p_{ij} = Pr\{\text{system is in state } (i, j)\}$.

The difference equations can be written by using the rate equality principle. Consider the state (0, 0). It can leave state (0, 0) through an arrival to channel 1 when the state becomes (1, 0). Thus, the rate of leaving state (0, 0) is $\lambda p_{0,0}$. It can enter the state (0, 0) either from the state (1, 0) through service completion of the unit under service in channel 1 or from the state (0, 1) through service completion of the unit under service in channel 2, the rate of entering state (0, 0) being $\mu p_{1,0} + \mu p_{0,1}$. Thus,

$$\lambda p_{0,0} = \mu p_{1,0} + \mu p_{0,1}. \tag{11.1}$$

Consider the state $(M, 0)$. It can leave this state though a service completion or through an arrival of a unit that will join channel 2 as channel 1 is full. Thus, the rate of leaving the state $(M, 0)$ is $\mu p_{M,0} + \lambda p_{M,0}$. It can enter the state $(M, 0)$ from either the state $(M - 1, 0)$ through an arrival or from the state $(M, 1)$ through a departure, the rate being $\lambda p_{M-1,0} + \mu p_{M,1}$. Thus,

$$\lambda p_{M,0} + \mu p_{M,0} = \lambda p_{M-1,0} + \mu p_{M,1}. \tag{11.2}$$

Consider the state (M, N). It can leave that state through a departure either from channel 1 or from channel 2, the rate of leaving being $2\mu p_{M,N}$. It can enter the state (M, N) either from the state $(M - 1, N)$ or from the state $(M, N - 1)$ through an arrival, the rate of entering being $\lambda p_{M-1,N} + \lambda p_{M,N-1}$. Thus,

$$2\mu p_{M,N} = \lambda p_{M-1,N} + \lambda p_{M,N-1}. \tag{11.3}$$

Consider the state $(0, N)$. It can leave that state either through an arrival to the channel 1 or from a departure from the channel 2, the rate being $(\lambda + \mu)p_{0,N}$. It can enter that state only from the state $(1, N)$ through a departure from the channel 1, the rate being $\mu p_{1,N}$. Thus,

$$(\lambda + \mu)p_{0,N} = \mu p_{1,N}. \tag{11.4}$$

Consider the state (M, L), $0 < L < N$. It can leave the state (M, L) either from the state $(M, L - 1)$ through an arrival who joins channel 2 or through a departure from either of the two channels, the rate being $(\lambda + 2\mu)p_{M,L}$. It can enter the state (M, L) from the state $(M - 1, L)$ through an arrival, from the state $(M, L - 1)$ through an arrival, or from the state $(M, L + 1)$ through a departure, the rate being $\lambda p_{M-1,L} + \lambda p_{M,L-1} + \mu p_{M,L+1}$. Thus,

$$(\lambda + 2\mu)p_{M,L} = \lambda p_{M-1,L} + \lambda p_{M,L-1} + \mu p_{M,L+1}. \tag{11.5}$$

Consider the state (K, N), $0 < K < M$. It can leave that state through an arrival who joins channel 1 or through a departure from either of the two channels, the rate being $(\lambda + 2\mu)p_{K,N}$. It can enter the state (K, N) either from the state $(K - 1, N)$ through an arrival or from the state $(K + 1, N)$ through a departure, the rate being $\lambda p_{K-1,N} + \mu p_{K+1,N}$. Thus,

$$(\lambda + 2\mu)p_{K,N} = \lambda p_{K-1,N} + \mu p_{K+1,N}. \tag{11.6}$$

Consider the state $(0, L)$, $0 < L < N$. It can leave that state through an arrival (to channel 1) or from a departure (from channel 2), the rate being $(\lambda + \mu)p_{0,L}$. It can enter the state $(0, L)$ from the state $(1, L)$ or from the state $(0, L + 1)$ through a departure, the rate being $\mu p_{1,L} + \mu p_{0,L+1}$. Thus,

$$(\lambda + \mu)p_{0,L} = \mu(p_{1,L} + p_{0,L+1}). \tag{11.7}$$

Consider the state $(K, 0)$, $0 < K < M$. It can leave that state through an arrival (to channel 1) or a departure (from channel 1), the rate being $(\lambda + \mu)p_{K,0}$. It can enter the state $(K, 0)$ either from the state $(K - 1, 0)$ through an arrival or from the states $(K + 1, 0)$ or $(K, 1)$ through a departure, the rate being $\lambda p_{K-1,0} + \mu p_{K+1,0} + \mu p_{K,1}$. Thus,

$$(\lambda + \mu)p_{K,0} = \lambda p_{K-1,0} + \mu p_{K+1,0} + \mu p_{K,1}. \tag{11.8}$$

Finally, consider the state (K, L), $0 < K < M$, $0 < L < N$. It can leave that state either through an arrival (to channel 1) or from a departure from either of the two channels, the rate being $(\lambda + 2\mu)p_{K,L}$. It can enter that state either from the state $(K - 1, L)$ through an arrival or from either of the states $(K + 1, L)$ or $(K, L + 1)$ through a departure, the rate being $\lambda p_{K-1,L} + \mu(p_{K+1,L} + p_{K,L+1})$. Thus,

$$(\lambda + 2\mu)p_{K,L} = \lambda p_{K-1,L} + \mu(p_{K+1,L} + p_{K,L+1}). \tag{11.9}$$

These nine sets of equations give the complete description of the system. Out of the preceding set of nine equations, one is dependent, implying that we solve all others and get p_{ij} in terms of one particular p_{kl}. Then using the normalizing condition $\sum_{i,j} p_{ij} = 1$, we can find this particular p_{kl} so that all

p_{ij} are completely determined. However, the equations of the system cannot be solved recursively as is usually done.

3.11.2. The Case $M = 1$, $N = N$

Consider the two-channel model with ordered entry and with no waiting space before channel 1 and with $N - 1$ waiting spaces before channel 2. Then we put $M = 1$, in the relevant equations. From (11.1), (11.7), and (11.4) and writing $\rho = \lambda/\mu$ we get

$$\begin{aligned} -\rho p_{0,0} + p_{0,1} + p_{1,0} &= 0 \\ -(1+\rho)p_{0,L} + p_{0,L+1} + p_{1,L} &= 0, \qquad 0 < L < N \qquad (11.10) \\ -(1+\rho)p_{0,N} + p_{1,N} &= 0. \end{aligned}$$

Denote

$$\mathbf{P}_0 = \begin{pmatrix} p_{00} \\ \vdots \\ p_{0i} \\ \vdots \\ p_{0N} \end{pmatrix}, \qquad \mathbf{P}_1 = \begin{pmatrix} p_{10} \\ \vdots \\ p_{1j} \\ \vdots \\ p_{1N} \end{pmatrix}. \qquad (11.11)$$

Writing

$$\mathbf{A}_{11} = \begin{pmatrix} -\rho & 1 & 0 & \cdots & 0 \\ 0 & -(1+\rho) & 1 & \cdots & 0 \\ \cdots & \cdots & \cdots & \cdots & \cdots \\ 0 & 0 & \cdots & \cdots & -(1+\rho) \end{pmatrix}, \qquad (11.12)$$

we can express the preceding equations as

$$\mathbf{A}_{11}\mathbf{P}_0 + \mathbf{P}_1 = \mathbf{O}. \qquad (11.13)$$

Putting $M = 1$, $N = N$ in equations (11.2), (11.3), and (11.5), we get

$$\begin{aligned} -(1+\rho)p_{1,0} + p_{1,1} + \rho p_{0,0} &= 0 \\ p_{1,L-1} - (\rho+2)p_{1,L} + p_{1,L+1} + p_{0,L} &= 0, \qquad 0 < L < N \qquad (11.14) \\ \rho p_{1,N-1} - 2p_{1,N} + \rho p_{0,N} &= 0. \end{aligned}$$

Writing

$$\mathbf{A}_{33} = \begin{pmatrix} -(1+\rho) & 1 & 0 & \cdots & 0 \\ \rho & -(2+\rho) & 1 & \cdots & 0 \\ 0 & 0 & -(2+\rho) & \cdots & 0 \\ \cdots & \cdots & \cdots & \cdots & \cdots \\ 0 & 0 & \cdots & \cdots & -2 \end{pmatrix}, \tag{11.15}$$

we can express the preceding equations (11.14) as

$$\mathbf{A}_{33}\mathbf{P}_1 + \rho\mathbf{I}\mathbf{P}_0 = \mathbf{O} \tag{11.16}$$

where $\mathbf{I}$ is the unit matrix of order $(N \times N)$. The equations (11.13) and (11.16) describe the system. From equation (11.13) we can immediately see that

$$\mathbf{P}_1 = -\mathbf{A}_{11}\mathbf{P}_0. \tag{11.17}$$

This relation allows $\mathbf{P}_1$ to be expressed in terms of $\mathbf{P}_0$.

Substituting in Eq. (11.16), we get

$$\begin{aligned} A_{33}(-\mathbf{A}_{11}\mathbf{P}_0) + \rho\mathbf{I}\mathbf{P}_0 &= \mathbf{O} \quad \text{or} \\ (-\mathbf{A}_{33}\mathbf{A}_{11} + \rho\mathbf{I})\mathbf{P}_0 &= \mathbf{O}, \end{aligned} \tag{11.18}$$

which is a set of homogeneous equations in terms of the unknown $\mathbf{P}_0$. The coefficient matrix is of rank N and hence this set of homogeneous equations can be solved and all terms of $\mathbf{P}_0$ can be solved in terms of any one of them. Using the relation

$$\sum_{i=0}^{1} \sum_{j=0}^{N} p_{ij} = 1, \tag{11.19}$$

we can find this term. Hence, the set of equations can be completely solved.

3.11.3. Particular Case: $M = N = 1$ (Overflow System)

Consider the two-channel model with ordered entry having no space for queue before any of the servers (except the ones in service, if any). Then putting $M = N = 1$ in Eqs. (11.1)–(11.4) we shall get the following equations

$$\lambda p_{0,0} = \mu p_{1,0} + \mu p_{0,1} \tag{11.20}$$

$$(\lambda + \mu) p_{1,0} = \lambda p_{0,0} + \mu p_{1,1} \tag{11.21}$$

$$2\mu p_{1,1} = \lambda p_{0,1} + \lambda p_{1,0} \tag{11.22}$$

$$(\lambda + \mu) p_{0,1} = \mu p_{1,1}. \tag{11.23}$$

We can use the matrix method outlined previously. Here we solve them recursively. Denote $\rho = \lambda/\mu$. From (11.23) we get

$$p_{1,1} = (1 + \rho)p_{0,1}. \tag{11.24}$$

Putting this value of $p_{1,1}$ in (11.22) we get

$$p_{1,0} = \left(\frac{1}{\rho}\right)[2(1 + \rho) - \rho]p_{0,1} = \left(\frac{2 + \rho}{\rho}\right)p_{0,1}. \tag{11.25}$$

From (11.21) we then get

$$\begin{aligned} p_{0,0} &= \left(\frac{1}{\rho}\right)[(1 + \rho)p_{1,0} - p_{1,1}] \\ &= \frac{2(1 + \rho)}{\rho^2} p_{0,1}. \end{aligned} \tag{11.26}$$

Equation (11.20) is the dependent equation. Using $\sum_{i,j=0}^{1} p_{i,j} = 1$, we get from Eqs. (11.24)–(11.26),

$$p_{0,1}\left[1 + (1 + \rho) + \frac{2 + \rho}{\rho} + \frac{2(1 + \rho)}{\rho^2}\right] = 1 \quad \text{or} \tag{11.27}$$

$$p_{0,1} = \frac{\rho^2}{(\rho + 1)(\rho^2 + 2\rho + 2)}$$

and so

$$\begin{aligned} p_{1,0} &= \frac{\rho(2 + \rho)}{(\rho + 1)(\rho^2 + 2\rho + 2)}, \\ p_{1,1} &= \frac{\rho^2}{\rho^2 + 2\rho + 2}, \quad \text{and} \\ p_{0,0} &= \frac{2}{\rho^2 + 2\rho + 2}. \end{aligned} \tag{11.28}$$

If N denotes the total number in the system (in both the channels combined) then

$$Pr(N = 0) = p_{0,0} = \frac{1}{\displaystyle\sum_{r=0}^{2} \frac{\rho^r}{r!}}.$$

$$Pr(N = 1) = p_{1,0} + p_{0,1} = \frac{\rho}{\sum_{r=0}^{2} \frac{\rho^r}{r!}}, \quad \text{and} \tag{11.29}$$

$$Pr(N = 2) = p_{1,1} = \frac{\rho^2/2}{\sum_{r=0}^{2} \frac{\rho^r}{r!}}$$

as is expected, these being the corresponding probabilities for the system $M/M/2/2$. Here $p_{1,1}$ is the probability that both the channels are busy, i.e., the probability that an arriving unit will be lost to the system. This gives the Erlang's loss formula for $M/M/2/2$.

The expected total number in the system equals

$$\sum_{k=0}^{2} kPr(N = k) = \frac{\rho(\rho + 1)}{\sum_{r=0}^{2} \frac{\rho^r}{r!}}. \tag{11.30}$$

Disney (1962, 1963) discusses further analysis of the working of such a system.

Notes.

(1) The system is called an overflow network system. (See Chapter on Networks.) The sojourn times in such a system are of simple structure and can be found trivially. For example, if S_m is the stationary sojourn time of the mth customer, then

$$Pr(S_m = 0) = p_{1,1} = \frac{\rho^2}{(\rho^2 + 2\rho + 2)}$$

$$Pr(S_m \le t) = \frac{2(1 + \rho)(1 - e^{-\mu t})}{(\rho^2 + 2\rho + 2)}.$$

(2) For the two-channel overflow model, (see Fig. 3.2). Consider an *overflow system*: The two-channel ordered-entry model with no waiting space, $M = N = 1$.

If $t_{m,j}$ is the time of the mth overflow from server j ($j = 1, 2$), then in equilibrium, the *interoverflow time* $\{t_{m+1,j} - t_{m,j}, m = 1, 2, \ldots\}$ is a sequence of IID random variables. The equilibrium overflow (interoverflow) distribution function $\Phi_j(x)$, from the jth server, $j = 1, 2$, satisfies the intergral equation (Palm's equation)

$$\Phi_j(x) = \int_0^x e^{-\mu_j v} d\Phi_{j-1}(v) + \int_0^x (1 - e^{-\mu_j v})\Phi_j(x - v) d\Phi_{j-1}(v)$$

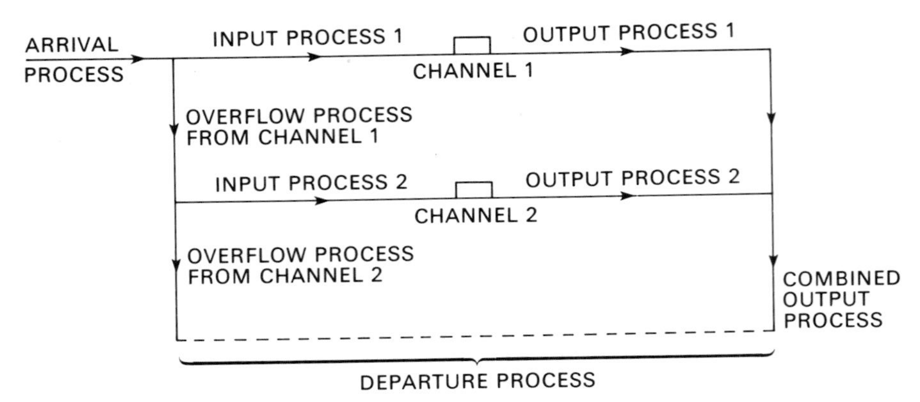

FIGURE 3.2. Diagram of a two-channel overflow ordered entry queue.
Figures 3.1 and 3.2 with kind permission from the authors, R. L. Disney and D. König, and the publishers of *SIAM Review*.

with

$$\Phi_0(x) = 1 - e^{-\lambda x}, \qquad x \geq 0.$$

The overflow distribution for the first server is given by

$$\Phi_1(x) = ae^{-bx} + (1 - a)e^{-cx}$$

where a, b, and c are functions of λ, μ_i (the rate of the ith server, $i = 1, 2$). The preceding shows that the overflow distribution is a sum of exponential terms and that the overflow process is not Poisson, but is a renewal process.

The overflow networks first appeared in telephony. Palm (1943) considered overflow distribution. The integral equation involving $\Phi_j(x)$ is known as *Palm's integral equation.* Khinchine (1960), considering Palm's treatment, introduced the idea of Palm functions into modern point process theory.

3.11.4. *Output Process*

If $t^0_{m,1}$ is the time of the mth output from server 1, then in equilibrium $\{t^0_{m+1,1} - t^0_{m,1}, m = 1, 2, \ldots\}$ is a sequence of IID random variables whose distribution is the sum of two independent RVs: the service-time distribution and the interarrival-time distribution. (See Disney and König, 1985.)

Remark. The queueing system with ordered entry has received considerable attention because of its importance in application, mainly in conveyor theory. (See Muth, and White 1979, for a survey.) For further work in this area, reference may be made, for example, to Elsayed (1983), Elsayed and Elayat (1976), Elsayed and Proctor (1977), Gregory and Litton (1975), Lin and

Elsayed (1978), Matsui and Fukuta (1977), Nawijn (1983, 1984), Newell (1984), Pourbabai (1987), Pourbabai and Sonderman (1986), Pritsker (1966), Proctor *et al.* (1977), Sonderman (1982), Yao (1987) and Shanthikumar, and Yao (1987).

Apart from conveyor systems, the ordered entry model also applies to communication networks, such as System Network Architecture (SNA) [see for example, Gray and Mcneill (1979)]. A third area of application is data-base systems, see, for example, Cooper and Solomon (1984).

Problems and Complements

3.1. (a) Combination of service channels: Consider that two identical $M/M/1$ queueing systems with the same rates λ, μ (intensity $\rho = \lambda/\mu$) are in operation side by side (with separate queues) in a premises. Show that the distribution of the total number N in the two systems taken together is given by

$$Pr(N = n) = (n + 1)(1 - \rho)^2\rho^n,\ n \geq 0.$$

(b) Consider (i) c number of $M/M/1$ queues each with rates λ and μ, (ii) a standard $M/M/c$ queue with rates $c\lambda$ and μ. Show that the average waiting time is less for the system (ii) for any $c > 1$. (This provides justification for having a common queue for a multi-server system in a bank, reservation counter and so on). [See Smith and Whitt (1981); also Rothkopf and Rech (1987)].

3.2. Dufcova and Zitek's (1975) model of an $M/M/1$ system with a class of queueing disciplines

Suppose that when the server becomes free, he accepts either the first in the queue with probability δ or the last in the queue with probability $1 - \delta$, so that $\delta = 1$ implies FCFS and $\delta = 0$ LCFS disciplines. Show the LST of the waiting-time distribution is given by

$$w_q^*(s) = (1 - \rho) + \rho\left[1 - \frac{sR}{\lambda(1 - R)}\right]$$

where $R = R(\delta)$ is the unique real root in (0, 1) of the equation

$$(1 - \delta)\mu x^2 - (\mu + \lambda - \delta\lambda + s)x + \lambda = 0;$$

show that the mean waiting time

$$E(W_q) = \frac{\lambda}{\mu(\mu - \lambda)}$$

is independent of R and of δ (as can be expected, it is independent of the queue discipline); and

$$\operatorname{var}(W_q) = \frac{2\lambda}{(\mu-\lambda)^2(\mu-\lambda+\delta\lambda)} - \frac{\rho^2}{(\mu-\lambda)^2}$$

and the extreme values of $\operatorname{var}(W_q)$ are for $\delta = 1$ and $\delta = 0$. Show that the results also hold good for the $M/M/c$ model.

3.3. An $M/M/1$ queue with control-limit policy and exponential start-up time (Baker, 1973)

Here the control policy is to turn off the system and withdraw the server when the system becomes empty and to turn on the system when the system size reaches $n(>0)$. When the system is turned on, it cannot immediately serve customers but requires some time to start up. It is assumed that the time required for start-up is exponential with mean $1/\gamma$. The interarrival- and service-time distributions are exponential with means $1/\lambda$ and $1/\mu$, respectively. Suppose that $P_{i,s}$ denotes the steady-state probability that there are i customers in the system and the server state is s (where $s = 0$ implies an idle state of the server in which the start-up has not begun or has not been completed) and $s = 1$ denotes the busy state of the server (in which service is being performed). Suppose that $P_i = P_{i,0} + P_{i,1}$ is the steady-state probability that there are i in the system.

Denote

$$\rho = \frac{\lambda}{\mu}\ (<1),$$

$$\theta = \frac{\lambda}{(\lambda+\gamma)}, \quad \text{and}$$

$$\omega = \frac{\gamma}{\mu}.$$

Show that the system satisfies the set of equations

$$P_{1,1} = \rho P_{0,0},$$

$$P_{i,0} = P_{i-1,0}, \qquad 1 \le i \le n-1,$$

$$P_{i,0} = \theta P_{i-1,0}, \qquad n \le i,$$

$$P_{i,1} = (1+\rho)P_{i-1,1} - \rho P_{i-2,1}, \qquad 2 \le i \le n,$$

$$P_{i,1} = (1+\rho)P_{i-1,1} - \rho P_{i-2,1} - \omega P_{i-1,0}, \qquad n+1 \le i.$$

Show further that

$$P_i = \alpha(1 - \rho^{i+1}), \qquad 0 \le i \le n-1,$$

$$P_i = \frac{\alpha[(1-\rho^{n-1})\rho^{i-n+2} + (1-\rho)(\rho^{i-n+2} - \theta^{i-n+2})]}{(\rho - \theta)}, \qquad n \le i,$$

where $\rho \neq \theta$, and $\alpha = (1-\theta)/[\theta + n(1-\theta)]$. Further show that the mean number in the system $E(N)$ equals

$$E(N) = \sum_{i=0}^{\infty} iP_i = \frac{\alpha n(n-1)}{2} + \frac{(\rho - 2\rho\theta + \theta)}{(1-\rho)(1-\theta)}.$$

See Borthakur *et al.* (1987) for a more general model, and Böhm and Mohanty (1990) for transient solution (through discrete-time analogue and use of combinatorial arguments) of $M/M/1$ queue under control limit policy and zero start-up time.

3.4. Naor's model for regulation of queue size (Naor, 1969)

Suppose that the cost to a customer of staying in a queue (i.e., for queueing) is c per unit time and that the reward collected at the end of service is R. Because of these costs a newly arrived customer weighs two alternatives: to join or not to join the queue by the net gains associated with them. For "individual optimization," show that the critical number n_s of customers is given by

$$n_s = \left[\frac{R\mu}{c}\right]$$

where $[\alpha]$ is the largest integer not exceeding α. For overall or "collective optimization," show that the critical number n_0 satisfies

$$\frac{n_0(1-\rho) - \rho(1-\rho^{n_0})}{(1-\rho)^2} \le \frac{R\mu}{c} < \frac{(n_0+1)(1-\rho) - \rho(1-\rho^{n_0+1})}{(1-\rho)^2},$$

that is, n_0 equals

$$n_0 = [v_0]$$

where v_0 is given by

$$\frac{v_0(1-\rho) - \rho(1-\rho^{v_0})}{(1-\rho)^2} = \frac{R\mu}{c};$$

further that $v_0 \le R\mu/c$, where the sign of equality holds if $R\mu/c$ equals unity. (λ, μ and ρ denote, respectively, the mean arrival rate, mean service rate, and traffic intensity in an $M/M/1$ queue with balking.)

3.5. De Vany's Model (1976)

De Vany uses the $M/M/1/K$ model to determine the effective demand function under the conditions: (1) the arrival stream is Poisson with rate $\lambda(p)$ (where p is price), (2) customers' orders are serviced on an FCFS basis with exponentially distributed service time with parameter μ, and (3) customers have a common balking value K that depends on the expected benefits of purchase at the firm relative to the expected benefits of purchase from an alternative supplier. Show that the effective demand for service equals

$$\begin{aligned}\lambda' &= \lambda[1 - p_K] \\ &= \lambda(p)[1 - B(\lambda(p), \mu, K)]\end{aligned}$$

where $B = p_K$ is the balking probability. Note that $\lambda' < \lambda$; this implies that there is a kind of excess demand of the firm's product in the sense that some potential customers arrive but turn away to the alternative firm as the queue they find on arrival is at the balking length K.

The equilibrium mean rate of demand λ' is a function of price, mean capacity μ, and the balking value K.

Taking derivatives shows that the price affects both the arrival rate and the proportion of those who stay on. A higher price reduces both the arrival rate and the balking value K, which causes a greater proportion of arrivals to balk.

Show that the demand curve is less elastic than the potential curve.

3.6. Show that for the $M/M/1$ queueing system starting with k customers at time 0, the joint distribution of the busy period T_k and the number $N(T_k)$ served during the busy period initiated by k customers is given by

$$P\{t \le T_k < t + dt, N(T_k) = n\} = e^{-(\lambda+\mu)t} \frac{k\lambda^{n-k}\mu^n t^{2n-k-1}}{n!(n-k)!} dt.$$

Hence, find the marginal distributions, that is, the PDF for T_1 (busy period initiated by a single customer) and the distribution of $N(T_1) = N$ (Prabhu, 1965).

3.7. Show that the expected busy period for an $M/M/1/K$ queueing system equals

$$E(T) = \frac{1 - \rho^{K+1}}{\mu(1 - \rho)}.$$

(Hints: Use $E(T)/E(I) = (1 - p_0)/p_0$. Here $E(I) = 1/\lambda'$ where λ' is the mean *effective* arrival rate.)

3.8. Show that for the $M/M/c$ system in steady state, the PGF of the number in the queue is given by

$$G(z) = \frac{1 - B^*[\lambda(1-z)/c]}{B^*[\lambda(1-z)/c] - z} p_{c-1} + \sum_{i=0}^{c-1} p_i$$

where $B^*(s) = \mu/(\mu + s)$ is the LST of service-time distribution.

3.9. $M/M/c$ queue with servers' vacations (Levy and Yechiali, 1976)

Consider an $M/M/c$ system in which a server proceeds on vacation when he has no unit to serve (the length of time he is in vacation being given by an exponential random variable with parameter θ) and in which the server proceeds on another vacation if he finds the queue empty on return. Suppose that the system is in steady state.

Find the joint distribution of the number of busy servers β and the number of customers N in the system

$$Pr(N = k, \beta = r), \qquad r = 0, 1, 2, \ldots, \qquad k \geq r.$$

Show that the average number of busy servers is λ/μ (which is the same as the average number of busy servers in an $M/M/c$ queue).

Show that the number of customers N in the system when all servers are on vacation has a geometric distribution given by

$$Pr(N = k) = \frac{c\theta}{\lambda + c\theta}\left(\frac{\lambda}{\lambda + c\theta}\right)^k.$$

Show that for $c = 2$, the average number of customers in the system is given by

$$L = \frac{\rho}{1-\rho} + \frac{\alpha\lambda[\lambda(1 - z_1) + \theta]}{\theta}, \qquad \rho = \frac{\lambda}{2\mu},$$

where

$$\alpha = [\lambda(1 - z_1^2) + 2\theta z_1]^{-1} \quad \text{and}$$

$$z_1 = \frac{(\lambda + \mu + \theta) - \{(\lambda + \mu + \theta)^2 - 4\lambda\mu\}^{1/2}}{2\lambda}.$$

(For queues with vacation, see Section 8.3.)

3.10. $M/M/1$ queue: waiting time in the system for an arrival at instant t, (virtual waiting time in the system).

Show that

$$F(x, t) = P\{W_s \le x | t\}$$
$$= 1 - e^{-\mu x} \sum_{n=0}^{\infty} \sum_{s=0}^{n} \frac{(\mu x)^s}{s!} p_n(t)$$

and the PDF of W_s is given by

$$w(x, t) = \frac{d}{dx} F(x, t)$$
$$= e^{-\mu x} \sum_{n=0}^{\infty} \sum_{s=0}^{n} \frac{(\mu x)^s}{s!} p_n(t) - e^{-\mu x} \sum_{n=0}^{\infty} \sum_{s=1}^{n} \frac{(\mu x)^{s-1}}{(s-1)!} p_n(t).$$

3.11. Suppose that a machine breaks down, independently of others, in accordance with a Poisson process, the average length of time for which a machine remains in working order being 36 hours. The duration of time required to repair a machine has an exponential distribution with mean 1 hour. Suppose that there are 10 machines and 1 mechanic. Find

(i) the probability that five or more machines will remain out of orders at the same time;
(ii) the average number of machines in working order;
(iii) the fraction of time, on the average, the mechanic will be busy; and
(iv) the average duration of time for which a machine is not in working order.

3.12. Consider a machine-repairperson problem with c repairpeople, m machines ($c < m$), and exponential working time and repair time having rates λ and μ, respectively. Suppose that m is very large. Show that (with notations as in Section 3.8.2)

(i) All the $a_0, a_1, \ldots, a_{c-1}$ approach zero; so also do $p_0, p_1, \ldots, p_c$ (so that all the c repairpeople are almost 100% busy at all times).
(ii) The distribution of the number of machines in working order is approximately given by the distribution of the number of busy servers in an $M/M/\infty$ queue with mean interarrival time $1/c\mu$ and mean service time $1/\lambda$.
(iii) The LST $w^*(s)$ approaches

$$\left[\exp\left(\frac{s}{\lambda}\right)\right]\left[\left(\frac{\mu}{s+\mu}\right)\right]\left[\frac{c\mu}{s+c\mu}\right]^{m-c}$$

and the response time is asymptotically and approximately normal with

$$\text{mean} = \frac{m}{c\mu} - \frac{1}{\lambda} \quad \text{and}$$

$$\text{variance} = \frac{m-c}{(c\mu)^2} + \frac{1}{\mu^2}$$

(Wong, (1979).

3.13. Transient solution of an $M/M/1$ queue; alternative approach (Parthasarathy, 1987). Define

$$\begin{aligned} q_k(t) &= \{\exp(\lambda + \mu)t\}[\mu p_k(t) - p_{k-1}(t)], \qquad k = 1, 2, \ldots \\ &= 0, \qquad k = 0, -1, -2, \ldots, \end{aligned}$$

$$\alpha = 2\sqrt{\lambda\mu}, \qquad \beta = \sqrt{\frac{\lambda}{\mu}} = \sqrt{\rho}$$

$I_n(t)$ is a modified Bessel function of order n. Assume that $Pr\{N(0) = a\} = 1$. Show that

$$\begin{aligned} q_k(t) = {} & \mu\beta^{k-a}(1 - \delta_{0a})[I_{n-a}(\alpha t) - I_{n+a}(\alpha t)] \\ & + \lambda\beta^{k-a-1}[I_{n+a+1}(\alpha t) - I_{n-a-1}(\alpha t)] \end{aligned}$$

and that, for $n = 1, 2, \ldots,$

$$\begin{aligned} p_n(t) &= Pr(\text{queue length at time } t \text{ is } n) \\ &= \left(\frac{1}{\mu}\right) \exp\{-(\lambda + \mu)t\} \sum_{k=1}^{n} q_k(t)\rho^{n-k} + \rho^n p_0(t) \quad \text{and} \end{aligned}$$

$$p_0(t) = \int_0^t q_1(y) \exp\{-(\lambda + \mu)y\} dy + \delta_{0a}.$$

Note: See also Syski (1988). The above solution involves one integral and one finite series. The result is shown (by Syski) to be equivalent to Cohen's result [1982, (4.31) p. 82] involving three integrals.

3.14. $M/M/1$: Two-dimensional state model

Let the state of the system by time t be given by the ordered pair (i, j), where i is the number of arrivals and j is the number of departures by time t, and let $p_{ij}(t)$ denote the probability that the system is in state (i, j) by time t. Let $f_{ij}(s)$ be the LT of $p_{ij}(t)$.

Show that $p_{ij}(t)$ satisfy the differential equations

$$p'_{00}(t) = -\lambda p_{00}(t) \tag{A}$$

$$p'_{i0}(t) = \lambda p_{i-1,0}(t) - (\lambda + \mu)p_{i0}(t), \qquad i \geq 1, \tag{B}$$

$$p'_{ii}(t) = \mu p_{i,i-1}(t) - \lambda p_{ii}(t), \qquad i \geq 1, \tag{C}$$

$$p'_{ij}(t) = \mu p_{i,j-1}(t) + \lambda p_{i-1,j}(t) - (\lambda + \mu)p_{ij}(t), \quad i \geq 2,\ 1 \leq j < i. \tag{D}$$

Show by induction that $f_{ij}(s)$ is given by

$$f_{0,0}(s) = \frac{1}{\lambda + s}$$

$$f_{ij}(s) = \left(\frac{\lambda}{\lambda + \mu + s}\right)^i \left(\frac{\mu}{\mu + s}\right)^j \sum_{k=0}^{j} \frac{(i-k)(i+k-1)!}{k!i!} \frac{(\lambda + s)^{k+1}}{(\lambda + \mu + s)^k}$$

$$\text{for} \quad i \geq 1,\ 0 \leq j \leq i. \tag{E}$$

Show further that

$$p_{0,0}(t) = e^{-\lambda t}$$

and, for $i \geq 1$, $0 \leq j \leq i$,

$$p_{ij}(t) = \left(\frac{\lambda}{\mu}\right)^i \frac{(\mu t)^j e^{-\lambda t}}{i!} \sum_{k=0}^{j} \frac{(i-k)}{k!}$$

$$\times \left\{ \sum_{m=0}^{j-k} \frac{(-1)^m (m+i+k-1)!}{m!(j-k-m)!(\mu t)^{m+k}} \left[1 - e^{-\mu t} \sum_{r=0}^{m+i+k-1} \frac{(\mu t)^r}{r!} \right] \right\}$$

$$= \frac{(\lambda t)^i (\mu t)^j \exp(-\lambda t - \mu t)}{i!j!} \left[\frac{j}{\mu t} + \left(1 - \frac{j}{\mu t}\right) \frac{i!}{(\mu t)^i} \sum_{r=i}^{\infty} \frac{(\mu t)^r}{r!} \right]. \tag{F}$$

Again $p_n(t) = P\{N(t) = n\}$ can be obtained from $p_{i,j}(t)$ as follows.

$$p_n(t) = P\{N(t) = n\} = \sum_{j=0}^{\infty} p_{n+j,j}(t). \tag{G}$$

Show further that the fraction of time the server is idle until time t is given by

$$I(t) = \frac{1}{t} \int_0^t \sum_{j=0}^{\infty} p_{jj}(\tau) d\tau$$

and the fraction of time the server is busy until time t is given by

$$B(t) = 1 - I(t)$$

(Pegden and Rosenshine (1982), Hubbard *et al.* (1986)).

3.15. Transient output distribution for an $M/M/c$ system

Let $R_t(k|n, m)$ etc., be defined as in Section 3.10.3, and let

$$R^*(k|n, m) = R_z^*(k|n, m) = \int_0^\infty e^{-zt} R_t(k|n, m)dt$$

be the LT of $R_t(k|n, m)$. Then (10.40) can be put as

$$[z + \mu \min(c, m + n - k)]R^*(k|n, m)$$
$$= R^*(k|n - 1, m) + \mu \min(c, m + n - k - 1)R^*(k - 1|n, m). \quad \text{(A)}$$

For the initial condition $P_t(0|0, m) = \exp\{-\mu \min(c, m)t\}$ show that the solution of (A) is given by

$$R^*(k|n, m) = \left[\frac{\prod_{j=n-k+1}\{z + \mu \min(c, m + j)\}}{\prod_{j=0}^{n-k}\{z + \mu \min(c, m + j)\}}\right]$$
$$\times \sum_{i_k=0}^{n} \sum_{i_{k-1}=0}^{i_k} \cdots \sum_{i_1=0}^{i_2} W(m - k + 1 + i_k) \quad \text{(B)}$$
$$\times W(m - k + 2 + i_{k-1}) \cdots W(m - 1 + i_2)W(m + i_1),$$

where $\prod_{j=p}^{q}(.)$ is defined to be 1 whenever $q < p$, (B1)

and

$$W(i) = \begin{cases} \dfrac{\mu \min(c, i)}{[z + \mu \min(c, i - 1)][z + \mu \min(c, i)]}, & i \geq 1 \\ 0, & i \leq 0. \end{cases} \quad \text{(B2)}$$

Further, show that

$$\lim_{t \to \infty} (\mu t)^{n-k} P_t(k|n, 0) = \lim_{t \to \infty} \left(\frac{n!\mu^{n-k}}{t^k}\right) R_t(k|n, 0)$$
$$= \frac{n!\mu^{n-k}}{k!} \lim_{z \to 0} z^{k+1} R_z^*(k|n, 0). \quad \text{(C)}$$

An explicit solution of $P_t(k|n, 0)$ when $N(0) = m = 0$ is also given by Everitt and Downs (1984).

3.16. Multiserver queue with balking and reneging

Consider a c-server queueing system with Poisson input with rate λ and exponential service time with rate μ for each of the c-servers. Suppose that (i) an arriving customer who finds all the c-servers busy on arrival may balk (leave without joining the system) with probability q or may join the system with probability $p(=1-q)$; (ii) after joining the queue, a customer may renege independently of others: he waits for a random length of time for service to begin, the length of time being an exponential random variable with parameter α, otherwise he departs; and (iii) a customer who balks or reneges and decides to return later is considered as a new arrival independent of his previous balking or reneging.

Find the differential-difference equations of the state $N(t)$ of the system. If $p_n(t) = P\{N(t) = n\}$ and $\lim_{t\to\infty} p_n(t) = p_n$, then show that

$$\begin{aligned} p_n &= \frac{1}{n!}\left(\frac{\lambda}{\mu}\right)^n p_0, \qquad n \le c, \\ &= \frac{(\lambda p)^{n-c}}{(c\mu+\alpha)(c\mu+2\alpha)\cdots[c\mu+(n-c)\alpha]}\, p_c, \qquad n > c. \end{aligned}$$

Find p_n when there is only balking and no reneging. Also deduce Erlang's loss formula. (Haghighi-Montazer *et al.*, 1986).

3.17. Two-channel model with ordered entry with $M = N = 1$

Suppose that the service rate at the two channels are different, being μ_1 and μ_2 at the first and second channel, respectively. Show that the steady-state probabilities are given by

$$p_{0,1} = \frac{\lambda^2\mu_1}{C},$$

$$p_{1,0} = \frac{\lambda\mu_2(\lambda+\mu_1+\mu_2)}{C},$$

$$p_{1,1} = \frac{\lambda^2(\lambda+\mu_2)}{C}, \quad \text{and}$$

$$p_{0,0} = \frac{\mu_1\mu_2(2\lambda+\mu_1+\mu_2)}{C},$$

where $C = (\lambda+\mu_1)[(\lambda+\mu_2)^2 + \mu_1\mu_2]$. Show further that the waiting time W in steady state has the distribution given by

$$\begin{aligned} P\{W \le t\} = [\{\mu_1\mu_2(2\lambda+\mu_1+\mu_2) + \lambda^2\mu_1\}\{1 - e^{-\mu_1 t}\} \\ + \lambda\mu_2(\lambda+\mu_1+\mu_2)(1 - e^{-\mu_2 t})]/C, \qquad t > 0 \end{aligned}$$

(Disney, 1962). Obtain the corresponding results for the particular case $\mu_1 = \mu_2$.

3.18. Two-channel model with ordered entry and with $M = 1$, $N = 3$

Let $P_{i,j}$ denote the probability that the number in channel 1 is i and that the number in channel 2 is j. (i, j include those in service, if any.) Show that

$$P_{0,0} = \frac{4(\rho+1)(\rho+2)}{A} \qquad p_{1,0} = \frac{\rho(\rho^2+8\rho+8)}{A}$$

$$P_{0,1} = \frac{\rho^2(3\rho+4)}{A} \qquad P_{1,1} = \frac{\rho^2(\rho+1)(\rho+4)}{A}$$

$$P_{0,2} = \frac{2\rho^3(\rho+1)}{A} \qquad P_{1,2} = \frac{\rho^3(\rho+1)(\rho+2)}{A}$$

$$P_{0,3} = \frac{\rho^4(\rho+1)}{A} \qquad P_{1,3} = \frac{\rho^4(\rho+1)^2}{A}$$

where

$$A = \rho^6 + 4\rho^5 + 8\rho^4 + 13\rho^3 + 20\rho^2 + 20\rho + 8, \qquad \rho = \frac{\lambda}{\mu}$$

(Disney, 1962).

3.19. $M/M/c$ queue ($c \geq 1$): transient state distribution.
Suppose that $k(\geq 1)$ customers are already present at time $t_0 = 0$ and that the nth new customer arrives at time t_n ($n \geq 1$).

Denote

X_n = number of customers present at time $t_n + 0$ (including the nth arrival at t_n).

$P_k(n, i) = Pr\{X_n = i \mid \text{number present at time } t_0\ (=0) \text{ is } k\}$

$\rho = \lambda/c\mu$ (which need not be <1).

[The first arriving customer at t_1 need not find k customers present but will find j ($0 \leq j \leq k$) customers.]

If $k \geq 1$ and $n \geq 1$, show that

$$P_k(n, k+n) = \left[\frac{\rho}{(\rho+1)}\right]^n \quad \text{if} \quad k \geq c$$

$$= \frac{\rho^n}{\prod_{j=1}^{n}\left[\rho + \dfrac{k+j-i}{c}\right]} \quad \text{if} \quad k+n \leq c$$

$$= \frac{\rho^n}{\left[(\rho+1)^{n-c+k}\prod_{j=1}^{c-k}\left\{\rho + \dfrac{k+j-1}{c}\right\}\right]} \quad \text{if} \quad k < c < k+n.$$

If $k = 0$, then for $n \geq 1$, show that

$$P_0(n, n) = \frac{\rho^n}{\prod_{j=1}^{n}\left[\rho + \frac{j-1}{c}\right]} \quad \text{if} \quad n \leq c$$

$$= \frac{\rho^n}{\left[(\rho+1)^{n-c}\prod_{j=1}^{c}\left\{\rho + \frac{(j-1)}{c}\right\}\right]} \quad \text{if} \quad n > c.$$

If $k \geq 1$, then for $2 \leq i \leq k$

$$P_k(1, i) = \left[\frac{\rho}{\left\{\rho + \frac{(i-1)}{c}\right\}} \prod_{j=1}^{k-i+1}\left\{1 - \frac{\rho}{\left\{\rho + \frac{(k-j+1)}{c}\right\}}\right\}\right] \quad \text{if} \quad k \leq c$$

$$= \frac{\rho}{(\rho+1)^{k-i+2}}, \qquad k > c, \qquad i > c$$

$$= \left[\frac{\rho}{\left\{(\rho+1)^{k-c+1}\left[\frac{\rho + (i-1)}{c}\right]\right\}}\right]\left[\prod_{j=1}^{c-i}\left\{1 - \frac{\rho}{[\rho + (c-j)/c]}\right\}\right],$$

$$i \leq c \leq k.$$

Let D_n = waiting time (or delay) in queue of the nth customer, and $G_q(x; n)$ = DF of an Erlang-q RV with mean q/η

$$= 1 - \{\exp(-\eta x)\}\sum_{j=0}^{q-1}\frac{(\eta x)^j}{j!}, \qquad x \geq 0.$$

Then show that

$$F_n(x) = P(D_n \leq x) = \sum_{i=1}^{s} P_k(n, i) + \sum_{i=c+1}^{k+n} G_{i-c}(x; c\mu)P_k(n, i) \quad \text{and}$$

$$E(D_n) = \left(\frac{1}{c\mu}\right)\left[\sum_{i=c+1}^{k+n}(i-c)P_k(n, i)\right].$$

(See Kelton and Law, 1985, who also consider the implication for steady-state simulation.)

3.20. Multichannel queue with ordered entry and heterogeneous servers (Mastsui and Fukuta, 1977)

Consider an ordered entry Poisson input queue (with rate λ) with c exponential servers, the ith server having rate μ_i. Suppose that the system is in steady state and that there is no waiting line before any of the servers. Denote

$$m_i = \frac{\mu_i}{\lambda}, \qquad i = 1, 2, \ldots, c$$

P_0 = probability that the system is idle

$P_{i_1, \ldots, i_k}$ = probability that i_1th, ..., i_kth channels are busy and others are idle ($i_k < c$)

$P_{1, 2, \ldots, c}$ = probability that the system is completely busy (with all the servers busy).

This gives the overflow probability.

Show that for $c = 2$

$$P_{1,2} = \frac{1 + m_2}{(1 + m_1)\{(1 + m_2)^2 + m_1 m_2\}}$$

and that the faster server should be assigned to the first channel to decrease the overflow probability (as should be intuitively clear).

Examine the case $c = 3$.

Note. See Yao (1987) for further results of such a system; also for comparison of various server arrangements and development of partial order.

3.21. Consider a three-channel Poisson queue with ordered entry having no waiting space (as in Section 3.11.3). Find the steady-state probabilities and verify the results with those of the corresponding loss system.

3.22. Multiserver Poisson queue with ordered entry (Nawijn, 1983). Consider the following two c-channel systems with ordered entry such that the c-channels are numbered $1, 2, \ldots, c$ and an arriving customer who finds a free channel joins the one with the lowest index:

(A) $M/M/c$ system (*queueing* or *delay* system)

(B) $M/M/c/c/$ system (*loss* system).

Denote

$p = \lambda/\mu$, $\rho = p/c < 1$

N = number in the $M/M/c$ system (A)

$B(k, p)$ = Erlang's loss formula for $M/M/k/k$, $k = 1, 2, \ldots, c$ (see Equation 7.3))

u_k = utilization factor of channel k in system (A)

v_k = utilization factor of channel k in system (B).

Show that, for $k = 1, 2, \ldots, c$,

$$v_k = p[B(k-1, p) - B(k, p)], \text{ and}$$

$$u_k = v_k Pr\{N \leq c\} + Pr\{N > c\}.$$

Verify that $\{v_k\}$ is monotone decreasing in k. Give an intuitive explanation. Find $\sum_{k=1}^{c} u_k$ and interpret the result.

References

Abate, J., and Whitt, W. (1987). Transient behavior of $M/M/1$ queue starting at the origin. *Queueing Systems* **2**, 42–66.

Abate, J., and Whitt, W. (1988). A transient behavior of the $M/M/1$ queue via Laplace transforms. *Adv. Appl. Prob.* **20**, 145–178.

Albin, S. L. (1984). Analyzing $M/M/1$ queues with perturbations in the arrival process. *J. Opnl. Res. Soc.* **35**, 303–309.

Bailey, N. T. J. (1954). A continuous time treatment of a simple queue using generating functions. *J.R.S.S.* **B16**, 288–291.

Baker, Kenneth, R. (1973). A note on the operating policies for the $M/M/1$ queue with exponential startups. *INFOR* **11**, 71–72.

Böhm, W., and Mohanty, S. G. (1990). On the transient solution of N-policy queues. *Statistics Research Report*, No. 11, McMaster University, Hamilton, Ontario.

Borthakur, A., Medhi, J., and Gohain, R. (1987). Poisson input queueing system with startup time and under control operating policy. *Comp. Opns. Res.* **14**, 33–40.

Bunday, B. D., and Scarton, R. E. (1980). The $G/M/r$ machine interference model. *Euro. J. Opnl. Res.* **4**, 399–402.

Burke, P. J. (1956). The output of a queueing system. *Opns. Res.* **4**, 699–704.

Burke, P. J. (1964). The dependence of delays in tandem queues. *Ann. Math. Stat.* **35**, 874–875.

Burke, P. J. (1968). The output process of a stationary $M/M/s$ queueing system. *Ann. Math. Stat.* **39**, 1144–1152.

Burke, P. J. (1972). Output processes and tandem queues in *Computer Communication Networks and Teletraffic* (J. Fox, ed.) Polytechnic Press, New York.

Chaudhry, M. L. and Templeton, J. G. C. (1973). A note on the distribution of a busy period for $M/M/c$ queueing system. *Math, Oper U. Stat.* **1**, 75–79.

Champernowne, D. G. (1956). An elementary method of solution of the queueing problem with a single server and constant parameters. *J.R.S.S.* **B18**, 125–128.

Cohen, J. W. (1982). *The Single Server Queue*, 2nd Ed., North-Holland, Amsterdam, The Netherlands.

Conolly, B. W. (1958). A difference equation technique applied to the simple queue with arbitrary arrival interval distribution. *J.R.S.S.* **B21**, 168–175.

Cooper, R. B. (1976). Queues with ordered servers that work at different rates. *Opsearch* **13**, 69–78.

Cooper, R. B. and Soloman, M. K. (1984). The average time until bucket overflow. *ACM Trans. Database Syst.* **9**, 392–398.

Descloux, A. (1962). *Delay Tables for Finite and Infinite Source Systems.* McGraw Hill, New York.

De Vany, A. (1976). Uncertainty, waiting time and capacity utilization: a stochastic theory of product quality. *J. Pol. Eco.* **84**, 523–541.

Dietrich, G., Krush, W., Michel, G., Ondra, F., Peter, E., and Wanger, G. (1966). *Teletraffic Engineering Manual.* Standard Electrik, Lorenz, Stutgart, Federal Republic of Germany.

Disney, R. L. (1962). Some multichannel queueing problems and ordered entry. *J. Industrial Eng.* **13**, 46–48.

Disney, R. L. (1963). Some multichannel queueing problems with ordered entry. An application to conveyor theory. *J. Industrial Eng.* **14**, 105–108.

Disney, R. L., and König, D. (1985). Queueing networks: a survey of their random processes. *SIAM Review* **27**, 335–403.

Dufkova, V., and Zitek, F. (1975). On a class of queue disciplines. *Aplikace Matmatiky Sv.* **20**, 345–358.

Elsayed, E. A. (1983). Miltichannel queueing systems with ordered entry and finite source. *Comp. & Opns. Res.* **10**, 213–222.

Elsayed, E. A., and Elayat, H. A. (1976). Analysis of closed-loop conveyor systems with multiple Poisson inputs and outputs. *Int. J. Prod. Res.* **14**, 99–107.

Elsayed, E. A., and Proctor, C. L. (1977). Ordered entry and random choice conveyors with multiple Poisson input. *Int. J. Prod. Res.* **15**, 439–451.

Everitt, D. E. and Downs, T. (1984). The output of the $M/M/s$ queue. *Opns. Res.* **32**, 796–808.

Feller, W. (1966). *Introduction to Probability Theory and its Applications*, **vol. 2**, Wiley, New York.

Feller, W. (1968). *Introduction to Probability Theory and its Applications*, **vol. I**, 3rd ed., Wiley, New York.

Goodman, J. B. and Massey, W. A. (1984). The non-ergodic Jackson network. *J. Appl. Prob.* **21**, 266-277.

Grassman, W. (1983). The convexity of the mean queue size of the $M/M/c$ queue with respect to the traffic intensity. *J. Appl. Prob.* **20**, 252–267.

Gray, J. P. and Mcneill, T. B. (1979). SNA multiple system networking. *IBM Syst. J.* **18**, 263–297.

Greenberg, H., and Greenberg, I. (1966). The number served in a queue. *Opns. Res.* **14**, 137–144.

Gregory, G., and Litton, C. D. (1975). A conveyor model with exponential service times. *Int. J. Prod. Res.* **13**, 1–7.

Gupta, S. K. (1966). Analysis of two-channel queueing system with ordered entry. *J. Industrial Engineering*, **17**, 54–55.

Haghighi-Montazer, A., Medhi, J., and Mohanty, S. G. (1986). On a multi-server Markovian queueing system with balking and reneging. *Comp. Opns. Res.* **13**, 421–425.

Haight, F. A. (1961). A distribution analogous to the Borel-Tanner. *Biometrika* **48**, 167-173.

Harel, A. (1987). Sharp bounds and simple approximations for the Erlang delay and loss formulas. *Mgmt. Sci.* (To appear).

Harel, A., and Zipkin, P. (1987). Strong convexity results for queueing systems. *Opns. Res.* **35**, 405–418.

Harel, A. and Zipkin, P. (1989). The convexity of a general performance measure for the $M/M/c$ queue. *J. Appl. Prob.* (To appear.)

Harris, R. (1974). The expected number of idle servers in a queueing system. *Opns. Res.* **22**, 1258–1259.

Hubbard, J. R., Pegden, C. D., and Rosenshine, M. (1986). The departure process of an $M/M/1$ queue. *J. Appl. Prob.* **23**, 249–255.

Jackson, R. R. P., and Henderson, J. C. (1966). The time dependent solution to the many server Poisson queue. *Opns. Res.* **14**, 720–723.

Jagerman, D. L. (1974). Some properties of the Erlang loss function. *Bell Syst. Tech. J.* **53**, 525–551.

Jordan, C. (1950). *The Calculus of Finite Differences.* Chelsea, New York.

Kaufman, J. S. (1979). The busy probability in $M/G/N/N$ loss systems. *Opns. Res.* **27**, 204–210.

Kelton, W. D., and Law, A. M. (1985). The transient behavior of $M/M/s$ queue with implications for steady state simulation. *Opns. Res.* **33**, 378–396.

Khinchin, A. Y. (1960). *Mathematical Models in the Theory of Queueing.* Griffin, London.

Kleinrock, L. (1975). *Queueing Systems*, Vol. 1, Wiley, New York.

Koenigsberg, E. (1980). Cycle queue models of semi-conductor noise and vehicle fleet operations. *Int. J. Electr.* **48**, 83-91.

Lemoine, A. J. (1976). On random walks and stable GI/G/I queues. *Math. Opns. Res.* **1**, 159–164.

Levy, Y., and Yechiali, U. (1976). An $M/M/s$ queue with servers vacations. *INFOR* **14**, 153–163.

Lin, B. W., and Elsayed, E. A. (1978). A general solution for multichannel queueing systems with ordered entry. *Int. J. Comp. Opns. Res.* **5**, 219–225.

Martins-Neto, A. F., and Wong, E. (1976). A martingale approach to queues, in: *Stochastic Systems: Modeling, Identification and Optimization.* (Wets, R. J-P., ed.), North-Holland, Amsterdam, The Netherlands.

Matsui, M., and Fukuta, J. (1977). On a multichannel queueing system with ordered entry and heterogeneous servers. *AIIE Trans.* **9**, 209–214.

Medhi, J. (1982). *Stochastic Processes.* Halsted Press, Wiley, New York and Wiley Eastern, New Dehli.

Messerli, E. J. (1972). Proof of a convexity property of the Erlang's B-formula. *Bell Syst. Tech. J.* **51**, 951–953.

Mohanty, S. G., and Jain, J. L. (1971). The distribution of the maximum queue length, the number of customers and the duration of the busy period for the queueing system $M/M/1$ involving batches. *INFOR* **9**, 161–166.

Mohanty, S. G. and Panny, W. (1990). A discrete time analogue of the $M/M/1$ queue and the transient solution, I & II. *Statistics Research Report* No. 9, McMaster University, Hamilton, Ontario.

Morisaku, T. (1976). Techniques for data-truncation in digital computer simulation. Ph.D. dissertation, U. of Southern California.

Morse, P. M. (1955). Stochastic properties of waiting lines. *Opns. Res.* **3**, 255–261.

Muth, E. J., and White, J. A. (1979). Conveyor theory: A survey. *AIIE Trans.* **11**, 270–277.

Naor, P. (1969). The regulation of queue size by levying tolls. *Econometrica* **37**, 15–23.

Nawijn, W. M. (1983). A note on many-server queueing systems with ordered entry, with an application to conveyor theory. *J. Appl. Prob.* **20**, 144–152.

Nawijn, W. M. (1984). On a two-server finite queueing system with ordered entry and deterministic arrivals. *Euro. J. Opnl. Res.* **18**, 388–395.

Neuts, M. F. (1974). The Markov renewal branching processes, in: *Math, methods in queueing theory* (A. B. Clarke, ed.), Lecture Notes in Ec. & Math. Systems. no. 98, Springer-Verlag, New York.

Neuts. M. F. (1976). Moment formulas for Markov renewal branching processes. *Adv. Appl. Prob.* **8**, 690–711.

Neuts, M. F. (1979). Queues solvable without Rouché's theorem. *Opns. Res.* **27**, 767–781.

Newell, G. F. (1984). *The $M/M/\infty$ Service System with Ranked Servers in Heavy Traffic.* Springer-Verlag, New York.

O'Brien, G. G. (1954). Some queueing problems. *J. Soc. Ind. Appl. Math.* **2**, 134.

Parthasarathy, P. R. (1987). A transient solution to an $M/M/1$ queue: a simple approach. *Adv. Appl. Prob.* **19**, 997–998.

Parthasarathy, P. R., and Sharafali, M. (1989). Transient solution to the many server Poisson queue: a simple approach *J. Appl. Prob.* **26**, 584–594.

Pegden, C. D., and Rosenshine, M. (1982). Some new results for the $M/M/1$ queue. *Mgmt. Sci.* **28**, 821–828.

Pourbabai, B. (1987). Approximation of the overflow process from a $G/M/N/K$ queueing system. *Mgmt. Sci.* **33**, 931–938.

Pourbabai, B., and Sonderman, D. (1986). Server utilization factors in queueing loss systems with ordered entry and heterogeneous servers. *J. Appl. Prob.* **23**, 236–242.

Prabhu, N. U. (1965). *Queues and Inventories.* Wiley, New York.

Pritsker, A. A. B. (1966). Applications of multichannel queueing results to the analysis of conveyor systems. *J. Ind. Engg.* **17**, 14–21.

Proctor, C. L., Elsayed, E. A., and Elayat, H. S. (1977). A conveyor system with homogeneous and heterogeneous servers with dual input. *Int. J. Prod. Res.* **15**, 73–85.

Reich, E. (1957). Waiting times when queues are in tandem. *Ann. Math. Stat.* **28**, 768–773.

Reich, E. (1965). Departure processes in *Proc. Symp. on Congestion Theory*, 439–457 (Smith, W. L. and Wilkinson, W. E. eds.), U. of North Carolina Press, Chapel-Hill, NC.

Rothpokf, M. H. and Rech, P. (1987). Perspective on queues. Combining queues is not always beneficial. *Opns. Res.* **35**, 906–909.

Rue, Robert, C., and Rosenshine, M. (1981). Optimal control for entry of many classes of customers to an $M/M/1$ queue. *Nav. Res. Log. Qrly.* **28**, 489–495.

Saaty, T. L. (1960). Time dependent solution of the many server Poisson queue. *Opns. Res.* **8**, 755–772.

Saaty, T. L. (1961). *Elements of Queueing Theory.* McGraw-Hill, New York.

Shanthikumar, J. G., and Yao, D. D. (1987). Comparing ordered entry queues with heterogeneous servers. *Queueing Systems* **2**, 235–244.

Sharma, O. P. and Gupta, U. C. (1982). Transient behavior of $M/M/1/N$ queue. *St. Proc. & Appl.* **13**, 327–331.

Smith, D. R., and Whitt, W. (1981). Resource sharing for efficiency in traffic system. *Bell System Tech. J.* **60**, 39–55.

Sonderman, D. (1982). An analytical model for recirculating conveyors with stochastic inputs and outputs. *Int. J. Prod. Res.* **20**, 591–605.

Sonderman, D. (1979). Comparing multi-server queues with finite waiting rooms. I. Same number of servers, II. Different number of servers. *Adv. Appl. Prob.* **11**, 439–447; 448–455.

Sphicas, G. P., and Shimshak, D. G. (1978). Waiting time variability in some single server queueing systems. *J. Opnl. Res. Soc.* **29**, 65–70.

Syski, R. (1986). *Introduction to Congestion Theory in Telephone Systems*, 2nd ed. North-Holland, Amsterdam, The Netherlands.

Syski, R. (1988). Further comments on the solution of the $M/M/1$ queue. *Adv. Appl. Prob.* **20**, 693.

Takács, L. (1962). *Introduction to the Theory of Queues.* Oxford University Press, New York.

Takács, L. (1969), On Erlang's formula. *Ann. Math. Stat.* **40**, 71–78.

Van Doorn, E. (1981). *Stochastic Monotonicity and Queueing Applications in Birth Death Processes.* Lecture Notes in Statistics, No. 4, Springer-Verlag. New York.

Whitt, W. (1981). Comparing counting processes and queues. *Adv. Appl. Prob.* **13**, 207–220.

Whitt, W. (1986). Deciding which queue to join: some counterexamples. *Opns. Res.* **34**, 55–62.

Weiss, E. N., and McClain, J. O. (1987). Administrative delays in acute health care facilities: a queueing analytic approach. *Opns. Res.* **35**, 35–44.

Wong, J. W. (1979). Response time distribution of the $M/M/m/N$ queueing model. *Opns. Res.* **27**, 1196–1202.

Yao, D. D. (1986). Convexity properties of the overflow in an ordered-entry system with heterogeneous servers. *Opns. Res. Lett.* **5**, 145–147.

Yao, D. D. (1987). The arrangement of servers in an ordered-entry system. *Opns. Res.* **35**, 759–763.

Zolotarev, V. M. (1977). General problems of the stability of mathematical models. *Proc.* 41*st Int. Stat. Ins.*, New Delhi, India.

4 Non–Birth-and-Death Queueing Systems: Markovian Models

4.1. Introduction

In the preceding chapter we discussed Markovian queueing processes that can be studied as birth-and-death processes. There the transitions occur to neighboring states: from state i to state $i-1$ $(i \geq 1)$ or from state i to state $i+1$ $(i \geq 0)$. We examine here some Markovian models that arise out of non–birth-and-death processes where transitions occur from a state to a state, not necessarily neighboring: from state i to, say, state $i-k$ $(i \geq k \geq 1)$ or from state i to, say, state $i+r$ $(i \geq 0, r \geq 1)$. The processes considered are Markovian and the Chapman–Kolmogorov equations pertaining to the model can be written down and the solutions can be obtained in a similar manner. We first consider systems where, of the two distributions (the interarrival-time distribution and the service-time distribution), one is exponential while the other is Erlangian.

4.1.1. The System $M/E_k/1$

We consider a single-channel system where the arrival process is Poisson with rate λ and the service-time distribution is E_k having density

$$b(t) = \frac{k\mu(k\mu t)^{k-1} e^{-k\mu t}}{(k-1)!}, \qquad t \geq 0,$$

with mean $1/\mu$.

The service time may be thought of as consisting of k independent exponential stages each with mean $(1/k\mu)$. As soon as a customer arrives in the system, he may be considered to possess k stages of service to be completed by him: until he has completed all the stages of service, the next arrival will wait in the queue. The system state at an instant may be studied in terms of stages of service remaining to be completed and the stages may be marked in the *reverse order*, the first stage corresponding to k, the second to $(k-1)$, and the last to stage 1, i.e., if he is at ith stage, he has yet to complete $k-(i-1)$ stages of service.

Thus, if at any instant there are $j(>0)$ customers in the system and the customer being served is at the ith stage $(1 \leq i \leq k)$ of service and if r denotes the number of stages contained in the total system (or number of stages that remain to be completed) at that time, then

$$\begin{aligned} r &= (j-1)k + [k-(i-1)] \\ &= jk - i + 1, \qquad r \geq 0. \end{aligned}$$

Here, a customer's arrival effects a transition from state i to state $i+k$ $(i \geq 0)$ in the system, whereas with a service completion (departure of a customer from the system), transition occurs from state i to $i-1$ $(i \geq 1)$; the process is non–birth-and-death, but may be described, following Keilson, as skip-free downward, since transitions from state i occur to a lower state only, to state $(i-1)$.

Let $N(t)$ be the number of stages in the total system at the epoch t and let

$$p_n(t) = Pr\{N(t) = n\}.$$

The Chapman–Kolmogorov equations can be written down as follows:

$$p_0(t+h) = p_0(t)[1-\lambda h] + p_1(h)[1-\lambda h]k\mu h + o(h),$$

and for $n \geq k$

$$\begin{aligned} p_n(t+h) = p_n(t)[1-\lambda h][1-k\mu h] \\ + p_{n+1}(t)[1-\lambda h][k\mu h] \\ + p_{n-k}(t)[\lambda h(1-k\mu h)] + o(h); \end{aligned}$$

for $1 \leq n \leq k$, the last term of the preceding will not occur.

Thus, we get, with $p_j(\cdot) = 0$ for $j < 0$,

$$p_0'(t) = -\lambda p_0(t) + k\mu p_1(t) \tag{1.1}$$

$$p_n'(t) = -(\lambda + k\mu)p_n(t) + k\mu p_{n+1}(t) + \lambda p_{n-k}(t), \qquad n \geq 1. \tag{1.2}$$

Assume that $\rho = \lambda/\mu < 1$; that the system is in steady state. Let $p_n = \lim_{t\to\infty} p_n(t)$ be the steady-state probability that there are n stages in the total system. Then we have, with $p_j = 0$ for $j < 0$,

$$\lambda p_0 = k\mu p_1 \tag{1.3}$$

$$(\lambda + k\mu)p_n = k\mu p_{n+1} + \lambda p_{n-k}, \qquad n \geq 1. \tag{1.4}$$

Let $P(s) = \sum_{n=0}^{\infty} p_n s^n$ be the PGF of $\{p_n\}$. Multiplying (1.4) by s^n and adding over all admissible values of n, we get

$$(\lambda + k\mu)\sum_{n=1}^{\infty} p_n s^n = k\mu \sum_{n=1}^{\infty} p_{n+1}s^n + \lambda \sum_{n=k}^{\infty} p_{n-k}s^k \quad \text{or}$$

$$(\lambda + k\mu)[P(s) - p_0] = \frac{k\mu}{s}[P(s) - p_0 - p_1 s] + \lambda s^k P(s),$$

whence we get

$$P(s) = \frac{p_0\left[(\lambda + k\mu) - \dfrac{k\mu}{s}\right] - k\mu p_1}{(\lambda + k\mu) - \dfrac{k\mu}{s} + \lambda s^k}.$$

Writing p_1 in terms of p_0 (from (1.3)) and simplifying, we get

$$P(s) = \frac{k\mu p_0(1-s)}{k\mu - (\lambda + k\mu)s + \lambda s^{k+1}}. \tag{1.5}$$

We can evaluate p_0 by using the relation $P(1) = 1$. Using L'Hôpital's rule we get

$$P(1) = \lim_{s\to 1} P(s) = \frac{-k\mu p_0}{-(\lambda + k\mu) + (k+1)\lambda}$$

so that

$$p_0 = \frac{(\mu - \lambda)}{\mu} = 1 - \frac{\lambda}{\mu} = 1 - \rho.$$

Thus, we have

$$P(s) = \frac{k\mu(1-\rho)(1-s)}{k\mu - (\lambda + k\mu)s + \lambda s^{k+1}} \tag{1.6}$$

$$= \frac{(1-\rho)(1-s)}{(1-s) - \dfrac{\lambda s}{k\mu}(1-s^k)}. \tag{1.6a}$$

To find p_n we have to expand $P(s)$ as a power series in s. The usual approach is by a partial fraction expansion of $P(s)$ for which one needs to find the zeros of the denominator; from (1.6a) we get

$$P(s) = \frac{(1-\rho)}{1 - \dfrac{\lambda}{k\mu}\{s + s^2 + \cdots + s^k\}}. \tag{1.6b}$$

Let the zeros of the denominator be $s_1, s_2, \ldots, s_k$; the zeros are unique, i.e., there is no multiple zero. The denominator of (1.6b) can be written as

$$-\frac{\lambda}{k\mu}\left\{s^k + \cdots + s - \frac{k\mu}{\lambda}\right\} = -\frac{\lambda}{k\mu}(s - s_1)\cdots(s - s_k)$$

$$\left(\text{where } (-1)^k(s_1 \cdots s_k) = -\frac{k\mu}{\lambda}\right)$$

$$= -\frac{\lambda}{k\mu}\{(-1)^k(s_1 \cdots s_k)\}$$

$$\times \left\{\left(1 - \frac{s}{s_1}\right)\left(1 - \frac{s}{s_2}\right)\cdots\left(1 - \frac{s}{s_k}\right)\right\}$$

$$= \left(1 - \frac{s}{s_1}\right)\cdots\left(1 - \frac{s}{s_k}\right).$$

Thus, we have

$$P(s) = \frac{(1-\rho)}{\left(1-\frac{s}{s_1}\right)\left(1-\frac{s}{s_2}\right)\cdots\left(1-\frac{s}{s_k}\right)}$$

$$= (1-\rho)\sum_{i=1}^{k}\frac{a_i}{\left(1-\frac{s}{s_i}\right)}, \tag{1.7}$$

where

$$a_i = \prod_{\substack{m=1\\ m\neq i}}^{k}\frac{1}{\left(1-\frac{s_i}{s_m}\right)}.$$

Thus,

$$P(s) = (1-\rho)\sum_{i=1}^{k} a_i\left(1-\frac{s}{s_i}\right)^{-1}$$

so that p_n, the coefficient of s^n, is given by

$$p_n = (1-\rho)\left[\sum_{i=1}^{k} a_i(s_i)^{-n}\right], \qquad n = 1, 2, \ldots. \tag{1.8}$$

Now p_n gives the probability of the number of *stages* in the system. If $p_r^{(c)}$ denotes the probability that the number of *customers* in the system is r, then we have the following relation between $p_n^{(c)}$ and p_n:

$$p_n^{(c)} = \sum_{m=(n-1)k+1}^{nk} p_m, \qquad n = 1, 2, 3, \ldots. \tag{1.9}$$

Thus, from p_n, $p_n^{(c)}$ can be obtained by using the preceding relation.

4.1.1.1. Particular Case: M/M/1. For $k = 1$, we have $E \equiv M$, then the system becomes $M/M/1$. Putting $k = 1$ in (1.6a) we have

$$P(s) = \frac{1-\rho}{1-s\rho} = (1-\rho)(1-s\rho)^{-1}$$

so that $p_n = (1-\rho)\rho^n$, $\quad n = 0, 1, 2, \ldots$. For $k = 1$, we get from (1.9)

$$p_r^{(c)} = \sum_{n=r}^{r} p_n = p_r$$

so that the steady-state distribution that there are n in the system is given by $p_n^{(c)} = (1-\rho)\rho^n$, $n \ge 0$.

Note 1: An alternative approach to find p_n is as follows: From (1.6a) we have

$$
\begin{aligned}
P(s) &= \frac{(1-\rho)}{1 - \dfrac{\lambda s}{\mu k}\left\{\dfrac{(1-s^k)}{1-s}\right\}} \\
&= (1-\rho)\left[1 - \frac{\lambda s}{\mu k}(1-s^k)(1-s)^{-1}\right]^{-1} \\
&= (1-\rho)\left[\sum_{m=0}^{\infty}\left(\frac{\lambda}{k\mu}\right)^m s^m(1-s^k)^m(1-s)^{-m}\right] \\
&= (1-\rho)\left[\sum_{m=0}^{\infty}\left(\frac{\lambda s}{k\mu}\right)^m \left\{\sum_{i=0}^{m}(-1)^i\binom{m}{i}(s^k)^i\right\}\right. \\
&\qquad \times \left.\left\{\sum_{j=0}^{\infty}\binom{m+j-1}{j}s^j\right\}\right] \quad (\text{where} \quad m \ge i,\ m+j-1 \ge j) \\
&= (1-\rho)\left[\sum_{m=0}^{\infty}\sum_{i=0}^{\infty}\sum_{j=0}^{\infty}(-1)^i\left(\frac{\lambda}{k\mu}\right)^m\binom{m}{i}\binom{m+j-1}{j}s^{m+ik+j}\right].
\end{aligned}
$$

Put $n = m + ik + j$ ($m \ge i$, $m \ge 1$), then

$$
\begin{aligned}
p_n &= \text{coefficient of } s^n \text{ in } P(s) \\
&= (1-\rho)\sum_{m,i,j}(-1)^i\left(\frac{\lambda}{k\mu}\right)^m\binom{m+j-1}{j} \qquad (1.10)
\end{aligned}
$$

where $m \ge i$, $m \ge 1$, $n = m + ik + j$. For example, for

$$
k = 2,\ p_7 = (1-\rho)(\rho_1^7 + 6\rho_1^6 + 10\rho_1^5 + 4\rho_1^4), \quad \text{where} \quad \rho_1 = \left(\frac{\lambda}{2\mu}\right) = \left(\frac{\rho}{2}\right).
$$

Note 2: Another model can be considered by specifying the state of the system by the number of customers in the queue (or system) and the number of the stage in which the customer is being served. Let $p_{q,i}(t)$ denote the probability that there are $q(\ge 0)$ in the queue and the customer being served, if any, is at stage $i(\ge 0)$, with $p_{0,0} = p_0$, and let $p_{q,i}$ be the corresponding steady-state probability. Then the differential-difference equations of the system can be easily written down. The equations are not easy to handle, however. The probability that the number in the system is n is given by

$$p_n = \sum_{i=1}^{k} p_{n-1,i}, \qquad (n \geq 1).$$

Note 3: We shall consider later the more general system $M/G/1$ from which results of $M/E_k/1$ can be deduced.

4.1.2. The System $E_k/M/1$

Now we consider a single-channel system where the interarrival time has Erlang-k distribution with mean $1/\lambda$ and the service time has exponential distribution with mean $(1/\mu)$. Here the arrival mechanism may be considered as consisting of k independent exponential stages each with mean $(1/k\lambda)$: an arriving customer has to pass through k successive independent stages and only when he has passed through all the stages is he finally admitted to the system; further, the next arrival after him, who arrives when the earlier arrival is at some stage of the arriving mechanism, cannot be admitted (to the first stage) and has to remain in queue unless the earlier arrival has completed all the stages and is admitted to the system. The system state at any epoch may be studied in terms of the arrival stage, the stage being marked in the order in which it occurs. Thus, if at any epoch, there are $j(>0)$ customers (already admitted) in the system and the arriving customer is at the ith stage $(1 \leq i \leq k)$ of the arriving mechanism, then the total number of (completed) stages in the total system r is given by

$$r = jk + (i - 1). \tag{1.11}$$

Here with the passage of a customer from stage i to stage $i + 1$ of the arrival mechanism the number of completed stages changes and transition occurs from state r to state $r + 1$, whereas with the completion of service of a customer, the transition occurs from state r to state $r - k$. Assume that $\rho = \lambda/\mu < 1$, i.e., the system is in steady state. Let p_n be the steady-state probability that the number of *stages* in the total system is n. If $p_r^{(c)}$ is the probability that the number of customers in the system is r, then

$$p_n^{(c)} = \sum_{m=nk}^{k(n+1)-1} p_m, \qquad n = 1, 2, \ldots. \tag{1.12}$$

The Chapman–Kolmogorov equations of the system can be easily written down. The differential-difference equations of the total system in steady state are given by

$$k\lambda p_0 = \mu p_k \tag{1.13}$$

$$k\lambda p_n = k\lambda p_{n-1} + \mu p_{n+k}, \qquad 1 \le n \le k-1 \tag{1.14}$$

$$(k\lambda + \mu)p_n = k\lambda p_{n-1} + \mu p_{n+k}, \qquad n \ge k. \tag{1.15}$$

Let $P(s) = \sum_{n=0}^{\infty} p_n s^n$ be the PGF of $\{p_n\}$. Multiplying (1.14) and (1.15) by s^n and adding over all admissible values of n, we get

$$\sum_{n=1}^{\infty} (k\lambda + \mu)p_n s^n - \sum_{n=1}^{k-1} \mu p_n s^n$$
$$= \sum_{n=1}^{\infty} k\lambda p_{n-1} s^n + \sum_{n=1}^{\infty} \mu p_{n+k} s^n.$$

Expressing the preceding in terms of $P(s)$ and the missing terms, we get

$$(k\lambda + \mu)[P(s) - p_0] - \sum_{n=1}^{k-1} \mu p_n s^n$$
$$= k\lambda s P(s) + \frac{\mu}{s^k}\left[P(s) - \sum_{n=0}^{k} p_n s^n\right]. \tag{1.16}$$

Multiplying by s^k/μ and collecting terms involving $P(s)$ on one side and noting that $\rho = \lambda/\mu$, we get

$$P(s)[k\rho s^{k+1} - (k\rho + 1)s^k + 1]$$
$$= -s^k(k\rho + 1)p_0 - s^k \sum_{n=1}^{k-1} p_n s^n + \sum_{n=0}^{k} p_n s^n$$
$$= -s^k \sum_{n=0}^{k-1} p_n s^n - k\rho s^k p_0 + \left(p_k s^k + \sum_{n=0}^{k-1} p_n s^n\right)$$
$$= (1 - s^k) \sum_{n=0}^{k-1} p_n s^n \quad \text{(using (1.13))}.$$

Thus,

$$P(s) = \frac{(1 - s^k) \sum_{n=0}^{k-1} p_n s^n}{k\rho s^{k+1} - (k\rho + 1)s^k + 1}. \tag{1.17}$$

The expression on the RHS involves $\sum_{n=0}^{k-1} p_n s^n$, which must be eliminated. The denominator has $(k + 1)$ zeros, of which 1 is one. By Rouché's theorem, we find that $(k-1)$ zeros are of modulus less than unity and one zero is of modulus greater than unity. Denote these zeros by $s_1, \ldots, s_{k-1}(|s_i| < 1, i = 1, 2, \ldots, k)$ and $s_0(|s_0| > 1)$. As $P(s)$ is analytic inside $|s| < 1$ and is bounded, the numerator must also have as zeros $s_1, \ldots, s_{k-1}$, which must come from one of the two factors in the numerator. The zeros of the factor $(1-s^k)$ have all modulus equal to unity so the factor $\sum_{n=0}^{k-1} p_n s^n$ must have as roots $s_1, \ldots, s_{k-1}$. Thus,

$$\frac{\sum_{n=0}^{k-1} p_n s^n}{k\rho s^{k+1} - (k\rho+1)s^k + 1} = \frac{\mathrm{A}(s-s_1)\cdots(s-s_{k-1})}{(1-s)(s-s_0)(s-s_1)\cdots(s-s_{k-1})}$$

$$= \frac{\mathrm{A}}{(1-s)(s-s_0)},$$

where A is a constant to be determined. Substituting in (1.17) we get

$$P(s) = \frac{(1-s^k)\mathrm{A}}{(1-s)(s-s_0)}. \tag{1.18}$$

Since $P(1) = 1$, we get by using L'Hôpital's rule

$$1 = P(1) = \lim_{s \to 1} P(s)$$

$$= \frac{-k\mathrm{A}}{-(1-s_0)}$$

so that

$$P(s) = \frac{(1-s^k)(1-s_0)}{k(1-s)(s-s_0)} = \frac{(1-s^k)\left(1-\dfrac{1}{s_0}\right)}{k(1-s)\left(1-\dfrac{s}{s_0}\right)}. \tag{1.19}$$

To find p_n we have to expand the RHS by resolving first into partial fractions. We have

$$P(s) = \frac{(1-s^k)}{k}\left[\frac{1}{1-s} - \frac{\dfrac{1}{s_0}}{1-\dfrac{s}{s_0}}\right]$$

$$= \frac{(1-s^k)}{k}\left[\sum_{r=0}^{\infty} s^r - \sum_{r=0}^{\infty} \frac{s^r}{(s_0)^{r+1}}\right]. \tag{1.20}$$

Thus, for $0 \le n < k$,

$$p_n = \frac{1}{k}\{1 - s_0^{-(n+1)}\}$$

and for $n \ge k$,

$$p_n = \frac{1}{k}\{1 - s_0^{-(n+1)} - 1 + (s_0)^{-(n-k+1)}\}$$

$$= \frac{1}{k}\{s_0^{k-n-1}(1 - s_0^{-k})\}.$$

Since s_0 is a root of $k\rho s^{k+1} - (k\rho + 1)s^k + 1 = 0$, we get

$$k\rho(s_0 - 1) = 1 - s_0^{-k}. \tag{1.21}$$

Thus, we have

$$p_n = \frac{1}{k}\{1 - s_0^{-(n+1)}\}, \qquad 0 \le n < k,$$

$$= \rho s_0^{k-n-1}(s_0 - 1), \qquad n \ge k. \tag{1.22}$$

Using (1.12), we get, for $n > 0$,

$$p_n^{(c)} = \sum_{m=nk}^{k(n+1)-1} \rho s_0^{k-m-1}(s_0 - 1)$$

$$= \rho s_0^{k-1}(s_0 - 1)\left[\frac{s_0^{-kn} - s_0^{-k(n+1)}}{1 - s_0^{-1}}\right]$$

$$= \rho s_0^{k-nk}(1 - s_0^{-k}) \tag{1.23}$$

$$= (\rho s_0^k)\{(1 - s_0^{-k})(s_0^{-k})^n\}; \tag{1.23a}$$

and, for $n = 0$,

$$p_0^{(c)} = \sum_{m=0}^{k-1} p_m = \sum_{m=0}^{k-1} \frac{1}{k}\{1 - s_0^{-(m+1)}\}$$

$$= 1 - \frac{1}{k}\frac{s_0^{-1}(1 - s_0^{-k})}{1 - s_0^{-1}}$$

$$= 1 - \frac{1}{k}\frac{(1 - s_0^{-k})}{(s_0 - 1)}$$

$$= 1 - \rho \quad \text{(using (1.21))}. \tag{1.24}$$

Thus, we find that the distribution of the number in the system is given by (1.23) for $n > 0$ and by (1.24) for $n = 0$. The form of the expression (1.23a) implies that the distribution of the number of customers in the system is *modified geometric* with a slightly modified first term. We shall take up further questions later on when we study more general $G/M/1$ systems.

Note: The arithmetic-geometric mean inequality to the characteristic equation (denominator of (1.17) equated to zero) gives the result that s_0 is real and that

$$1 < s_0 \leq \rho^{-2/(k+1)}.$$

4.2. Bulk Queues

Bailey (1954) introduced the concept of bulk queues; he considered a situation where service can be effected in a batch of up to C customers, i.e., all waiting customers up to a fixed capacity C are taken for service in a batch. Gaver (1959) introduced bulk-arrival queues, where arrival could be in bulk or batch. The literature on bulk queues, with bulk arrival and/or with bulk service is now quite vast. Chaudhry and Templeton (1983) discuss this subject at great length.

We shall first discuss bulk queues that can be modeled as non–birth-and-death processes. The resulting queueing processes will be Markovian. The Chapman–Kolmogorov equations or the balance equations can be written down easily and the analysis can be done in a straightforward manner similar to that of birth-and-death queueing processes.

4.2.1. Markovian Bulk-Arrival System: $M^X/M/1$

Let us consider a single-server queueing system in which customers arrive in batches in accordance with a time-homogeneous Poisson process with parameter λ. Assume that the batch size X is a RV with PF

$$Pr(X = k) = a_k, \qquad k = 1, 2, 3, \ldots .$$

that is, the probability that a batch of k arrives in an infinitesimal interval $(t, t + h)$ is $\lambda a_k h + o(h)$.

Let $A(s) = \sum_{k=1}^{\infty} a_k s^k$ be the PGF of X and $\bar{a} = A'(1) = E(X)$ be the mean of X. The arrival process is a compound Poisson process with mean arrival rate $\lambda\bar{a}$. Assume that service takes place singly and that the distribution of service time is exponential with mean $1/\mu$. The process is still Markovian but is a non–birth–death, as transitions, in case of arrival of a batch of size $k(k \geq 1)$, will occur to a state differing by k, and will not necessarily be to a neighboring one. The system is denoted by $M^X/M/1$. The traffic intensity is $\rho = \lambda E(X)/\mu = \lambda\bar{a}/\mu$. The balance equations can be written down as:

$$p_0'(t) = -\lambda p_0(t) + \mu p_1(t) \tag{2.1}$$

$$p_n'(t) = -(\lambda + \mu)p_n(t) + \mu p_{n+1}(t) + \sum_{k=1}^{n} \lambda a_k p_{n-k}(t), \qquad n \geq k \geq 1. \tag{2.2}$$

This queue is considered also in Stadje (1989).

4.2.1.1. Steady-State Solution. Assume that the steady-state solution exists. Then we get

$$0 = -\lambda p_0 + \mu p_1 \tag{2.3a}$$

$$0 = -(\lambda + \mu)p_n + \mu p_{n+1} + \lambda \sum_{k=1}^{n} a_k p_{n-k}, \qquad n \geq k \geq 1. \tag{2.3b}$$

Let $P(s) = \sum_{n=0}^{\infty} p_n s^n$ be the PGF of $\{p_n\}$. Multiplying (2.3b) by s^n for $n = 1, 2, 3, \ldots$ and adding them to (2.3a), we get

$$0 = -\lambda P(s) - \mu[P(s) - p_0] + \left(\frac{\mu}{s}\right)[P(s) - p_0] + \lambda \sum_{n=1}^{\infty} \sum_{k=1}^{n} a_k p_{n-k} s^n. \tag{2.4}$$

The last term on the RHS of Eq. (2.4) can be written as

$$\lambda \sum_{k=1}^{\infty} a_k s^k \left\{ \sum_{n=k}^{\infty} p_{n-k} s^{n-k} \right\} = \lambda A(s)P(s).$$

Then from Eq. (2.4) we get

$$P(s) = \frac{\mu(1-s)p_0}{\mu(1-s) - \lambda s[1 - A(s)]}. \tag{2.5a}$$

Since $P(1) = 1$, we get, applying L'Hôpital's rule,

$$1 = \lim_{s \to 1} P(s) = \frac{-\mu p_0}{-\mu + \lambda\bar{a}}$$

whence

$$p_0 = 1 - \lambda\frac{\bar{a}}{\mu} = 1 - \rho.$$

The condition $\rho < 1$ is sufficient for existence of steady state. From (2.5a) we get

$$P(s) = \frac{\mu(1-s)(1-\rho)}{\mu(1-s) - \lambda s[1 - A(s)]}$$

$$= \frac{(1-\rho)}{1 - \dfrac{\lambda s}{\mu(1-s)}\{1 - A(s)\}}. \tag{2.5b}$$

4.2.1.2. Expected Number in the System. The expected number in the system $E(N)$ is given by $E(N) = P'(1)$. To find $P'(1)$, we can proceed in a straightforward manner or, better still, we can use a method described shortly where $P(s)$ is indeterminate at $s=1$. To find $P(1)$ we used L'Hôpital's rule and it was easy. To find $P(1)$ and also $P'(1)$, proceed as follows.

The power series for $P(s)$ is convergent in the full unit interval and, thus, by Abel's theorem the functions $P(s)$ and $P'(s)$ are continuous at $s = 1$. The indeterminancy in $P(s)$ can be eliminated by expanding the denominator as a power series about the point $s = 1$ and then canceling the factor $(1 - s)$ from the numerator and the denominator of $P(s)$.

Considering the expansion of $f(s) = 1 - C(s)$ about $s = 1$ by Taylor's theorem. We get

$$f(s) = f(1) + (s-1)f'(1) + \frac{(s-1)^2}{2}f''(1) + \text{higher powers of } (s-1).$$

Now

$$f(1) = 0, \qquad f'(1) = -C'(1) = -\bar{a} \quad \text{and}$$

$$f''(1) = -C''(1) = -\sigma^2 + \bar{a}^2 - \bar{a},$$

where $\bar{a}$ is the mean and σ^2 is the variance of the bulk distribution $\{a_k\}$. Thus,

$$P(s) = \frac{\mu(1-s)(1-\rho)}{\mu(1-s) - \lambda s\left[(s-1)f'(1) + \dfrac{(s-1)^2}{2!}f''(1) + \cdots\right]}$$

$$= \frac{\mu(1-\rho)}{\mu + \lambda s f'(1) + \dfrac{\lambda s(s-1)}{2!}f''(1) + \cdots}. \tag{2.6}$$

It readily follows that

$$P(1) = \frac{\mu(1-\rho)}{\mu - \lambda\bar{a}} = 1$$

$$\begin{aligned} P'(1) &= \frac{\mu(1-\rho)}{(\mu - \lambda\bar{a})^2}\left[-\lambda f'(1) - \frac{\lambda}{2} f''(1)\right] \\ &= \frac{1}{\mu - \lambda\bar{a}}\left[\lambda C'(1) + \frac{\lambda}{2} C''(1)\right] \\ &= \frac{1}{\mu - \lambda\bar{a}}\left[\lambda\bar{a} + \frac{\lambda}{2}(\sigma^2 + \bar{a}^2 - \bar{a})\right]. \end{aligned}$$

Thus,

$$E(N) = \frac{\lambda(\sigma^2 + \bar{a}^2 + \bar{a})}{2(\mu - \lambda\bar{a})}, \tag{2.7}$$

which gives the expected number in the system. In the same manner $P''(1)$ can be found and the variance of the number in the system can be obtained.

4.2.1.3. Particular Cases

(1) By considering that $a_k = 0$, $k \neq 1$, $a_1 = 1$, we get the corresponding result for the $M/M/1$ queue.

(2) Let X have geometric (decapitated) distribution

$$a_k = P(X = k) = c(1-c)^{k-1}, \qquad 0 < c < 1, \qquad k = 1, 2, \ldots,$$

with PGF $A(s) = cs/[1 - (1-c)s]$ and $\bar{a} = E(X) = A'(1) = 1/c$. Then

$$\begin{aligned} P(s) &= \frac{\mu(1-\rho)(1-s)}{\mu(1-s) - \{s\lambda(1-s)\}/\{1-(1-c)s\}} \\ &= \frac{(1-\rho)\{1-(1-c)s\}}{1 - \{(1-c) + c\rho\}s}, \qquad \text{since } \rho = \lambda/c\mu \\ &= (1-\rho)\left[\frac{1}{1 - \{(1-c) + c\rho\}s} \; \frac{(1-c)s}{1 - \{(1-c) + c\rho\}s}\right] \\ &= (1-\rho)\left[\sum_{n=0}^{\infty}\{(1-c) + c\rho\}^n s^n - (1-c)\sum_{n=0}^{\infty}\{(1-c) + c\rho\}^n s^{n+1}\right] \end{aligned}$$

so that

$$p_n = (1-\rho)[\{(1-c)+c\rho\}^n - (1-c)\{(1-c)+c\rho\}^{n-1}]$$
$$= c\rho(1-\rho)[(1-c)+c\rho]^{n-1}, \qquad n \geq 1.$$

This, together with $p_0 = 1 - \rho$, gives $\{p_n\}$.

(3) Let the batch be of size 1 or 2 with equal probability

$$a_1 = a_2 = \tfrac{1}{2}, \qquad a_k = 0,\ k > 2;$$

$$A(s) = \frac{1}{2}s(1+s), \qquad \rho = \frac{2}{3}\frac{\lambda}{\mu}.$$

Then

$$P(s) = \frac{\mu(1-\rho)(1-s)}{\mu(1-s) - \lambda s\{1 - \frac{1}{2}s(1+s)\}} = \frac{3(1-\rho)}{6 - 4\rho s - 2\rho s^2}$$
$$= \frac{3(1-\rho)}{(s_2 - s_1)}\left[\frac{1}{s-s_1} - \frac{1}{s-s_2}\right],$$

where s_1 and s_2 are the roots of the equation $(3/\rho) - 2s - s^2 = 0$. The RHS can be expanded in powers of s and from there an expression for p_n can be obtained.

A more general case where X has a Bernoulli distribution with $a_1 = p$, $a_2 = 1 - p$, $0 < p < 1$, can be similarly treated. Jensen, *et al.* (1977) consider the case where X has a multinomial distribution.

(4) For fixed batch size $M^r/M/1$, suppose each batch consists of exactly r arrivals. Then $a_r = 1$, $a_k = 0$, $k \neq r$ and $A(s) = s^r$. The PGF of the number of customers in the system $M^r/M/1$ is [from (2.5a)],

$$P(s) = \frac{\mu(1-s)(1-\rho)}{\mu(1-s) - \lambda s(1-s^r)}, \tag{2.8}$$

where $\rho = r\lambda/\mu$.

4.2.2. Equivalence of $M^r/M/1$ and $M/E_r/1$ Systems

In the system $M/E_r/1$, arrivals occur singly in accordance with a Poisson process, while the service consists of r stages, each exponential with mean $(1/r\mu)$ (parameter $r\mu$) and the completion of the service requires total service in r stages with mean $r(1/r\mu) = 1/\mu$. The system can be considered as one in which "each customer" arrival amounts to arrival of "r customers," such that each of these would require a single stage of exponential service with mean $(1/r\mu)$. This latter system is $M^r/M/1$ with mean service time $(1/r\mu)$. Thus, the distribution of the number of customers in the system $M^r/M/1$ (with μ replaced by μr) is the same as the distribution of the number of stages in the system $M/E_r/1$. The equivalence of the two systems is thus established. Replacing μ by $r\mu$ in (2.8) we get the PGF of the number of stages in $M/E_r/1$, which is given by

$$P(s) = \frac{r\mu(1-s)(1-\rho)}{r\mu(1-s) - \lambda s(1-s^r)}, \qquad (\rho = \lambda/\mu). \tag{2.9}$$

4.2.3. Waiting Time Distribution in an $M^X/M/1$ Queue

The waiting time of a test unit consists of two parts: (1) the time to complete service of all the units in the system found by an arriving group and (2) the time to serve all units of the group served before the test unit. Burke (1975) has obtained the distribution under (2) for the more general service-time distribution (as we shall discuss later on), from which the distribution for the case of the exponential service time can be easily deduced.

We shall confine our discussion here to (1), which is the waiting time in the queue W_q of the *first customer* (to be served) of an arriving group. The LST of the distribution of W_q and its moment can be easily obtained as discussed next. Let ${}_1w_Q(t)$ be its density and ${}_1w_q^*(\alpha)$ its LT. We have

$$w_q(x) = p_0 = 1 - \rho, \qquad x = 0$$

and for $x \geq 0$

$$w_q(x) = (1-\rho)\delta(t) + \sum_{n=1}^{\infty} \frac{\mu^n x^{n-1} e^{-\mu x}}{\Gamma(n)} p_n.$$

The LT is given by

$$w_q^*(\alpha) = (1-\rho) + \sum_{n=1}^{\infty} \left(\frac{\mu}{\alpha+\mu}\right)^n p_n$$

$$= P\left(\frac{\mu}{\alpha+\mu}\right)$$

where $P(s)$, the PGF of $\{p_n\}$, is given by (2.5b). Thus, $w_q^*(\alpha)$ can be written as

$$w_q^*(\alpha) = \frac{(1-\rho)}{1 - \left(\frac{\lambda}{\alpha}\right)\left\{1 - A\left[\frac{\mu}{(\alpha+\mu)}\right]\right\}} \tag{2.10}$$

where

$$A\left[\frac{\mu}{(\alpha + \mu)}\right] = \sum_{k=1}^{\infty} a_k \left(\frac{\mu}{\mu + \alpha}\right)^k$$

is the LST of a batch service, i.e., the LST of the time required by a server to serve all customers of a batch (also termed *supercustomer*, the batch being treated as a supercustomer). The preceding expression is in terms of λ, $\rho = \lambda\bar{a}/\mu$, and the PGF A(.) of the bulk distribution X with PF, $P(X = k) = a_k$ and $\bar{a} = E(X)$.

To find the moment we consider the limit as $\alpha \to 0$ of

$$f(\alpha) = \frac{\lambda\left[1 - A\left(\frac{\mu}{\mu + \alpha}\right)\right]}{\alpha}.$$

Using L'Hôpital's rule, we find

$$\lim_{\alpha \to 0} f(\alpha) = \lim_{\alpha \to 0} \frac{\lambda\left[A'\left(\frac{\mu}{\mu + \alpha}\right)\right]\left[\frac{\mu}{(\mu + \alpha)^2}\right]}{1}$$

$$= \frac{\lambda\bar{a}}{\mu} = \rho$$

so that

$$w_q^*(0) = 1.$$

To find the expected waiting time we first find $\lim_{\alpha \to 0} f'(\alpha)$.

$$\lim_{\alpha \to 0} f'(\alpha) = \lambda \frac{\alpha\left[A'\left(\frac{\mu}{\mu + \alpha}\right)\right]\left[\frac{\mu}{(\mu + \alpha)^2}\right] - \left[1 - A\left(\frac{\mu}{\mu + \alpha}\right)\right]}{\alpha^2}$$

$$= \lambda \frac{\alpha\left[-A''\left(\frac{\mu}{\mu + \alpha}\right)\frac{\mu^2}{(\mu + \alpha)^4} - \frac{2\mu}{(\mu + \alpha)^3} A'\left(\frac{\mu}{\mu + \alpha}\right)\right]}{2\alpha}$$

$$= -\frac{\lambda}{2\mu^2}\left[(\sigma^2 + \bar{a}^2 - \bar{a}) + 2\bar{a}\right]$$

$$= -\left(\frac{\lambda}{2\mu^2}\right)(\sigma^2 + \bar{a}^2 + \bar{a}).$$

Thus,

$$\frac{d}{d\alpha} w_q^*(\alpha)\bigg|_{\alpha=0} = \frac{1-\rho}{[1-f(0)]^2} f'(0) = -\frac{\lambda(\sigma^2 + \bar{a}^2 + \bar{a})}{2\mu^2(1-\rho)}$$

so that the expected waiting time of the first customer is

$$\frac{\lambda(\sigma^2 + \bar{a}^2 + \bar{a})}{2\mu^2(1-\rho)}. \tag{2.11}$$

4.2.3.1. Particular Cases

(1) $M^X/M/1$: where X is geometric with $A(s) = cs/[1-(1-c)s]$.

Then,

$$\begin{aligned} w_q^*(\alpha) &= \frac{1-\rho}{1-\left(\dfrac{\lambda}{\alpha}\right)\left\{1 - A\left[\dfrac{\mu}{(\mu+\alpha)}\right]\right\}} \\ &= \frac{(1-\rho)(\alpha + c\mu)}{\alpha + c\mu - \lambda}, \qquad \rho = \frac{\lambda}{c\mu} \end{aligned}$$

and the expected delay of the first customer of an arriving group is

$$\frac{\lambda\left\{\dfrac{1-c}{c^2} + \dfrac{1}{c^2} + \dfrac{1}{c}\right\}}{2\mu^2(1-\rho)} = \frac{\rho}{c\mu(1-\rho)}. \tag{2.12}$$

(2) $M/M/1$: Here $a = 1, \sigma = 0, A(s) = s$, and the first customer of an arriving group is the test customer. The LT of the PDF of the waiting time distribution (of a customer) is

$$\begin{aligned} w_q^*(\alpha) &= \frac{1-\rho}{1-\left(\dfrac{\lambda}{\alpha}\right)(1-\mu)(\mu+\alpha)} \\ &= \frac{(1-\rho)(\alpha+\mu)}{\alpha - \lambda + \mu} \end{aligned} \tag{2.13}$$

and the expected waiting time of a customer is

$$\frac{\rho}{\mu(1-\rho)}. \tag{2.14}$$

4.2.4. Transient-State Behavior

Let us consider now the transient state of the system. Assume that $p_0(0) = 1$, i.e., at time 0 there is no customer in the system. Let

$$p_n^*(\alpha) = \text{LT} \quad \text{of} \quad p_n(t) = \int_0^\infty e^{-\alpha t} p_n(t)dt \quad \text{and}$$

$$P^*(s, \alpha) = \sum_{n=0}^{\infty} p_n^*(\alpha)s^n$$

be the GF of $\{p_n^*(\alpha)\}$. Taking the LT of (2.1) and (2.2) we get

$$\alpha p_0^*(\alpha) - 1 = -\lambda p_0^*(\alpha) + \mu p_1^*(\alpha) \tag{2.15}$$

$$\alpha p_n^*(\alpha) = -(\lambda + \mu)p_n^*(\alpha) + \mu p_{n+1}^*(\alpha) + \sum_{k=1}^{\infty} \lambda a_k p_{n-k}^*(\alpha), \qquad n \geq k \geq 1. \tag{2.16}$$

Proceeding in the same manner, we get on simplification

$$P^*(s, \alpha) = \frac{s + \mu(s-1)p_0^*(\alpha)}{s(\alpha + \lambda + \mu) - \mu - \lambda s A(s)}. \tag{2.17}$$

It can be easily verified by taking limits that the corresponding steady-state result follows. We have

$$\begin{aligned} P(s) &= \sum_{n=0}^{\infty} p_n s^n = \sum_{n=0}^{\infty} \left\{ \lim_{t \to \infty} p_n(t) \right\} s^n \\ &= \sum_{n=0}^{\infty} \lim_{\alpha \to 0} \alpha p_n^*(\alpha) s^n \\ &= \lim_{\alpha \to 0} \alpha P^*(s, \alpha), \end{aligned} \tag{2.18}$$

as can be verified; $p_0^*(\alpha)$ can be evaluated by the method discussed by Luchak (1958).

Note: The initial condition $p_i(0) = 1$ will yield the same form as (2.17) with s^{i+1} in place of the first term s in the numerator of (2.17).

4.2.4.1. Busy Period Distribution. Let $N^*(t)$ be the number in the system during a busy period starting with one customer at $t = 0$. Let

$$q_n(t) = P\{N^*(t) = n \mid N^*(0) = 1\}. \tag{2.19}$$

Then $q_1(0) = 1$, $q_n(0) = 0$, $n \geq 2$. The equations governing the zero-avoiding state probabilities will be

$$q_n'(t) = -(\lambda + \mu)q_n(t) + \mu q_{n+1}(t) + \lambda \sum_{k=2}^{n} a_k q_{n-k}(t), \qquad n \geq k \geq 2$$

$$q_1'(t) = -(\lambda + \mu)q_1(t) + \mu q_2(t). \tag{2.20}$$

The GF of the LT of $q_n(t)$ can be obtained, and from there the LT $q_1^*(s)$ of the busy-period density can be found.

4.2.5. The System $M^X/M/\infty$

The transient state was considered by Reynolds (1968). Let $N(t)$ be the number in the system (or the number of busy servers) at time t. We suppose that $\{N(0) = i\}$, that is, there are i in the system at time 0. Let

$$p_{i,j}(t) = P\{N(t) = j \mid N(0) = i\} \quad \text{and} \tag{2.21}$$

$$Q(s, t) = \sum_{j=0}^{\infty} p_{i,j}(t)s^j \tag{2.22}$$

be the PGF of $\{p_{i,j}(t)\}$. We have $Q(s, 0) = s^i$. Now $p_{i,j}(t)$ satisfies the following differential-difference equations:

$$p_{i,j}'(t) = \lambda \sum_{r=1}^{j} a_r p_{i,j-r}(t) - (\lambda + j\mu)p_{i,j}(t)$$

$$+ (j + 1)\mu p_{i,j+1}(t), \qquad j = 0, 1, 2, \ldots. \tag{2.23}$$

Multiplying (2.23) by s^j and summing over all j we get

$$\frac{\partial Q}{\partial t} = \mu(1 - s)\frac{\partial Q}{\partial s} - \lambda[\{1 - A(s)\}Q]. \tag{2.24}$$

This is an equation of the form

$$p\frac{\partial Q}{\partial t} + q\frac{\partial Q}{\partial s} = r$$

where p, q, and r are functions of t, s, and Q. Equation (2.24) can be solved by the method of Lagrange, which uses the subsidiary equation

$$\frac{dt}{1} = \frac{ds}{-\mu(1 - s)} = \frac{dQ}{\lambda\{1 - A(s)\}Q}. \tag{2.25}$$

From the first two, we get, by integration,

$$(s-1)e^{-\mu t} = C_1, \tag{2.26}$$

where C_1 is a constant; and from the last two, by integration, we get

$$\int \frac{dQ}{Q} = \frac{\lambda}{\mu} \int \left(\frac{1-A(s)}{1-s} \right) ds \tag{2.27}$$

$$= \frac{\lambda}{\mu} \int A_1(s) ds, \tag{2.27a}$$

where

$$A_1(s) = \frac{1-A(s)}{1-s} = \sum_{r=0}^{\infty} Pr(X > r) s^r \tag{2.27b}$$

is the generating function of $\{Pr(X > r)\}$. Integration of (2.27) gives

$$Q \exp\left[-\frac{\lambda}{\mu} \int_0^s A_1(s) ds \right] = C_2 \tag{2.28}$$

where C_2 is a constant. Two constants C_1 and C_2 can be functionally related by writing $C_2 \equiv F(C_1)$ where F is an arbitrary function. Thus, from (2.26) and (2.28) we get

$$Q(s, t) \exp\left[-\frac{\lambda}{\mu} \int_0^s A_1(s) ds \right] = F[(s-1)e^{-\mu t}]. \tag{2.29}$$

Thus, the general solution of (2.24) is given by

$$Q(s, t) = \left\{ \exp\left[\frac{\lambda}{\mu} \int_0^s A_1(y) dy \right] \right\} \{F[(s-1)e^{-\mu t}]\}. \tag{2.29a}$$

To determine F, we put $t = 0$; LHS becomes $Q(s, 0) = s^i$, so that

$$s^i = \left[\exp\left\{ \frac{\lambda}{\mu} \int_0^s A_1(y) dy \right\} \right] F(s-1)$$

whence

$$F(s) = (s+1)^i \exp\left\{ -\frac{\lambda}{\mu} \int_0^{s+1} A_1(y) dy \right\}. \tag{2.30}$$

Put

$$e^{-\mu t} = p, \qquad 1 - e^{-\mu t} = q; \tag{2.31}$$

then

$(s-1)e^{-\mu t}+1 = sp+q$ and

$$F\{(s-1)e^{-\mu t}\} = F\{(sp+q-1)\} = (sp+q)^i \exp\left\{-\frac{\lambda}{\mu}\int_0^{sp+q} A_1(y)dy\right\}. \tag{2.32}$$

Thus, from (2.29a) and (2.32) we get

$$Q(s,t) = (sp+q)^i \exp\left\{\frac{\lambda}{\mu}\left[\int_0^s A_1(y)dy - \int_0^{sp+q} A_1(y)dy\right]\right\} \tag{2.33}$$

$$= (sp+q)^i \exp\left\{-\frac{\lambda}{\mu}\int_s^{sp+q} A_1(y)dy\right\}. \tag{2.33a}$$

Thus, we obtain the PGF $Q(s,t)$ of $\{p_{i,j}(t)\}$. When $i=0$, i.e.,

$$P\{N(0)=0\}=1,$$

then

$$Q(s,t) = \exp\left\{-\frac{\lambda}{\mu}\int_s^{sp+q} A_1(y)dy\right\}. \tag{2.34}$$

4.2.5.1. Steady-State Result. As $t\to\infty$, $p=e^{-\mu t}\to 0$ and $q=1-e^{-\mu t}\to 1$, $Q(s,t)\to Q(s)$ so that from (2.33a) we get

$$Q(s) = \sum_{j=0}^{\infty} p_{i,j} = \exp\left\{-\frac{\lambda}{\mu}\int_s^1 A_1(y)dy\right\} \tag{2.35}$$

which is independent of the initial state i, so that $p_{i,j}=p_j$, $j=0,1,2,\ldots$. Writing

$$\int_s^1 A_1(y)dy = \int_0^1 A_1(y)dy - \int_0^s A_1(y)dy$$

$$= k(1-G(s)) \tag{2.36}$$

where

$$k=\int_0^1 A_1(y)dy, \qquad G(s)=\frac{1}{k}\int_0^s A_1(y)dy,$$

we get

$$G(1)=\frac{1}{k}\int_0^1 A_1(y)dy = \frac{1}{k}k=1,$$

so that $G(s)$ is a PGF. It is the PGF of the distribution

$$\{f_r\} \quad \text{where} \quad f_r = \frac{P(X > r - 1)}{kr}, \qquad r = 1, 2, 3, \ldots .$$

Thus, from (2.35) and (2.36) we get that

$$Q(s) = \exp\left\{-\frac{\lambda k}{\mu}[1 - G(s)]\right\} \tag{2.37}$$

so that the distribution of $\{p_j\}$ is compound Poisson.

4.2.5.2. Particular Case M/M/∞: Transient State. Let arrivals occur singly; then $P(X = 1) = 1$, $A(s) = s$, and $A_1(s) = 1$. Then from (2.33) we get

$$\begin{aligned} Q(s, t) &= (sp + q)^i \exp\left\{-\frac{\lambda}{\mu}(sp + q - s)\right\} \\ &= (sp + q)^i \exp\left\{-\frac{\lambda q}{\mu}(1 - s)\right\} \\ &= (sp + q)^i \exp\left\{\frac{\lambda}{\mu}(1 - e^{-\mu t})(s - 1)\right\}. \end{aligned} \tag{2.38}$$

With $i = 0$, we get

$$Q(s, t) = \exp\left\{\frac{\lambda}{\mu}(1 - e^{-\mu t})(s - 1)\right\} \tag{2.39}$$

so that the distribution of $\{p_{0,j}(t)\}$ is Poisson with mean $(\lambda/\mu)(1 - e^{-\mu t})$. Thus,

$$p_j(t) = \left\{\exp\left[-\frac{\lambda}{\mu}(1 - e^{-\mu t})\right]\right\}\frac{\left[\frac{\lambda}{\mu}(1 - e^{-\mu t})\right]^j}{j!}. \tag{2.40}$$

This result is obtained later as a particular case of $M/G/\infty$. Evidently, in steady state, the distribution of $\{p_j\}$ is Poisson with mean λ/μ and this happens irrespective of the magnitude of λ/μ.

4.2.5.3. Regression of N(t) on N(0). Reynolds (1968) gives another interesting result for $M^X/M/\infty$ system. Suppose $i > 0$. From (2.33) we get the expected number in the system, given $N(0) = i$. We have

$$E\{N(t) \mid N(0) = i\} = \frac{\partial}{\partial s} Q(s, t)\bigg|_{s=1}.$$

Noting that

$$\frac{d}{ds}\int_0^s A_1(y)dy = A_1(s),$$

$$\frac{d}{ds}\int_0^{sp+q} A_1(y)dy = \left[\frac{d}{ds}(sp+q)\right]A_1(sp+q)$$

$$= pA_1(sp+q), \quad \text{and}$$

$$\frac{d}{ds}\left\{\exp\frac{\lambda}{\mu}\left[\int_0^s A_1(y)dy - \int_0^{sp+q} A_1(y)dy\right]\right\}\Bigg|_{s=1}$$

$$= \frac{\lambda}{\mu}[A_1(1) - pA_1(1)]\exp\left\{\frac{\lambda}{\mu}\left[\int_0^1 A_1(y)dy - \int_0^1 A_1(y)dy\right]\right\}$$

$$= \frac{\lambda}{\mu}qA_1(1)$$

$$= \frac{\lambda}{\mu}qE(X),$$

we get

$$E\{N(t)|N(0)=i\} = ip + \frac{\lambda}{\mu}qA_1(1)$$

$$= \frac{\lambda}{\mu}(1 - e^{\mu t})E(X) + N(0)e^{-\mu t}.$$

This shows that the regression of $N(t)$ on $N(0)$ is linear being of the form $E\{Y|x\} = a + bx$.

4.3. Queueing Models with Bulk (Batch) Service

We have so far considered models with individual service. Now we shall consider systems where services are offered in batches, instead of personalized service of one at a time. Bailey (1954) was the first to consider bulk service. The literature on bulk service has grown over the years. Such models find applications in several situations, such as transportation; mass transit vehicles and carriers are natural batch servers.

A number of policies of bulk service are considered in literature. These are discussed below.

(1) Bailey (1954) considered that the server serves in batches of size not more than, say, b, the maximum capacity of the server. If the server, on

completion of a batch service, finds not more than b waiting, then he takes all of them in a batch for service; if he finds more than b waiting, then he takes for service a batch of size b (in order of arrival or in any other order), while others, in excess of b units, wait and join the queue. An example of this type of server is an elevator.

(2) A service batch may be of a fixed size, say, k. The server waits till there are k in the queue and starts service as soon as the queue reaches this size; if, on completion of a batch service, he finds more than k waiting, the server takes a batch of size k (in order of arrival or in any other order), while others, in excess of k units, wait in the queue.

(3) A server may take in a batch a minimum number of units, say, a less than or equal to his capacity, say, b. The server adopts the following policy. If, on completion of a batch service, he finds q units waiting and if

(i) $0 \le q < a$, then he waits till the queue size grows to a,

(ii) $a \le q \le b$, then he takes a batch of size q for service, and

(iii) $q > b$, then he takes a batch of size b for service (in order of arrival or in random order), while those in excess of b units, wait in the queue.

We shall call this the rule the *general bulk service rule*, as rules under (1) and (2) above can be covered as particular cases of this rule. Neuts (1967) considered this rule; this has been further investigated, for example, by Medhi, Borthakur, Sim, Templeton, Chaudhry, Powell, Humblet, and others. Kosten earlier considered such a situation with infinite server capacity.

(4) The size of a batch may be a random variable; it may depend on the unfilled capacity of the server. This rule has been considered, for example, by Cohen, Prabhu, Bhat, Teghem, and others. Newell considers a model such that batch service may be extended to accommodate additional units during the course of the service to the extent of the unfilled capacity (of the server or service channel), if there be any. This may be called accessible batch service. For references, see Medhi (1984).

We shall examine here the general bulk service, which, though considered in literature, has not been treated in standard textbooks.

4.3.1. The System $M/M(a, b)/1$

We assume that the input is Poisson (with single arrivals at each epoch of Poisson occurrence) with rate λ. The service is in batches under general bulk service rule as discussed above; service commences only when the queue size reaches or exceeds a, the capacity being $b(\ge a \ge 1)$. The service time distribution of a batch is assumed to be exponential with parameter μ; for simplicity, the service time is taken to be independent of the batch size. We denote the system by the notation $M/M(a, b)/1$. Denote the states of the system by (i, n), where i is an indicator variable, $i = 1$ implies that the server is busy in serving a batch of size s $(a \le s \le b)$, and $i = 0$ implies that the server is idle, with n being the number of units in the queue. Denote $p_{i,n}(t) = Pr\{\text{at time } t, \text{ the system is at state } (i, n)\}$. $p_{i,n}(t)$ is non-zero only for $i = 1$, $n \ge 0$, and $i = 0$, $0 \le n \le a - 1$.

It is clear that the system is Markovian, arising out of non–birth–death processes. The Chapman–Kolmogorov equations lead to:

$$p'_{1,n}(t) = -(\lambda + \mu)p_{1,n}(t) + \lambda p_{1,n-1}(t) + \mu p_{1,n+b}(t), \quad n = 1, 2, \ldots \tag{3.1}$$

$$p'_{1,0}(t) = -(\lambda + \mu)p_{1,0}(t) + \lambda p_{0,a-1}(t) + \mu \sum_{r=a}^{b} p_{1,r}(t) \tag{3.2}$$

$$p'_{0,0}(t) = -\lambda p_{0,0}(t) + \mu p_{1,0}(t) \tag{3.3}$$

$$p'_{0,q}(t) = -\lambda p_{0,q}(t) + \lambda p_{0,q-1}(t) + \mu p_{1,q}(t), \quad q = 1, 2, \ldots, a - 1. \tag{3.4}$$

It is to be noted that (3.4) will not occur when $a = 1$.

4.3.1.1. *Steady-State Solution.*

Assume that steady state exists. Let

$$p_{0,q} = \lim_{t \to \infty} p_{0,q}(t)$$

$$p_{1,n} = \lim_{t \to \infty} p_{1,n}(t).$$

The steady-state equations of the system become:

$$0 = -(\lambda + \mu)p_{1,n} + \lambda p_{1,n-1} + \mu p_{1,n+b}, \quad n = 1, 2, \ldots, \tag{3.5}$$

$$0 = -(\lambda + \mu)p_{1,0} + \lambda p_{0,a-1} + \mu \sum_{r=a}^{b} p_{1,r} \tag{3.6}$$

$$0 = -\lambda p_{0,0} + \mu p_{1,0} \tag{3.7}$$

$$0 = -\lambda p_{0,q} + \lambda p_{0,q-1} + \mu p_{1,q}, \quad q = 1, 2, \ldots, a - 1. \tag{3.8}$$

Equation (3.8) will not occur when $a = 1$.

Solution of the preceding difference equations would give the probabilities $p_{0,q}$, $p_{1,n}$. Denoting the displacement operator by E (i.e., $E\{p_{1,r}\} = p_{1,r+1}$), Eq. (3.5) can be written as

$$-(\lambda + \mu)Ep_{1,n-1} + \lambda p_{1,n-1} + \mu E^{b+1}\{p_{1,n-1}\} = 0 \quad \text{or}$$

$$h(E)\{p_{1,n}\} = 0, \qquad n = 0, 1, 2, \ldots,$$

with characteristic equation

$$h(z) \equiv \mu z^{b+1} - (\lambda + \mu)z + \lambda = 0. \tag{3.9}$$

Suppose that $f(z) = -(\lambda + \mu)z$ and $g(z) = \mu z^{b+1} + \lambda$. Consider the circle $|z| = 1 - \delta$, where δ is arbitrarily small; writing $z = (1 - \delta)e^{i\theta}$, it can be shown that on the contour of the circle,

$$|g(z)| < |\ (z)|.$$

Hence from Rouché's theorem it follows that $f(z)$ and $f(z) + g(z)$ will have the same number of zeros inside $|z| = 1 - \delta$. Since $f(z)$ has only one zero inside this circle, $f(z) + g(z) \equiv h(z)$ will also have only one zero inside $|z| = 1 - \delta$. This root of $h(z) = 0$ is real and unique if and only if $\rho = \lambda/b\mu < 1$. Denote this real root by r $(0 < r < 1)$ and other b roots by $r_1, \ldots, r_b, |r_i| \geq 1$.

Then r satisfies the equation

$$b\rho = \frac{\lambda}{\mu} = \frac{r(1 - r^b)}{1 - r} = r + r^2 + \cdots + r^b. \tag{3.10}$$

We have from (3.10), when $0 < \rho < 1$

$$\rho \leq r \leq \rho^{2/(b+1)}$$

which is useful in solving Eq. (3.10) for r. The solution of (3.5) can now be written as

$$p_{1,n} = \mathrm{A}r^n + \sum_{i=1}^{b} \mathrm{A}_i r_i^n, \qquad n = 0, 1, 2, \ldots$$

where the A's are constants. Since $\sum_{n=0}^{\infty} p_{1,n} < 1$ we must have $\mathrm{A}_i = 0$ for all i, so that, for $n = 0, 1, 2, \ldots,$

$$\begin{aligned} p_{1,n} &= Ar^n \\ &= p_{1,0}r^n \\ &= \left(\frac{\lambda}{\mu}\right)p_{0,0}r^n \quad \text{from (3.7)} \\ &= \left(\frac{1 - r^b}{1 - r}\right)p_{0,0}r^{n+1}. \end{aligned} \tag{3.11}$$

Then from (3.6), using (3.11) and on simplifying, we get

$$p_{0,a-1} = \frac{1 - r^a}{1 - r}\, p_{0,0}. \tag{3.12}$$

Finally, putting $q = a - 1, a - 2, \ldots, 1$ and using (3.11) we get, recursively for $q = a - 2, a - 3, \ldots, 1$,

$$p_{0,q} = \frac{1 - r^{q+1}}{1 - r} p_{0,0} \tag{3.13}$$

and because (3.12) holds, (3.13) holds for $q = 1, 2, \ldots, a - 1$. Using

$$\sum_{q=0}^{a-1} p_{0,q} + \sum_{n=0}^{\infty} p_{1,n} = 1,$$

we get

$$p_{0,0} = \left[\frac{a}{1 - r} + \frac{r^{a+1} - r^{b+1}}{1 - r}\right]^{-1}. \tag{3.14}$$

Thus, (3.11), (3.13), and (3.14) give the steady-state probabilities. The expected number in the queue $E(Q)$

$$\begin{aligned} E(Q) &= \sum_{q=0}^{a-1} q p_{0,q} + \sum_{n=0}^{\infty} n p_{1,n} \\ &= \frac{p_{0,0}}{1 - r} \sum_{q=0}^{a-1} \{q - rq^{r+1}\} + \frac{p_{0,0}(1 - r^b)}{1 - r} \sum_{n=0}^{\infty} n r^{n+1} \\ &= \frac{p_{0,0}}{1 - r} \left\{\frac{a(a - 1)}{2} + \frac{r^2[ar^{a-1}(1 - r) - (1 - r^a)]}{(1 - r)^2}\right\} \\ &\quad + \frac{p_{0,0}(1 - r^b)}{1 - r} \frac{r^2}{(1 - r)^2}. \end{aligned} \tag{3.15}$$

4.3.1.2. *Particular Cases*

(1) Fixed batch size $M/M(k, k)/1$.

Here $a = b = k$, $\rho = \lambda/k\mu$, and r is the real root lying in (0, 1) of

$$\frac{\lambda}{\mu} = r + r^2 + \cdots + r^k.$$

Let p_n be the probability that the number in the system N is n. Then

$$p_0 = p_{0,0} = \frac{1 - r}{k}$$

$$p_q = p_{0,q} = \frac{1 - r^{q+1}}{1 - r} p_0 = \frac{1 - r^{q+1}}{k}, \qquad q = 1, 2, \ldots, k-1,$$

$$p_{m+k} = p_{1,m} = \left(\frac{\lambda}{\mu}\right) p_0 r^m = \rho(1-r) r^m = p_k r^m, \qquad m = 0, 1, 2, \ldots .$$

Or, in compact form,

$$p_n = \begin{cases} \dfrac{1 - r^{n+1}}{k}, & n = 0, 1, 2, \ldots k-1 \\ \rho(1-r) r^{n-k} = p_k r^{n-k}, & n = k, k+1, \ldots . \end{cases} \tag{3.16}$$

(2) $M/M/1$: The distribution of the number in the system N is obtained by putting $a = b = 1$, then $r = \rho$ and

$$p_0 = P(N = 0) = p_{0,0} = 1 - \rho$$

$$p_n = P(N = n) = p_{1,n-1} = (1 - \rho)\rho^n, \qquad n = 1, 2, \ldots .$$

(3) $M/M(1, b)/1$: Usual bulk service rule.

(4) $M/M(a, \infty)/1$: Here the server has infinite capacity. As $b \to \infty$, $r^b \to 0$ and $\lambda/\mu \to r/(1-r)$, so that $r = \lambda/(\lambda + \mu)$ and

$$p_{0,0} = \left[\frac{a}{1-r} + \frac{r^{a+1}}{1-r}\right]^{-1}.$$

Note. Neuts and others after him used analytic methods to study systems under this rule. Later Neuts considered an altogether different approach—an algorithmic approach for queues that have a modified matrix-geometric structure. This method uses a matrix method as an alternative to closed-form analytic methods. For a description of this method, refer to Neuts (1979, 1981).

4.3.2. Distribution of the Waiting Time for the System M/M(a, b)/1

Medhi (1975) obtained the waiting-time distribution. Assume that the system is in steady state. Let the random variable W_q denote the waiting time in the queue for an arriving unit. Denote

$$\omega(t) = \text{the PDF of } W_q$$

$$\begin{aligned} f(\alpha, k; t) &= \text{the PDF of gamma distribution with parameters } \alpha, k \\ &= \alpha^k t^{k-1} \frac{\exp(-\alpha t)}{\Gamma(k)}, \qquad t > 0, \qquad k = 1, 2, \ldots . \end{aligned}$$

$$\Gamma_x(\alpha, k) = \int_0^x f(\alpha, k; t)dt$$

$$= 1 - \sum_{r=0}^{k-1} e^{-\alpha x} \frac{(\alpha x)^r}{r!} \tag{3.17}$$

$$e(m, z) = \sum_{k=0}^{m-1} \frac{z^k}{k!} \quad \text{and}$$

$$E(m, z) = e^{-z} e(m, z) = \sum_{k=0}^{m-1} \frac{e^{-z} z^k}{k!};$$

$E(m, z) \equiv Pr(X \leq m - 1)$ is the DF of Poisson RV X with mean $z(>0)$. We have then

$$\sum_{q=0}^{a-2} r^q f(\lambda, a - q - 1; t) = \lambda r^{a-2} \exp(-\lambda t) e\left(a - 1, \frac{\lambda t}{r}\right). \tag{3.18}$$

The states (sets of) in which a test unit, on arrival, may find the system are:

$$\begin{array}{lll} \text{(i)} & (0, a - 1) & \\ \text{(ii)} & (0, q), 0 \leq q \leq a - 2 & \\ \text{(iii)} & (1, n), a - 1 \leq m \leq b - 1 & n = kb + m \\ \text{(iv)} & (1, n), 0 \leq m \leq a - 2 & k = 0, 1, 2, \ldots. \end{array} \tag{3.19}$$

If he (or she) finds the system in state (i), then the unit does not wait. Thus, $Pr\{W_q = 0\} = p_{0,a-1}$, the probability of blocking (or delay) equals $Pr\{W_q > 0\} = 1 - p_{0,a-1}$.

If he finds the system in a state of (ii), the unit has to await for the arrival of $a - 1 - q$ units after him. The waiting time, i.e., the time required for $a - 1 - q$ arrivals, has a gamma distribution with parameters λ, $a - 1 - q$.

If he finds the system in a state of (iii), then he has to wait for the completion of services of $k + 1$ batches, including that of the one under service. The time required for completion of services of $k + 1$ batches has a gamma distribution with parameters μ, $k + 1$.

Consider finally, that the test unit finds the system in a state of (iv); then he has to wait until either the services of $k + 1$ groups are completed or $a - 1 - m$ units arrive, whichever occurs later. The waiting time in this case is a RV Z, which is the maximum of two gamma variates, i.e.,

$$Z = \max\{\text{gamma variate with parameters } \lambda, a - 1 - m;$$
$$\text{gamma variate with parameters } \mu, k + 1\}.$$

We have

$$
\begin{aligned}
F_Z(t) &= Pr\{Z \le t\} = \Gamma_t(\lambda, a-1-m)\Gamma_t(\mu, k+1), \\
u(t) &= F'_Z(t) \\
&= f(\lambda, a-1-m; t)\Gamma_t(\mu, k+1) \\
&\quad + \Gamma_t(\lambda, a-1-m) f(\mu, k+1; t) \\
&= u_1(t) + u_2(t), \quad \text{say.}
\end{aligned}
\tag{3.20}
$$

Conditioning on the state in which a test unit finds the system, on arrival, we have $Pr\{t \le W_q < t + dt\} = \omega(t)dt$, where

$$
\begin{aligned}
\omega(t) &= \sum_{q-0}^{a-2} f(\lambda, a-1-q; t) p_{0,q} \\
&\quad + \sum_{k=0}^{\infty} \sum_{m=0}^{b-1} f(\mu, k+1; t) p_{1,kb+m} \\
&\quad + \sum_{k=0}^{\infty} \sum_{m=0}^{a-2} u(t) p_{1,kb+m}, \quad 0 < t < \infty.
\end{aligned}
\tag{3.21}
$$

Substituting the values of $p_{0,q}$ and $p_{1,n}$ [as given in (3.13) and (3.11), respectively] and the expressions for f and u [given in (3.20)] in (3.21) we get an explicit expression for $\omega(t)$ as follows.

On simplification, the first term on the RHS of (3.21) becomes

$$
p_{0,0} \frac{\left[E(a-1, \lambda t) - r^{a-1} \exp(-\lambda t) e\left(a-1, \frac{\lambda t}{r}\right)\right]}{(1-r)}. \tag{3.22}
$$

The second member on the RHS of (3.21) reduces to

$$
\begin{aligned}
&\left(\frac{\lambda}{\mu}\right) p_{0,0} \left[\sum_{m=a-1}^{b-1} r^m\right]\left[\mu \exp(-\mu t) \sum_{k=0}^{\infty} \frac{(\mu t r^b)^k}{k!}\right] \\
&\qquad = \frac{[\lambda p_{0,0}(r^{a-1} - r^b)\exp\{-\mu t(1-r^b)\}]}{(1-r)}.
\end{aligned}
$$

We break up the third term into two parts u_1 and u_2 (corresponding to $u(t) = u_1(t) + u_2(t)$). The part corresponding to u_1 equals

$$\left(\frac{\lambda}{\mu}\right)p_{0,0}\left[\sum_{r=0}^{a-2} r^m f(\lambda, a-1-m; t)\right]\int_0^t \left\{\sum_{k=0}^{\infty}(r^b)^k f(\mu, k+1; x)\right\}dx$$

$$= p_{0,0}r^{a-1}\exp(-\lambda t)e\left(a-1, \frac{\lambda t}{r}\right) \times \frac{[1-\exp\{-\mu(1-r^b)t\}]}{(1-r)}, \tag{3.23}$$

and the part corresponding to u_2 equals

$$\left(\frac{\lambda}{\mu}\right)p_{0,0}\left[\sum_{k=0}^{\infty}(r^b)^k f(\mu, k+1; t)\right]\left[\sum_{m=0}^{a-2} r^m \Gamma_t(\lambda, a-1-m)\right]$$

$$= \lambda p_{0,0}[\exp\{-\mu(1-r^b)t\}]\left[\sum_{m=0}^{a-2} r^m\left\{1-\sum_{q=0}^{a-2-m}\frac{e^{-\lambda t}(\lambda t)^q}{q!}\right\}\right]$$

$$= \lambda p_{0,0}[\exp\{-\mu(1-r^b)t\}]\left[\frac{1-r^{a-1}}{1-r} - \frac{E(a-1, \lambda t)}{1-r} + \frac{r^{a-1}\exp(-\lambda t)e\left(a-1, \lambda\frac{t}{r}\right)}{1-r}\right]. \tag{3.24}$$

Now, adding the expressions on the RHS of (3.22) to (3.24), we get

$$\omega(t) = \left[\frac{\lambda p_{0,0}}{(1-r)}\right]\left((1-r^b)\exp\{-\mu(1-r^b)t\} + E(a-1, \lambda t)\{1-\exp[-\mu(1-r^b)t]\}\right), \qquad 0 < t < \infty. \tag{3.25}$$

(We may write $\lambda(1-r)/r$ in place of $\mu(1-r^b)$.)

The PDF of the conditional waiting-time distribution given that an arrival has to wait (i.e., an arriving unit who finds the service channel busy or idle with less than $a-1$ waiting) is given by

$$v(t) = \frac{\omega(t)}{[1-p_{0,a-1}]}. \tag{3.26}$$

4.3.2.1. Moments of the Distribution of W_q. To find the moments, we note that, for positive integral values of k and for $a \geq 2$,

$$J_1(k) = \int_0^\infty t^k \exp\{-\mu(1-r^b)t\}dt = \frac{\Gamma(k+1)}{\{\mu(1-r^b)\}^{k+1}} \tag{3.27}$$

$$\begin{aligned} J_2(k) = \int_0^\infty t^k E(a-1, \lambda t)dt &= \sum_{s=0}^{a-2} \int_0^\infty \frac{t^k (\lambda t)^s e^{-\lambda t}\, dt}{s!} \\ &= \frac{1}{\lambda} \prod_{i=1}^{k} \left(\frac{s+i}{\lambda}\right) \end{aligned} \tag{3.28}$$

$$\begin{aligned} J_3(k) &= \int_0^\infty t^k E(a-1, \lambda t) \exp\{-\mu(1-r^b)t\}dt \\ &= \frac{1}{\lambda} \sum_{s=0}^{a-2} \left(\frac{1}{r^{s+1}}\right) \prod_{i=1}^{k} \left(\frac{s+i}{\lambda}\right). \end{aligned} \tag{3.29}$$

We then put $E(W_q)$ in terms of the above quantities as

$$\begin{aligned} E(W_q) &= \int_0^\infty t\omega(t)dt = \left[\frac{\lambda p_{0,0}}{1-r}\right][(1-r^b) J_1(1) + J_2(1) - J_3(1)] \\ &= \frac{\lambda p_{0,0}}{1-r} \left[\frac{1}{\mu^2(1-r^b)} + \frac{a(a-1)}{2\lambda^2} + \frac{ar^{a+1}(1-r) - r^2(1-r^a)}{\lambda^2(1-r)^2}\right]. \end{aligned} \tag{3.30}$$

Using (3.10), we get

$$\mu^2(1-r^b) = \frac{\lambda^2(1-r)^2}{r^2(1-r^b)}$$

and, thus,

$$E(W_q) = \frac{p_{0,0}}{\lambda(1-r)} \left[\frac{r^2(1-r^b)}{(1-r)^2} + \frac{a(a-1)}{2} + \frac{r^2\{ar^{a-1}(1-r) - (1-r^a)\}}{(1-r)^2}\right]. \tag{3.31}$$

Comparing (3.15) and (3.31) we at once verify that Little's formula $L_q = \lambda W_q$ holds for such a bulk-service system also. Since (3.27)–(3.30) hold for all positive integral values of k, higher moments $E\{W_q^k\}$, $k = 2, 3, \ldots$ can be obtained easily.

4.3.2.2. *Particular Case: M/M(1, b)/1.* We have for $a = 1$,

$$p_{0,0} = \frac{(1-r)}{1-r+\left(\dfrac{\lambda}{\mu}\right)} = \frac{(1-r)^2}{(1-r)^2 + r(1-r^b)}.$$

$$p_{1,n} = \frac{(1-r)(1-r^b)}{(1-r)^2 + r(1-r^b)} r^{n+1}, \qquad n = 0, 1, 2, \ldots .$$

The waiting-time density is given by

$$\omega(t) = \left[\frac{\lambda p_{0,0}}{(1-r)}\right][(1-r^b)\exp\{-\mu(1-r^b)t\}], \tag{3.32}$$

and the expected waiting time by

$$E(W_q) = \frac{r}{\mu[(1-r)^2 + r(1-r^b)]}. \tag{3.33}$$

The conditional waiting-time distribution of only those who wait is exponential with mean $1/\mu\{1-r^b\}$.

When $b = 1$, then $r = \rho$ and the corresponding results for an $M/M/1$ queue follow.

For the model $M/M(a, \infty)/1$, $r = \lambda/(\lambda + \mu)$; a model with large b has been considered by Sim and Templeton (1983).

Notes.

(1) The unique positive root $r(<1)$ of the equation $h(z) = 0$ is independent of a. Values of r for certain values of b and ρ are given by Cromie and Chaudhury (1976).

(2) Medhi (1979) has given the values of $E(W_q)/b$ and $\mathrm{var}(W_q)/b^2$ for certain values of a, b, and p, and also their limiting values for large b and for $\theta = a/b = 0(0.1)(1.0)$.

4.3.3. *Service Batch-Size Distribution*

This has been obtained by Sim and Templeton (1985) for the more general c-server $M/M(a, b)/c$ system.

Let Y be the service batch size for a randomly chosen customer. Then Y is a RV that assumes values between a and b. Let k_j be the probability that j customers arrive during an interval T having gamma distribution with parameters μ, $k+1$. Then

$$\begin{aligned}
k_j &= \int_0^\infty \frac{e^{-\lambda t}(\lambda t)^j}{j!} \frac{(\mu)^{k+1}(t)^k e^{-\mu t}}{\Gamma(k+1)}\, dt \\
&= \frac{\lambda^j \mu^{k+1}}{j!k!} \int_0^\infty e^{-(\lambda+\mu)t} t^{(k+j)} dt \\
&= \frac{\lambda^j \mu^{k+1}}{j!k!} \frac{(k+j)!}{(\lambda+\mu)^{k+j+1}} \\
&= \binom{k+j}{j} \left(\frac{\lambda}{\lambda+\mu}\right)^j \left(\frac{\mu}{\lambda+\mu}\right)^{k+1}, \qquad j = 0, 1, 2, \ldots .
\end{aligned} \tag{3.34}$$

Consider the four states in which an arriving customer may find the system.

In case he finds the system in state (i) $(0, a-1)$ or (ii), $(0, q)$, with $0 \le q \le a-2$, he will eventually be served in a batch of size exactly a.

In case he finds the system in state (iii), $(1, n)$, with $n = kb + m$, $a - 1 \le m \le b - 1$, he will have to wait for the completion of services of $(k+1)$ batches the duration of which equals T. The size of the service depends on the number of customers who arrive during the interval T. Thus, for $k = 0, 1, 2, \ldots, a-1 \le m \le b-1$,

$$Pr\{Y = y | \text{state } (1, kb+m)\} = \begin{cases} 0, & y \le m \\ k_{y-m-1}, & m < y < b-1 \\ \sum_{i=b-m-1}^{\infty} k_i, & y = b. \end{cases} \tag{3.35}$$

In case he finds the system in state (iv), $(1, n)$, $n = kb + m$, $0 \le m \le a-2$, the size of the batch will depend on the number of customers who arrive during the interval T. Thus, for

$$k = 0, 1, \ldots, \qquad 0 \le m \le a-2,$$

$$Pr\{Y = y | \text{state } (1, kb+m)\} = \begin{cases} \sum_{i=0}^{a-m-1} k_i, & y = a \\ k_{y-m-1}, & a < y \le b-1 \\ \sum_{i=b-m-1}^{\infty} k_i, & y = b. \end{cases} \tag{3.36}$$

It follows that the service batch-size distribution for a *randomly chosen customer* is given by

$$g_a = Pr\{Y = a\} = \sum_{j=0}^{a-1} p_{0,j} + \sum_{k=0}^{\infty} \sum_{m=0}^{a-2} \sum_{i=0}^{a-m-1} k_i p_{1,kb+m} \tag{3.37}$$

$$g_y = Pr\{Y = y\} = \sum_{k=0}^{\infty} \sum_{m=0}^{y-1} k_{y-m-1} p_{1,kb+m}, \qquad a < y \le b-1 \tag{3.38}$$

$$g_b = Pr\{Y = b\} = \sum_{k=0}^{\infty} \sum_{m=0}^{b-1} \sum_{i=b-m-1}^{\infty} k_i p_{1,kb+m}. \tag{3.39}$$

Using the expressions for the state probabilities and simplifying, one can get explicit expressions for g_a, g_y, and g_b (Sim and Templeton, 1985).

Note. Neuts and Nadarajan (1982) consider the same case. The distribution of the size of a randomly chosen batch has not been considered here.

4.4. *M/M(a, b)/1*: Transient-State Distribution

Let us assume that time is reckoned from the instant the server has taken a batch for service, leaving none in the queue, i.e., $p_{1,0}(0) = 1$. Let $p^*_{1,n}(s)$, $p^*_{0,q}(s)$ denote, respectively, the LT of $p_{1,n}(t)$ and $p_{0,q}(t)$. Taking LT of Eqs. (3.1)–(3.4) we get

$$(s + \lambda + \mu)p^*_{1,n}(s) = \lambda p^*_{1,n-1}(s) + \mu p^*_{1,n+b}(s), \qquad n \ge 1 \tag{4.1}$$

$$(s + \lambda + \mu)p^*_{1,0}(s) - 1 = \lambda p^*_{0,a-1}(s) + \mu \sum_{r=a}^{b} p^*_{1,r}(s) \tag{4.2}$$

$$(s + \lambda)p^*_{0,0}(s) = \mu p^*_{1,0}(s) \tag{4.3}$$

$$(s + \lambda)p^*_{0,q}(s) = \lambda p^*_{0,q-1}(s) + \mu p^*_{1,q}(s), \qquad q = 1, 2, \ldots, a-1 (>0). \tag{4.4}$$

and the last equation will not occur for $a = 1$. The equations can now be solved in the same manner as in the case of steady-state solutions. Equation (4.1) can be written as

$$h_1(E)\{p^*_{1,n}(s)\} = 0, \qquad n = 1, 2, \ldots,$$

where

$$h_1(z) = \mu z^{b+1} - (s + \lambda + \mu)z + \lambda = 0. \tag{4.5}$$

Equation (4.5) will have only one real zero inside $|z| = 1$, if and only if $\rho = \lambda/b\mu < 1$. Denote this real boot by $R \equiv R(s)$ and the other roots by $R_1, \ldots, R_b$. Note that as $s \to 0$, $R \to r$, $R_i \to r_i$. Thus

$$p_{1,n}^*(s) = AR^n + \sum_{i=1}^{b} A_i R_i^n, \qquad n = 0, 1, 2, \ldots.$$

Since

$$\sum_n p_{1,n}^*(s) + \sum_q p_{0,q}^*(s) = \frac{1}{s}, \tag{4.6}$$

we have $A_i = 0$ for all i, so that

$$\begin{aligned} p_{1,n}^*(s) &= AR^n, \qquad n = 1, 2, \ldots, \\ &= p_{1,0}^*(s)R^n \\ &\qquad\qquad \text{(choosing } A \text{ such that (4.1) is satisfied for } n = 1\text{)}, \\ &= \left(\frac{s+\lambda}{\mu}\right) p_{0,0}^*(s)R^n \quad \text{from (4.3).} \end{aligned} \tag{4.7}$$

From (4.2), using (4.7) and simplifying, we get

$$\begin{aligned} p_{0,a-1}^*(s) = &\frac{(s+\lambda)(s+\lambda+\mu)}{\lambda\mu} \\ &-\left(\frac{s+\lambda}{\lambda}\right)\frac{R^a(1-R^{b-a+1})}{1-R}\,p_{0,0}^*(s) - \frac{1}{\lambda}. \end{aligned} \tag{4.8}$$

Then putting $q = a - 1, \ldots, 1$, we get for $q = a - 2, \ldots, 1$,

$$p_{0,q}^*(s) = \left(\frac{\lambda}{s+\lambda}\right)^q \left\{\frac{\lambda}{\lambda - (s+\lambda)R}\right\}\left[1 - \left\{\frac{(s+\lambda)R}{\lambda}\right\}^{q+1}\right] p_{0,0}^*(s). \tag{4.9}$$

Finally, using (4.6), we get

$$\begin{aligned} [p_{0,0}^*(s)]^{-1} = &\frac{\lambda s}{\mu(1-R)} - \frac{\lambda s(R - R^a)}{(1-R)(\lambda - sR - \lambda R)} + \frac{s(s+\lambda+\mu)}{\mu} \\ &- \frac{s(R^a - R^{b+1})}{1-R} + \frac{\lambda^2}{\lambda - sR - \lambda R}\left\{1 - \left(\frac{\lambda}{s+\lambda}\right)^{a-1}\right\}. \end{aligned} \tag{4.10}$$

This can also be obtained from the relation (4.4) using expressions for $p_{0,a-1}^*(s)$, $p_{0,a-2}^*(s)$, $p_{1,a-1}^*(s)$ in terms of $p_{0,0}^*(s)$. Equations (4.7)–(4.10) give the LT's of the state probabilities.

4.4.1. Steady-State Solution

The steady-state probabilities can be obtained easily by taking limits. When steady state exists, we have

$$p_{1,n} = \lim_{t\to\infty} p_{1,n}(t) = \lim_{s\to 0} sp^*_{1,n}(s).$$

Thus, $\lim_{s\to 0} R = r$, where r is the *unique* real root in (0, 1) of the equation $h_1(z) = 0$ as $s \to 0$, i.e., of $h(z) = 0$ (Eq. (3.9)) and r satisfies Eq. (3.10). The steady-state probabilities can thus be found and these agree with those given in Eqs. (3.11)–(3.14).

4.4.2. Busy-Period Distribution

The busy-period distribution of the $M/M/(a, b)/1$ system can be obtained in a manner similar to that used for the simple queue $M/M/1$ in Section 3.9.2. We consider a process that avoids the states $0, 1, \ldots, a-1$, immediately after completion of a service. Here a busy period T commences with the start of a service (a batch with a units) and lasts until, for the first time on completion of a service, there are X units ($X = 0, 1, \ldots, a-1$) in the queue. Let the joint distribution of T and X be given by

$$F_j(t) = Pr\{T \le t, X = j\}, \tag{4.11}$$

and $$f_j(t)dt = Pr\{t \le T < t + dt, X = j\}, \qquad j = 0, 1, \ldots, a-1, \tag{4.12}$$

$f^*_j(s)$ be the LT of $f_j(t)$.

Let us consider the process that avoids the states j, $j = 0, 1, \ldots, a-1$, when the server becomes free. Then

$$f_j(t) = \mu p_{1,j}(t), \qquad j = 0, 1, \ldots, a-1, \tag{4.13}$$

where $p_{1,j}(t)$ is determined by the following equations:

$$p'_{1,n}(t) = -(\lambda + \mu)p_{1,n}(t) + \lambda p_{1,n-1}(t) + \mu p_{1,n+b}(t), \qquad n \ge 1 \tag{4.14}$$

$$p'_{1,0}(t) = -(\lambda + \mu)p_{1,0}(t) + \mu \sum_{r=a}^{b} p_{1,r}(t). \tag{4.15}$$

From the definition of busy period, we have $p_{1,0}(0) = 1$. Taking the LT of

(4.14) and (4.15) and using $p_{1,0}(0) = 1$, we get

$$(s + \lambda + \mu)p^*_{1,n}(s) = \lambda p^*_{1,n-1}(s) + \mu p^*_{1,n+b}(s), \qquad n \geq 1 \tag{4.16}$$

$$(s + \lambda + \mu)p^*_{1,0}(s) = 1 + \mu \sum_{r=a}^{b} p^*_{1,r}(s). \tag{4.17}$$

From (4.16) we get as before

$$p^*_{1,n}(s) = R^n p^*_{1,0}(s) \tag{4.18}$$

and using it, we get from (4.17)

$$p^*_{1,0}(s) = \frac{1 - R}{\mu - \mu R^a + s}. \tag{4.19}$$

Finally

$$p^*_{1,n}(s) = \frac{R^n(1 - R)}{s + \mu - \mu R^a}, \qquad n \geq 0. \tag{4.20}$$

Thus, for $0 \leq j \leq a - 1$

$$f^*_j(s) = \frac{\mu(1 - R)R^j}{s + \mu - \mu R^a} \tag{4.21}$$

and the LT $b^*(s)$ of the busy-period PDF $b(t)$ of T is given by

$$\begin{aligned} b^*(s) = \sum_{j=0}^{a-1} f^*_j(s) &= \frac{\mu(1 - R)}{s + \mu - \mu R^a} \sum_{j=0}^{a-1} R^j \\ &= \frac{\mu(1 - R^a)}{s + \mu - \mu R^a}. \end{aligned} \tag{4.22}$$

Inverting (4.22) one gets the PDF $b(t)$ of T

$$b(t) = \sum_{j=0}^{a-1} f_j(t). \tag{4.23}$$

We have

$$\begin{aligned} b^*(0) &= \sum_{j=0}^{a-1} f^*_j(0) \\ &= \int_0^\infty b(t)dt = 1. \end{aligned} \tag{4.24}$$

Now $f_j^*(0)$ is the probability that the busy period terminates leaving j $(0 \le j \le a-1)$ in the queue. The corresponding idle periods are distributed as gamma with parameters λ, j. We have

$$\begin{aligned} f_j^*(0) &= \lim_{s\to 0} f_j^*(s) \\ &= \frac{(1-r)r^j}{(1-r^a)}, \ j = 0, 1, \ldots, a-1, \end{aligned} \tag{4.25}$$

where r is the unique root in (0, 1) of (3.9) ($r = \lim R$ as $s \to 0$).

4.4.2.1. *Mean and Variance of the Busy Period.* Moments of T can be obtained by differentiating $b^*(s)$. From (4.22),

$$b^*(s) = 1 - \frac{s}{s + \mu - \mu R^a}. \tag{4.26}$$

We have

$$\begin{aligned} E(T) &= -\frac{d}{ds} b^*(s)\bigg|_{s=0} \\ &= \frac{(s+\mu-\mu R^a) - s(1-\mu a R^{a-1}R')}{(s+\mu-\mu R^a)^2}\bigg|_{s=0} \\ &= \frac{\mu(1-r^a)}{[\mu(1-r^a)]^2} = \frac{1}{\mu(1-r^a)} \end{aligned} \tag{4.27}$$

where $R' = (d/ds)R = R'(s)$. As R satisfies (4.5), we have

$$\mu R^{b+1} - (s + \lambda + \mu)R + \lambda = 0.$$

Differentiating, we get

$$\mu(b+1)R^b R' - R - (s+\lambda+\mu)R' = 0 \quad \text{or}$$

$$R' = \frac{R}{\mu(b+1)R^b - (s+\lambda+\mu)}$$

and at $s = 0$, we get

$$R_0' = \frac{r}{\mu(b+1)r^b - (\lambda + \mu)} = \frac{1}{\mu} \frac{r}{(b+1)r^b - 1 - \dfrac{r(1-r^b)}{1-r}} \tag{4.28}$$

$$= \frac{1}{\mu} \frac{(1-r)r}{br^b(1-r) - (1-r^b)}, \quad \text{and} \tag{4.28a}$$

$$E(T^2) = \frac{d^2}{ds^2} b^*(s)\bigg|_{s=0} = \frac{2[1 - a\mu r^{a-1} R_0']}{[\mu(1-ra)]^2} \tag{4.29}$$

where R_0' is given by (4.28). We have

$$\operatorname{var}(T) = \frac{1}{[\mu(1-r^a)]^2} - \left(\frac{2a\mu r^{a-1}}{[\mu(1-r^a)]^2}\right)\left(\frac{1}{\mu}\right)\left(\frac{(1-r)r}{br^b(1-r) - (1-r^b)}\right) = E(T)^2 + \frac{2ar^a(1-r)}{[\mu(1-ra)]^2[(1-r^b) - br^b(1-r)]} \tag{4.30}$$

$$= \frac{2ar^a(1-r) + (1-r^b) - br^b(1-r)}{[\mu(1-r^a)]^2[(1-r^b) - br^b(1-r)]}. \tag{4.30a}$$

4.4.2.2. Particular Cases.

M/M(1, b)/1

When $a = 1$, the busy period always terminates with none in the queue (i.e., termination of a busy period leaves the system empty) and in this case the busy and idle periods alternate. We can then use the result

$$\frac{E(T)}{E(I)} = \frac{1 - p_{0,0}}{p_{0,0}}$$

to find the mean busy period $E(T)$. We have $E(I) = 1/\lambda$, and when $a = 1$.

$$p_{0,0} = \frac{(1-r)^2}{1 - r + r^2 - r^{b+1}}$$

so that

$$E(T) = \frac{1}{\lambda} \frac{1 - p_{0,0}}{p_{0,0}} = \frac{1}{\lambda} \frac{r(1 - r^b)}{(1 - r)^2} = \frac{1}{\mu(1 - r)}.$$

We get the same result as given by (4.27) with $a = 1$. Putting $a = 1$ in (4.30)

$$\text{var}(T) = \left[\frac{1}{\mu(1 - r)}\right]^2 + \frac{2r}{\mu^2(1 - r)[(1 - r^b) - br^b(1 - r)]}.$$

M/M(k, k)/1

Putting $a = b = k$ in (4.27) and (4.30) we get

$$E(T) = \frac{1}{\mu(1 - r^a)} \quad \text{and}$$

$$\text{var}(T) = \frac{kr^k(1 - r) + (1 - r^k)}{[\mu(1 - r^k)]^2[(1 - r^k) - kr^k(1 - r)]}$$

where r is the unique root in (0, 1) of

$$\mu z^{k+1} - (\lambda + \mu)z + \lambda = 0.$$

M/M/1

Putting $k = 1$ (and noting that $r = \rho$), we get

$$E(T) = \frac{1}{\mu(1 - \rho)} \quad \text{and}$$

$$\text{var}(T) = \frac{1 - \rho^2}{[\mu(1 - \rho)]^2[(1 - \rho) - \rho(1 - \rho)]}$$

$$= \frac{1 + \rho}{\mu^2(1 - \rho)^3}.$$

4.5. Two-Server Model: *M/M(a, b)/2*

Consider that there are two parallel (homogeneous) servers and service is under a general bulk-service rule. When both the servers are free, each is equally likely to take a batch for service. Let

$p_{r,n}(t) = Pr\{$at time t, r channels are busy and there are n waiting in the queue$\}$; $p_{r,n}(t)$ is nonzero for $r = 0, 1,\ 0 \le n \le a - 1;\ r = 2, n \ge 0$.

$$p_{r,n} = \lim_{t \to 0} p_{r,n}(t)$$

$$\rho = \frac{\lambda}{2\mu b} = \text{the traffic intensity.}$$

We have the following differential-difference equations governing $p_{r,n}(t)$:

$$p'_{2,n}(t) = -(\lambda + 2\mu)p_{2,n}(t) + \lambda p_{2,n-1}(t) + 2\mu p_{2,n+b}(t), \qquad n \geq 1 \tag{5.1}$$

$$p'_{2,0}(t) = -(\lambda + 2\mu)p_{2,0}(t) + \lambda p_{1,a-1}(t) + 2\mu \sum_{k=a}^{b} p_{2,k}(t) \tag{5.2}$$

$$p'_{1,0}(t) = -(\lambda + \mu)p_{1,0}(t) + \lambda p_{0,a-1}(t) + 2\mu p_{2,0}(t) \tag{5.3}$$

$$p'_{1,q}(t) = -(\lambda + \mu)p_{1,q}(t) + \lambda p_{1,q-1}(t) + 2\mu p_{2,q}(t), \qquad 1 \leq q \leq a-1 \tag{5.4}$$

$$p'_{0,q}(t) = -\lambda p_{0,q}(t) + \lambda p_{0,q-1}(t) + \mu p_{1,q}(t), \qquad 1 \leq q \leq a-1 \tag{5.5}$$

$$p'_{0,0}(t) = -\lambda p_{0,0}(t) + \mu p_{1,0}(t). \tag{5.6}$$

It is to be noted that (5.4) and (5.5) will not occur when $a = 1$. Assume that steady-state solutions exist. Then $p_{r,n}$ will be the solutions of the following equations:

$$0 = -(\lambda + 2\mu)p_{2,n} + \lambda p_{2,n-1} + 2\mu p_{2,n+b}, \qquad n \geq 1 \tag{5.7}$$

$$0 = -(\lambda + 2\mu)p_{2,0} + \lambda p_{1,a-1} + 2\mu \sum_{k=a}^{b} p_{2,k} \tag{5.8}$$

$$0 = -(\lambda + \mu)p_{1,0} + \lambda p_{0,a-1} + 2\mu p_{2,0} \tag{5.9}$$

$$0 = -(\lambda + \mu)p_{1,q} + \lambda p_{1,q-1} + 2\mu p_{2,q}, \qquad 1 \leq q \leq a-1 \tag{5.10}$$

$$0 = -\lambda p_{0,q} + \lambda p_{0,q-1} + \mu p_{1,q}, \qquad 1 \leq q \leq a-1 \tag{5.11}$$

$$0 = -\lambda p_{0,0} + \mu p_{1,0}. \tag{5.12}$$

As before, (5.10) and (5.11) will not occur for $a = 1$ and either (5.9) or (5.12) may be used. These equations can be solved very much the same way. Equation (5.7) can be written as

$$g(E)[p_{2,n}] = 0, \qquad n = 0, 1, 2, \ldots,$$

where

$$g(z) = 2\mu z^{b+1} - (\lambda + 2\mu)z + \lambda = 0. \tag{5.13}$$

Note that (5.13) is obtained by writing 2μ in place of μ in (3.9). $g(z)$ has a unique real root in (0, 1) if and only if $\rho = \lambda/2b\mu < 1$. Denote this real root by r; r satisfies

$$p = b\rho = \frac{\lambda}{2\mu} = \frac{r(1 - r^b)}{1 - r} \tag{5.14}$$

(the same as (3.11), but with μ replaced by 2μ). Thus, it follows that

$$p_{2,n} = p_{2,0}r^n, \qquad n \geq 0. \tag{5.15}$$

Substituting this value of $p_{2,n}$ in (5.8), one gets $p_{1,a-1}$; then from (5.10) $p_{1,a-2}, p_{1,a-3}, \ldots, p_{1,0}$. Using the value of $p_{1,0}$ one gets $p_{0,a-1}$ from (5.9) and from (5.11), $p_{0,a-2}, \ldots, p_{0,0}$, and finally $p_{1,0}$ from (5.12). Thus, all the probabilities can be found in terms of $p_{2,0}$. Using the normalizing condition

$$\sum_{n=0}^{\infty} p_{2,n} + \sum_{q=0}^{a-1} p_{1,q} + \sum_{q=0}^{a-1} p_{0,q} = 1,$$

one gets $p_{2,0}$. Thus, all the probabilities can be completely obtained.

Note. $p_{2,0}$ can also be obtained by using (4.11) with $q = a - 1$ and expressing $p_{0,a-1}$, $p_{0,a-2}$, and $p_{1,a-1}$ all in terms of $p_{2,0}$.

See Medhi and Borthakur (1972) for further results.

4.5.1. *Particular Case: M/M(1, b)/2*

Two-server Markovian queue under usual bulk service.

Using (5.15) in (5.8) with $a = 1$, we get

$$\begin{aligned} 0 &= -(\lambda + 2\mu)p_{2,0} + \lambda p_{1,0} + 2\mu \sum_{k=1}^{b} p_{2,k} \\ &= -(\lambda + 2\mu)p_{2,0} + \lambda p_{1,0} + 2\mu p_{2,0} \frac{r(1 - r^b)}{1 - r} \end{aligned}$$

and using (5.14), we get

$$\begin{aligned} p_{1,0} &= \frac{1}{\lambda}\left[(\lambda + 2\mu) - 2\mu \frac{\lambda}{2\mu}\right]p_{2,0} \\ &= \frac{2\mu}{\lambda} p_{2,0}. \end{aligned}$$

Finally from (5.12) (or from (5.9)) we get

$$p_{0,0} = \frac{1}{2}\left(\frac{2\mu}{\lambda}\right)^2 p_{2,0}.$$

Writing in terms of $p_{0,0}$, and $p = \lambda/2\mu$

$$p_{2,n} = p_{2,0}r^n = 2p^2 p_{0,0} r^n, \qquad n \geq 0 \tag{5.15a}$$

$$p_{1,0} = 2\left(\frac{\lambda}{2\mu}\right)p_{0,0} = 2p p_{0,0}. \tag{5.16}$$

Using $\sum_{n=0}^{\infty} p_{2,n} + p_{1,0} + p_{0,0} = 1$, we get

$$p_{0,0} = \left[1 + 2p + \frac{2p^2}{1-r}\right]^{-1} = \frac{(1-r)}{2p^2 + (2p+1)(1-r)}. \tag{5.17}$$

The expected number in the queue $E(Q)$ is given by

$$\begin{aligned} E(Q) &= \sum_{n=0}^{\infty} n p_{2,n} \\ &= p_{2,0} \sum_{n=0}^{\infty} n r^n = p_{2,0}\left[\frac{r}{(1-r)^2}\right] \\ &= \frac{2p^2 r}{(1-r)[2p^2 + (2p+1)(1-r)]}. \end{aligned} \tag{5.18}$$

When $b = 1$, $r = \rho = \lambda/2\mu = p$; we have for an $M/M/2$ queue:

$$p_{0,0} = \frac{1-\rho}{2\rho^2 + (2\rho+1)(1-\rho)} = \frac{1-\rho}{1+\rho},$$

$$p_{2,n} = \frac{(2\rho^{n+2})(1-\rho)}{1+\rho},$$

$$p_{1,0} = \frac{(2\rho)(1-\rho)}{1+\rho}, \quad \text{and}$$

$$E(Q) = \frac{2\rho^3}{(1-\rho)(1+\rho)} = \frac{2\rho^3}{1-\rho^2}. \tag{5.19}$$

Note that with our earlier notation for $M/M/2$

$$p_0 = p_{0,0}, \qquad p_{1,0} = p_1,$$

$$p_{2,n} = p_{n+2}, \qquad n \geq 0$$

where p_m = Prob {number in the system is m} in the usual $M/M/2$ notation.

4.6. The $M/M(1, b)/c$ Model

Ghare (1968) considered this model. We discuss it with some modification of the procedure adopted by him. Let

$$p_{r,n}(t) = Pr\{r \text{ channels are busy, } n \text{ units in queue at time } t\};$$

$p_{r,n}(t)$ is nonzero only when $r \le c$, $0 \le n \le \infty$, and $(c - r)n = 0$. Let $p^*_{r,n}(s)$ be the LT of $p_{r,n}(t)$ and $p_{r,n}$ be the corresponding steady-state probabilities. The equations involving $p_{r,n}$ satisfy the following equations

$$p'_{c,n}(t) = -(\lambda + c\mu)p_{c,n}(t) + \lambda p_{c,n-1}(t) + c\mu p_{c,n+b}(t), \qquad n > 0 \tag{6.1}$$

$$p'_{c,0}(t) = -(\lambda + c\mu)p_{c,0}(t) + \lambda p_{c-1,0}(t) + c\mu \sum_{k=1}^{b} p_{c,k}(t) \tag{6.2}$$

$$\begin{aligned} p'_{r,0}(t) = -(\lambda + r\mu)p_{r,0}(t) + \lambda p_{r-1,0}(t) + (r+1)\mu p_{r+1,0}(t), \\ 1 \le r \le c-1 \end{aligned} \tag{6.3}$$

$$p'_{0,0}(t) = -\lambda p_{0,0}(t) + \mu p_{1,0}(t) \tag{6.4}$$

Let the initial condition be $p_{c,0}(0) = 1$, i.e., that time is reckoned from the instant that all the channels become busy, leaving none in the queue. Taking LTs, we get

$$(s + \lambda + c\mu)p^*_{c,n}(s) = \lambda p^*_{c,n-1}(s) + c\mu p^*_{c,n+b}(s), \qquad n > 0 \tag{6.5}$$

$$(s + \lambda + c\mu)p^*_{c,0}(s) = 1 + \lambda p^*_{c-1,0}(s) + c\mu \sum_{k=1}^{b} p^*_{c,k}(s) \tag{6.6}$$

$$(s + \lambda + r\mu)p^*_{r,0}(s) = \lambda p^*_{r-1,0}(s) + (r+1)\mu p^*_{r+1,0}(s), \quad 1 \le r \le c-1 \tag{6.7}$$

$$(s + \lambda)p^*_{0,0}(s) = \mu p^*_{1,0}(s). \tag{6.8}$$

Consider the equation

$$h(z) \equiv \lambda - (\lambda + s + c\mu)z + c\mu z^{b+1} = 0. \tag{6.9}$$

The equation has a unique root $R = R(s)$ in (0, 1). Then from (6.5) we get, as before

$$\begin{aligned} p^*_{c,n}(s) &= AR^n \\ &= p^*_{c,0}(s)R^n, \qquad n \ge 0. \end{aligned} \tag{6.10}$$

Then using (6.10) we can find $p^*_{c-1,0}(s)$ from (6.6) and then recursively $p^*_{r,0}(s)$, $r = 1, 2, \ldots, c-2$, from (6.7). Thus, all the probabilities can be obtained in terms of $p^*_{0,0}(s)$ and this can be evaluated as before. Instead, the general solution of the set of equations (6.7) and (6.8) can be written (see our treatment of $M/M/c$ queue in Section 3.10 by the method of Jackson and Henderson, 1966) as

$$p^*_{r,0}(0) = p^*_{0,0}(s) \sum_{j=0}^{r} \frac{\left(\frac{\lambda}{\mu}\right)^{r-j} \Gamma\left(j + \frac{s}{\mu}\right)}{\Gamma\left(\frac{s}{\mu}\right) j!(r-j)!} \tag{6.11}$$

$$= p^*_{0,0}(s)\phi_r(s), \qquad 0 \le r < c \tag{6.11a}$$

where

$$\phi_r(s) = \left[\frac{\left(\frac{\lambda}{\mu}\right)^r}{r!}\right] {}_2F_0\left(-r, \frac{s}{\mu}; -; -\frac{\mu}{\lambda}\right), \tag{6.12}$$

and ${}_2F_0$ is a confluent hypergeometric series [see Abramowitz and Stegun (1968)]. Using (6.11a) and (6.6) we get

$$(s + \lambda + c\mu)p^*_{c,0}(s) = 1 + \lambda p^*_{0,0}(s)\phi_{c-1}(s) + c\mu p^*_{c,0}(s)\frac{R(1 - R^b)}{1 - R} \quad \text{or}$$

$$\left[(s + \lambda + c\mu) - \frac{c\mu R(1 - R^b)}{1 - R}\right] p^*_{c,0}(s) = 1 + \lambda p^*_{0,0}(s)\phi_{c-1}(s).$$

Using the fact that R satisfies (6.9), i.e., the relation

$$\lambda - (\lambda + s + c\mu)R + c\mu R^{b+1} = 0,$$

we get

$$(s + \lambda + c\mu) - \frac{c\mu R(1 - R^b)}{1 - R} = \frac{s + c\mu(1 - R)}{1 - R}.$$

Thus,

$$p^*_{c,0}(s) = \frac{1 - R}{s + c\mu(1 - R)}[1 + \lambda p^*_{0,0}(s)\phi_{c-1}(s)]. \tag{6.13}$$

Thus, we get $p^*_{c,n}(s)$, $p^*_{r,0}(s)$ $(r = 1, \ldots, c - 1)$, all in terms of $p^*_{0,0}(s)$. Using the relation

$$\sum_{n=0}^{\infty} p^*_{c,n}(s) + \sum_{r=1}^{c-1} p^*_{r,0}(s) + p^*_{0,0}(s) = \frac{1}{s}, \tag{6.14}$$

we get

$$p_{c,0}^*(s)\left(\frac{1}{1-R}\right) + \sum_{r=1}^{c-1} \phi_r(s) p_{0,0}^*(s) + p_{0,0}^*(s) = \frac{1}{s}.$$

Thus, we have

$$p_{0,0}^*(s)\left[1 + \sum_{r=1}^{c-1} \phi_r(s) + \frac{\lambda \phi_{c-1}(s)}{s + c\mu(1-R)}\right] = \frac{1}{s} - \frac{1}{s + c\mu(1-R)}$$
$$= \frac{c\mu(1-R)}{s\{s + c\mu(1-R)\}}.$$

and finally,

$$p_{0,0}^*(s) = \frac{c\mu(1-R)}{s}\left[\lambda \phi_{c-1}(s) + \{s + c\mu(1-R)\} \sum_{r=0}^{c-1} \phi_r(s)\right]^{-1} \tag{6.15}$$

noting that $\phi_0(s) = 1$.

4.6.1. *Steady-State Results M/M(1, b)/c*

Now

$$\lim_{s \to 0} R = r$$

$$\lim_{s \to 0} \phi_m(s) = \frac{\left(\frac{\lambda}{\mu}\right)^m}{m!} + \lim_{s \to 0} \sum_{j=1}^{m} \frac{\left(\frac{\lambda}{\mu}\right)^{m-j}}{j!(m-j)!}\left[\left(\frac{s}{\mu}\right)\left(\frac{s}{\mu}+1\right)\cdots\left(\frac{s}{\mu}+j-1\right)\right]$$
$$= \frac{\left(\frac{\lambda}{\mu}\right)^m}{m!}, \tag{6.16}$$

since

$$\frac{\Gamma(a+n)}{\Gamma(a)} = \begin{cases} 1, & \text{for } n = 0, \quad a = 0 \\ a(a+1)\cdots(a+n-1), & n = 1, 2, 3, \ldots. \end{cases}$$

Thus,

$$p_{0,0} = \lim_{s \to 0} sp^*_{0,0}(s) = c\mu(1-r)\left[\lambda \frac{\left(\frac{\lambda}{\mu}\right)^{c-1}}{(c-1)!} + c\mu(1-r)\sum_{m=0}^{c-1} \frac{\left(\frac{\lambda}{\mu}\right)^m}{m!}\right]^{-1}$$

$$= \left[\frac{\left(\frac{\lambda}{\mu}\right)^c}{c!(1-r)} + \sum_{m=0}^{c-1} \frac{\left(\frac{\lambda}{\mu}\right)^m}{m!}\right]^{-1} \tag{6.17}$$

$$p_{m,0} = \lim_{s \to 0} sp^*_{m,0}(s) = \lim_{s \to 0}[\{sp^*_{0,0}(s)\}\{\phi_m(s)\}]$$

$$= p_{0,0} \frac{\left(\frac{\lambda}{\mu}\right)^m}{m!}, \qquad m = 1, 2, \ldots, c-1 \tag{6.18}$$

$$p_{c,n} = \lim_{s \to 0} sp^*_{c,n}(s) = \lim_{s \to 0} sp^*_{c,0}(s) R^n$$

$$= r^n \left[\frac{\lambda}{c\mu} \cdot \frac{\left(\frac{\lambda}{\mu}\right)^{c-1}}{(c-1)!} p_{0,0}\right]$$

$$= \frac{\left(\frac{\lambda}{\mu}\right)^c}{c!} r^n p_{0,0}, \qquad n \geq 0. \tag{6.19}$$

We have

$$E(Q) = \frac{\left(\frac{\lambda}{\mu}\right)^c}{c!(1-r)} p_{0,0}. \tag{6.20}$$

4.6.1.1. *Particular Cases*

(1) By putting $c = 2$, we obtain the results for $M/M(1, b)/2$.

(2) $M/M/c$: Putting $b = 1$, we get $r = \lambda/c\mu = \rho$, so that

$$p_0 = p_{0,0} = \left[\sum_{m=0}^{c-1} \frac{\left(\frac{\lambda}{\mu}\right)^m}{m!} + \frac{\left(\frac{\lambda}{\mu}\right)^c}{c!(1-\rho)}\right]^{-1}$$

$$p_n = p_{n,0} = \left[\frac{\left(\frac{\lambda}{\mu}\right)^n}{n!} \right] p_{0,0}, \qquad n \le c$$

$$p_n = p_{c,n-c} = \frac{\left(\frac{\lambda}{\mu}\right)^c}{c!} \rho^{n-c} p_{0,0}, \qquad n \ge c.$$

where $p_n = Pr\{\text{number in systems is } n\}$ in an $M/M/c$ queue.

M/M(a, b)/c. This model has been investigated by Neuts and Nadarajan (1982) and Sim and Templeton (1985). We mention here a simple result relating to its busy period.

Suppose that the busy period T starts with the instant when all the servers become busy and lasts until the instant when any one of the servers becomes free for the first time. Then the LT of the busy period can be the obtained from that of $M/M(a, b)/1$ by replacing μ by $c\mu$. Thus, the LST of the busy period T is given by

$$b^*(s) = \frac{c\mu(1 - R^a)}{s + c\mu(1 - R^a)}$$

and the probability that the busy period terminates with j in the queue is given by

$$f_j^*(0) = \frac{(1 - r)r^j}{(1 - r^a)}, \qquad 0 \le j \le a - 1.$$

The mean busy period is $E(T) = 1/c\mu(1 - r^a)$.

Problems and Complements

4.1. The system $M/E_k/1$: transient behavior. Define

$$P(s, t) = \sum_{n=0}^{\infty} p_n(t)s^n$$

$$P^*(s, \alpha) = \sum_{n=0}^{\infty} p_n^*(\alpha)s^n$$

where $p_n^*(\alpha)$ is the LT of $p_n(t)$. Assume that $p_0(0) = 1$. Show that

$$P^*(s, \alpha) = \frac{s - k\mu(1 - s)p_0^*(\alpha)}{s(\alpha + \lambda + k\mu) - k\mu - \lambda s^{k+1}}, \qquad |s| < 1,$$

and that

$$\lim_{\alpha \to 0} p_0^*(\alpha) = \frac{c}{k\mu(1 - c)}$$

where $c \equiv c(s, \alpha)$ is the only zero of $(\alpha + \lambda + k\mu)s - k\mu - \lambda s^{k+1}$ inside the unit circle $|s| = 1$. Noting that

$$\lim_{\alpha \to 0} \alpha p_n^*(\alpha) = \lim_{t \to \infty} p_n(t) = p_n \quad \text{and}$$

$$\lim_{\alpha \to 0} \alpha P^*(s, \alpha) = \lim_{t \to \infty} P(s, t) = P(s) = \sum_{n=0}^{\infty} p_n s^n,$$

derive the expression for the PGF $P(s)$ [in the form given in (1.6)].

4.2. The system $E_k/M/1$

Write down the Chapman–Kolmogorov equations and obtain an expression for $P^*(s, \alpha)$.

Obtain the relevant equations for finding the distribution of the busy period T (the interval from the arrival of a customer in an empty system to the first subsequent instant when the system becomes empty) of an $E_k/M/1$ system. Show the LT $f^*(s)$ of the PDF of the busy period T is given by

$$f^*(s) = \frac{\mu(\alpha - 1)}{\alpha(s + \mu) - \mu}$$

where $\alpha \equiv \alpha(s)$ is the root lying outside $|z| = 1$ of the equation

$$g(z, s) \equiv z - \left[\frac{z(s + \mu + k\lambda) - \mu}{k\lambda z}\right]^k = 0.$$

Show that the mean busy period is given by

$$E(T) = \frac{\alpha_0}{\mu(\alpha_0 - 1)}$$

where α_0 is the unique root outside $|z| = 1$ of the equation $g(z, 0) = 0$.

4.3. The System $M^X/M/1$

(a) Find the variance of the number in the system for the $M^X/M/1$ system.

(b) Find the mean and variance of the number in the system for the $M^X/M/1$ system where X has a geometric distribution.

(c) Find the mean and variance of the number in the system for the $M^r/M/1$ system.

4.4. $M^X/E_k/1$ system (Restrepo, 1965).
Let the state of the system be denoted by (n, s), $n = 0, 1, 2, \ldots, s = 1, 2, \ldots, k$ where n is the number of customers in the system and s is the number of phases that remain to be completed by the person in service, if any. Let $p_{n,s}$ be the steady-state probability that the system is in state (n, s), $n = 1, 2, \ldots, s = 1, \ldots, k$; for $n = 0$, $p_{0,0} = p_0$. Writing $\theta = \lambda/k\mu$, show that the system of equations can be obtained as follows:

$$0 = p_{1,1} - \theta p_0 \tag{c.1}$$

$$0 = p_{1,s+1} - (1+\theta)p_{1,s} \tag{c.2}$$

$$0 = p_{2,1} - (1+\theta)p_{1,k} + \theta a_1 p_0 \tag{c.3}$$

$$0 = p_{n,s+1} - (1+\theta)p_{n,s} + \theta \sum_{m=1}^{n-1} a_m p_{n-m,s} \tag{c.4}$$

$$0 = p_{n+1,1} - (1+\theta)p_{n,k} + \theta a_n p_0 + \theta \sum_{m=1}^{n-1} a_m p_{n-m,k} \tag{c.5}$$

with (c.2), (c.4), (c.5) restricted to $s < k$ and $n > 1$. How would you proceed to solve the preceding equations in terms of p_0 recursively starting from Eq. (c.1)? [$A(z)$ is the PGF of X, and $a_k = Pr\{X = k\}$.]

4.5. For a continuation of Problem 4.4, show that the generating function

$$F(z) = p_0 + \sum_{n=1}^{\infty} \sum_{s=1}^{k} p_{n,s} z^n \tag{c.6}$$

is given by the expression

$$F(z) = \frac{p_0(1-z)}{1 - z\{1 + \theta - \theta A(z)\}^k}. \tag{c.7}$$

Show that

$$p_0 = 1 - \frac{\lambda\bar{a}}{\mu} = 1 - \rho, \qquad (E(X) = \bar{a},\ \sigma = \mathrm{var}(X))$$

$$E\{N\} = \rho + E(Q)$$

$$= \rho + \frac{k+1}{2k\mu}\left\{\frac{(\lambda\bar{a})^2}{\mu - \lambda\bar{a}}\right\} + \frac{\lambda}{2(\mu - \lambda\bar{a})}\{\sigma^2 + \bar{a}^2 - \bar{a}\}. \tag{c.8}$$

Deduce that for $E_k \equiv E_1 \equiv M$, i.e., $M^X/M/1$

$$E\{N\} = \frac{\lambda(\sigma^2 + \bar{a}^2 + \bar{a})}{2(\mu - \lambda\bar{a})}$$

and for $M^X/D/1$,

$$E\{N\} = \frac{\lambda(\sigma^2 + \bar{a}^2 + \bar{a})}{2(\mu - \lambda\bar{a})} - \frac{(\lambda\bar{a})^2}{2\mu(\mu - \lambda\bar{a})}$$

(Restrepo, 1965).

4.6. Multiple Poisson bulk arrival (MPBA) system (Jensen *et al.* 1977) Consider that groups of size j arrive according to a Poisson process with intensity λ_j, $\{N_j(t), t \geq 0\}, j = 1, 2, \ldots, m$, and that these processes are mutually independent. Note that the group size is restricted to m. Then $M(t) = \sum_{j=1}^{m} N_j(t)$ denotes the number of groups that arrive in $(0, t)$, while the compound Poisson process $\{N(t), t \geq 0\}$ where

$$N(t) = \sum_{j=1}^{m} jN_j(t)$$

gives the number of arrivals in $(0, t)$. The mean arrival rate of customers is $\sum_{j=1}^{m} j\lambda_j$. Then the arrival distribution X is given by

$$a_j = Pr(X = j) = \frac{\lambda_j}{\lambda}\,\delta(m - j), \qquad j = 1, 2, \ldots, m,$$

where

$$\begin{aligned} \delta(x) &= 1 \quad \text{for } x \geq 0 \\ &= 0 \quad \text{for } x < 0. \end{aligned}$$

For such a compound Poisson arrival process and exponential service, show that the steady-state probabilities p_n of the distribution of number in the system satisfy

$$p_{k+1} = \sum_{j=1}^{m} \rho_j p_{k+1-j}, \qquad k=0, 1, 2, \ldots,$$

where

$$\rho_j = \sum_{k=j}^{m} \frac{\lambda_k}{\mu}, \qquad j = 1, \ldots, m, \qquad p_r = 0, \qquad r < 0.$$

(This relation is deducible from (2.3a) and (2.3b).) Further show that

PGF $P(s) = \sum_{n=0}^{\infty} p_n s^n$ is given by

$$P(s) = \frac{p_0}{1 - \sum_{j=1}^{m} p_j s^j}$$

where

$$p_0 = 1 - \sum_{j=1}^{m} p_j = 1 - \sum_j \frac{\lambda_j}{\mu} = 1 - \rho.$$

4.7. $M^X/M/\infty$
Show that given $N(0) = i$, the correlation coefficient ρ between $N(t)$ and $N(0)$ equals $e^{-\mu t}$ and is independent of the arrival rate λ and of the batch size X. As $t \to \infty$, $\rho \to 0$, as is expected (Reynolds, 1968).

4.8. $M^r/M/c$ (Kabak, 1968) with rates λ and $\mu = 1$
Consider (i) the delay system $M^r/M/c/\infty$, where customers are allowed to wait, and (ii) the loss system $M^r/M/c/c$, where customers are lost when all the channels are busy. If $\{p_i\}$ is the steady-state probability distribution of the system size, show that p_j satisfies the following equations.

For the *delay* system,

$$\begin{aligned} p_i &= \frac{\lambda}{i} \sum_{j=\max(0,i-r)}^{i-1} p_j, \qquad 1 \le i \le c-1, \\ &= \frac{\lambda}{c} \sum_{j=\max(0,i-r)}^{i-1} p_j, \qquad i \ge c, \quad \text{and} \end{aligned}$$

$$\sum_{i=0}^{\infty} p_i = 1.$$

For the *loss* system,

$$\begin{aligned} p_i &= \frac{\lambda}{i} \sum_{j=0}^{i-1} p_j, \qquad 1 \le i \le c, \qquad i \le r \\ &= \frac{\lambda}{i} \sum_{j=i-r}^{i-1} p_j, \qquad 1 \le i \le c, \qquad i \ge r, \quad \text{and} \end{aligned}$$

$$\sum_{i=0}^{c} p_i = 1.$$

4.9. For a continuation of Problem 4.8, consider waiting-time distribution in the delay system. Show that the distribution of waiting time W in the queue

$$P(W > t) = \left(\frac{1}{r}\right) \sum_{d=1}^{\infty} \sum_{i=\max(0, d+c-r)}^{d+c-1} p_i F$$

where

$$F = \frac{\Gamma(d, ct)}{\Gamma(d)}, \quad \text{and}$$

$$\Gamma(a, x) = \int_x^{\infty} e^{-x} x^{a-1}\, dx$$

(Kabak, 1968).

4.10. Model $M/M(1, b; \mu_k)/1$, with mean service time $1/\mu_k$ for a batch of k

Distribution of queue size and number in the batch being served in Poisson queue with usual bulk service has been considered by Cosmetatos (1983c). Show that the steady-state difference equations for $p(n; k)$, where $p(n; k)$ denotes the steady-state probability of n customers being in the queue and k being served $(0 \le k \le b)$, satisfy the following equations:

$$0 = -p(0; 0) + \sum_{k=1}^{b} \mu_k p(0; k)$$

$$0 = -(\lambda + \mu_1)p(0; 1) + \lambda p(0; 0) + \sum_{k=1}^{b} \mu_k p(1; k)$$

$$0 = -(\lambda + \mu_m)p(0; m) + \sum_{k=1}^{b} \mu_k p(m; k), \qquad 2 \le m \le b$$

$$0 = -(\lambda + \mu_k)p(n; k) + \lambda p(n-1; k), \qquad n \ge 1, \qquad 1 \le k \le b-1$$

$$0 = -(\lambda + \mu_b)p(n; b) + \lambda p(n-1; b) + \sum_{k=1}^{b} \mu_k p(n+b; k), \qquad n \ge 1$$

$$0 = p(n; 0), \qquad n \ge 1.$$

Indicate how the probabilities $p(0; k)$, $0 \le k \le b-1$ and the marginal probabilities $p(n) = \sum_{k=0}^{b} p(n; k)$ can be obtained from the previous equations. Examine the special case $b = 2$, $1/\mu_2 = c/\mu_1$, $c = 1, 2$.

Cosmetatos examines the effect of such a server-sharing rule on performance measures such as the average server utilization, average queue size, and average number of customers in the system and the benefits of the server-sharing scheme (Cosmetatos, 1983c).

4.11. Batch size in systems with general service rule

For an $M/M(a, b)/1$ system, let $\omega = \lambda/(\lambda + \mu)$, $p = \lambda/\mu$. The distribution of service batch size Y is given by

$$g_a = Pr\{Y = a\}$$
$$= 1 - \frac{pP_{0,0}}{\omega(1-r)^2}[-(r-\omega) + (1-\omega)r^{a+1}(a+1-ar)],$$

$$g_y = Pr\{Y = y\}$$
$$= \left(\frac{(1-\omega)}{\omega}\right) yr^y pP_{0,0}, \qquad a+1 \le y \le b-1,$$

$$g_b = Pr\{Y = b\}$$
$$= \frac{b(r-\omega)}{\omega r(1-r)} pP_{0,0}, \quad \text{and}$$

$$E\{Y\} = a\left\{1 + \frac{(r-\omega)pP_{0,0}}{\omega(1-r)^2} + \frac{(1-\omega)r^{a+1}pP_{0,0}}{(1-r)^3}[1+r+a-ar]\right.$$
$$\left. - \frac{(r-\omega)pP_{0,0}}{\omega(1-r)^3}[(1+2b)+(1-2b)r]\right.$$

Replacing ω by $\omega_c = \lambda/(\lambda + c\mu)$, and p by $p_c = \lambda/c\mu$, one gets the corresponding expressions for the $M/M(a, b)/c$ model (Sim and Templeton, 1985).

Note. This gives the distribution of the size of the batch in which a randomly chosen customer is served. This is to be distinguished from the distribution of the size of a randomly chosen batch.

4.12. The model $M/M(1, b)/2$

Show that

$$p_{1,0}^*(s) = \left[\frac{(s+\lambda)}{\mu}\right] p_{0,0}^*(s)$$

$$p_{2,n}^*(s) = \left[\frac{\{(s+\lambda)^2 + s\mu\}}{2\mu^2}\right] p_{0,0}^*(s), \qquad n \ge 0.$$

Find $p^*_{0,0}(s)$. Suppose that $(\rho = \lambda/2b\mu < 1)$ and that the system is in steady state and r denotes the unique real root in (0, 1) of Eq. (5.13).

Show that the steady-state probabilities are as given in (5.15a) to (5.17).

4.13. The model $M/M(a, b)/2$: busy period

The busy period may be defined as the interval during which (i) both the servers remain busy or (ii) at least one server remains busy.

Case (i): Show that the busy-period distribution can be obtained from that of the one-channel case by replacing μ by 2μ wherever μ occurs in the one-channel result. Find $b^*(s)$, $f^*_j(0)$, $0 \le j \le a-1$ and the mean and the variance of the busy period.

Case (ii): Here $p_{1,0}(0) = 1$; the busy period is the interval between commencement of service in one of the channels with none in the other channel or queue and the first subsequent epoch when both the channels become free. Write down the equations involving

$$p^*_{ij}(s), \qquad i=2, \qquad j \ge 0; \qquad i = 1, \qquad j = 0, 1, \ldots, a-1,$$

and solve them. Denote R as the real root in (0, 1) of Eq. (5.13). Denote

$$C(s) = \frac{2\mu p^*_{2,0}(s)}{(s+\lambda+\mu)R - \lambda}$$

$$= \frac{[2\mu\lambda^a(1-R)]}{\{(s+\lambda+\mu)^a[(s+\lambda+\mu)R-\lambda][s+2\mu(1-R^a)] - 2\lambda\mu(1-R)[(s+\lambda+\mu)^a R^a - \lambda^a]\}}.$$

Show that, for $0 \le q \le a-1$,

$$p^*_{1,q}(s) = \frac{1}{\lambda}\left(\frac{\lambda}{s+\lambda+\mu}\right)^{q+1} + C(s)\left[R^{q+1} - \left(\frac{\lambda}{s+\lambda+\mu}\right)^{q+1}\right]$$

and hence find the LST of the busy-period distribution. Find the probability that the busy period ends with $q (0 \le q \le a-1)$ in the queue, and verify that the total probability equals 1.

4.14. The model $M/M(1, b)/2$

Define the busy period as the interval during which at least one of the servers remains busy with $p_{1,0}(0) = 1$. Show that the LST $b^*(s)$ of the busy period is given by

$$b^*(s) = \frac{\mu[(s+2\mu) - 2\mu R]}{s(s+\lambda+\mu) + 2\mu(s+\mu)(1-R)}$$

and that the average busy period equals

$$\frac{(1-r)+\left(\dfrac{\lambda}{2\mu}\right)}{\mu(1-r)}.$$

(R and r refer to the corresponding quantities of the two-channel model.) Find the corresponding results for an $M/M/2$ queue as a particular case.

4.15. The system $M/M(a, b)/1$

(a) Let Y be the number of customers that arrive during the period the server is idle with $q (0 \le q \le a-1)$ in the queue, and let Z be the number of customers present in the queue at the epoch the server's idle period commences. Let $P_Y(s)$ and $P_Z(s)$ be the PGF of Y and Z, respectively. Show that

$$P_Y(s) = \frac{s(1-r)(s^a - r^a)}{(1-r^a)(s-r)} \quad \text{and}$$

$$P_Z(s) = \frac{(1-r)\{1-(rs)^a\}}{(1-r^a)(1-rs)}.$$

Find $E(Y)$ and $E(Z)$ and verify that $E(Y) + E(Z) = a$.

(b) Show that the LST of the server's idle period I is given by

$$I^*(s) = \frac{\lambda(1-r)}{(1-r^a)} \frac{1}{(s+\lambda)^a} \frac{\lambda^a - \{r(s+\lambda)\}^a}{\lambda - r(s+\lambda)}.$$

Show further that

$$E(I) = \frac{a(1-r) - r + r^{a+1}}{(1-r)(1-r^a)}.$$

Verify that when $a = b = 1$,

$$I^*(s) = \frac{\lambda}{(\lambda + s)}.$$

(c) Denote

$$\begin{aligned} p_0 &= Pr\{\text{the server is idle}\} \\ &= \sum_{q=0}^{a-1} p_{0,q}. \end{aligned}$$

Using the relation

$$p_0 = \frac{E(I)}{E(I) + E(B)},$$

find the fraction of time the server is idle. Show that the average number of batches served during a server busy period is

$$\frac{1}{(1 - r^a)}.$$

Deduce the corresponding results for an $M/M/1$ queue.

References

Abramowitz, M., and Stegun, I. A. (Eds.) (1968). *Handbook of Mathematical Functions.* National Bureau of Standards, Washington, D.C.

Bailey, N. T. J. (1954). On queueing processes with bulk service. *J. Roy. Stat. Soc.* **B16**, 80–97.

Borthakur, A. (1971). A Poisson queue with a general bulk service. *J. Ass. Sc. Soc.* **Xiv**, 162–167.

Burke, P. J. (1975). Delays in single server queues with batch input. *Opns. Res.* **23**, 830–833.

Chaudhry, M. L., and Templeton, J. G. C. (1983): *A First Course in Bulk Queues*, Wiley, New York.

Chaudhry, M. L., Medhi, J., Sim, S. H., and Templeton, J. G. C. (1987). On a two heterogeneous-server Markovian queue with general bulk service rules. *Sankhyā*, 49 Series B, 36-50.

Cosmetatos, G. P. (1983a). Closed form equilibrium results for the $M/M\ (\alpha, \infty)/N$ queue. *Euro. J. Opns. Res.* **12**, 203–204.

Cosmetatos, G. P. (1983b). A steady state approximation for the $E_R^X/M/1$ queue. *J. Opnl. Res. Soc.* **34**, 899–902.

Cosmetatos, G. P. (1983c). Increasing productivity in exponential queues by server-sharing. *Omega* **11**, 187–193.

Cromie, M. V., and Chaudhry, M. L. (1976). Analytically explicit results for the queueing system $M^X/M/C$ with charts and tables for certain measures of efficiency. *Oper. Res. Qrtly.* **27**, 733–745.

Easton, G. D., and Choudhry, M. L. (1982). The queueing system $E_k/M(a, b)/1$ and its numerical analysis. *Comp. & Opns. Res.* **9**, 197–205.

Gaver, D. P. (1959). Imbedded Markov chain analysis of a waiting time process in continuous time. *Ann. Math. Stat.* **30**, 698–720.

Ghare, P. M. (1968). Multichannel queueing system with bulk service. *Opns. Res.* **16**, 189–192.

Jackson, R. R. P., and Henderson, J. C. (1966). The time dependent solution to the many server Poisson queue. *Opns. Res.* **14**, 720–722.

Jensen, G. L., Paulson, A. S., and Sullo, P. (1977). Steady state solution for a particular $M^{(k)}/M/1$ queueing system. *Nav. Res. Log. Qrly.* **24**, 651–659.

Kabak, I. W. (1968). Blocking and delay in $M^{(n)}/M/c$ bulk queueing systems. *Opns. Res.* **16**, 830–839.

Luchak, G. (1958). The continuous time solution of the equations of the single channel queue with a general class of service-time distribution by the method of generating functions. *J. Roy. Stat. Soc. B* **20**, 176–181.

Medhi, J. (1975). Waiting time distribution in a Poisson queue with general bulk service rule. *Management Sci.* **21**, 777–782.

Medhi, J. (1979). Further results in a Poisson queue under a general bulk service rule. *Cahiers Centre d'Etudes de Rech. Opér.* **21**, 183–189.

Medhi, J. (1982). *Stochastic Processes*, Halsted Press, Wiley, New York.

Medhi, J. (1984). *Recent Developments in Bulk Queueing Models*, Wiley Eastern, New Dehli.

Medhi, J., and Borthakur, A. (1972). On a two server Markovian queue with a general bulk service. *Cahiers Centre d'Etudes de Rech. Opér.* **14**, 151–158.

Neuts, M. F. (1967). A general class of bulk queues with Poisson input. *Ann. Math. Stat.* **38**, 759–770.

Neuts, M. F. (1979). Queues solvable without Rouchés theorem. *Opns. Res.* **27**, 767–781.

Neuts, M. F. (1981). *Matrix-Geometric Solutions to Stochastic Models—An Algorithmic Approach*, The Johns Hopkins University Press, Baltimore, Md.

Neuts, M. F., and Nadarajan, R. (1982). A multiserver queue with thresholds for the acceptance of customers into service. *Opns. Res.* **30**, 948–960.

Restrepo, R. A. (1965). A queue with simultaneous arrivals and Erlang service distributions. *Opns. Res.* **13**, 375–381.

Reynolds, J. F. (1968). Some results for the bulk arrival infinite server Poisson queue. *Opns. Res.* **16**, 186–189.

Sim, S. H., and Templeton, J. G. C. (1983). Computational procedures for steady-state characteristics of unscheduled multi-carrier shuttle systems. *Euro. J. Opl. Res.* **12**, 190–202.

Sim, S. H., and Templeton, J. G. C. (1985). Steady state results for the $M/M(a, b)/c$ batch service system. *Euro. J. Opnl. Res.* **21**, 260–267.

Stadje, W. (1989). Some exact expressions for the bulk arrival queue $M^X/M/1$. *Queueing Systems* **4**, 85–92.

5 Network of Queues

5.1. Network of Markovian Queues

The queueing models that we have been examining so far are such that every customer or unit demands *one* service and leaves the system as soon as it is obtained. Very often we come across situations where a customer may need more than one service (or different kinds of service) from different servers and may be required to wait before different service channels. We can consider two broad types of models. For example, customers at a store may require services in a number of *successive* stages: they may be served at a counter and *then* go to the cash counter for payment; a bank customer may have first to go to a manager with his check and then go to the cashier to receive payment. Here each customer has to receive service from two different servers *one after another* and may have to queue up for service before each of the servers. This is how this kind of model differs from the model with service in several stages, e.g., Erlang-k service. To model such types of queueing situations, one is led to consider queues *in series* or queues *in tandem.* There may be situations in which customers may not be required to go from each service counter to the next one. For example, in maintenance and repair facilities with a number of counters, a particular job may not be required to receive service and thus may not have to pass through each service channel. An outpatient at a hospital may require some, though not each one, of the service facilities, but on the other hand may be required to go to one service again after passing through some other service counter. To model systems of the preceding types, one is required to consider instead of one service channel a *network* of service channels with a separate queue before each of the channels. Modeling of complex systems often involves such a network of queues. Queueing network models have applications in diverse areas, such as production and assembly lines, maintenance and repair operations, airport terminals, communication networks, computer-sharing and multiprogramming systems, and health-care centers. Before considering general network models, we examine systems in which services are required to be provided in a number of *successive* stages.

5.2. Channels in Series or Tandem Queues

Jackson (1954) was the first to consider queues in series as a model of a queueing system for the overhaul of aircraft engines, where stages of overhaul involve successive operations such as stripping, inspecting, repairing, assembling, and testing. Every unit or customer is served at each of the stages (called phases) one after another and may have to queue up before each service facility, which may have one or more similar parallel servers. The input to each stage (or phase), after the first, is the output from the preceding stage. It is assumed that the queueing space before each service facility is large enough to accommodate any number of waiting customers.

For the sake of simplicity we consider at first a system with two phases.

Assume that units arrive in accordance with a Poisson process with rate λ; these units constitute the input to the first stage. Units receive service at the first counter, the service-time distribution being an independent exponential with mean $1/\mu_1$.

The units that emerge from the first stage of service constitute the input to the second stage of service. A unit joins a queue (if needed) before the second service channel, receives service in turn, and then leaves the system. It is assumed that the service-time distribution at the second stage is also independent exponential, with mean $1/\mu_2$. The state of the system at an instant t is given by (n_1, n_2) where n_1 is the number of units in the first phase or before the first service channel and n_2 that before the second channel; let

$p(n_1, n_2, t)$ denote the corresponding probability. The system can be denoted by the notation $M/M/1 \to \cdot /M/1$.

The balance equations can be easily written down as follows:

$$p'(n_1, n_2, t) = \lambda p(n_1 - 1, n_2, t) + \mu_1 p(n_1 + 1, n_2, t) + \mu_2 p(n_1, n_2 + 1, t) - (\lambda + \mu_1 + \mu_2)p(n_1, n_2, t), \quad n_1 \geq 1, \quad n_2 \geq 1, \tag{2.1}$$

$$p'(n_1, 0, t) = \lambda p(n_1 - 1, 0, t) + \mu_2 p(n_1, 1, t) - (\lambda + \mu_1)p(n_1, 0, t), \quad n_1 \geq 1, \tag{2.2}$$

$$p'(0, n_2, t) = \mu_1 p(1, n_2, t) + \mu_2 p(0, n_2 + 1, t) - (\lambda + \mu_2)p(0, n_2, t), \quad n_2 \geq 1, \tag{2.3}$$

$$p'(0, 0, t) = -\lambda p(0, 0, t) + \mu_2 p(0, 1, t). \tag{2.4}$$

Assume that $\rho_1 = \lambda/\mu_1 < 1$, $\rho_2 = \lambda/\mu_2 < 1$; then the steady state is reached for large t and $\lim_{t\to\infty} p(n_1, n_2, t)$ exists. Denote $\lim_{t\to\infty} p(n_1, n_2, t) = p(n_1, n_2)$. Then it can be shown that

$$p(n_1, n_2) = \rho_1^{n_1}\rho_2^{n_2}p(0, 0) \quad \text{and}$$

$$p(0, 0) = (1 - \rho_1)(1 - \rho_2).$$

Thus,

$$p(n_1, n_2) = [(1 - \rho_1)\rho_1^{n_1}][(1 - \rho_2)\rho_2^{n_2}]. \tag{2.5}$$

The relation (2.5) indicates that in steady state each phase behaves *independently* of the other; further that the second phase behaves like a system with an input process that is Poisson with rate λ. This also follows from Burke's theorem on the output process of an $M/M/1$ queue in steady state. (See Section 3.2.3.) The input process to the second phase is an *independent* Poisson process with rate λ. The second phase behaves as an $M/M/1$ queue independent of the behavior of the first phase. Hence it follows that in steady state (when $\rho_1 < 1, \rho_2 < 1$), $p(n_1, n_2)$ is given by (2.5). The result can be extended to a finite number of simple $M/M/1$ queues in series or in tandem.

Example 5.1. Suppose that k $M/M/1$ queues in tandem are in steady state, with λ as the arrival rate and μ_i as the service rate of the ith phase. Find

(1) the probability $P(n_1, \ldots, n_k)$ that there are n_1 in the first phase, n_2 in the second phase, and so on;

(2) the expected number in ith phase; and

(3) the expected number in the complete system.

For $\mu_1 = \mu_2 = \cdots = \mu_k$ find the probability that there are n units in the complete system.

Since each of the phases behaves independently of the others

$$P(n_1, \ldots, n_k) = \prod_{i=1}^{k} (1 - \rho_i)\rho_i^{n_i}, \quad \text{where } \rho_i = \frac{\lambda}{\mu_i}.$$

The ith phase is an $M/M/1$ queue with rates λ and μ_i and so the expected number in the system in ith phase is given by

$$\frac{\rho_i}{1 - \rho_i}$$

and the expected number in the complete system is given by

$$\sum_{i=1}^{k} \frac{\rho_i}{1 - \rho_i}.$$

Suppose that $\mu_1 = \cdots = \mu_k$, i.e., $\rho_1 = \cdots = \rho_k = \rho$. The number of distinguishable arrangements in which the total number of n (indistinguishable) things can be put in k cells is

$$\binom{n + k - 1}{k - 1}.$$

Hence the probability that there are (a total of) n units in the complete system is given by

$$\binom{n + k - 1}{k - 1}(1 - \rho)^k \rho^n. \tag{2.6}$$

5.2.1. *Queues in Series with Multiple Channels at Each Phase*

Suppose that there are k service channels in series in steady state. The arrivals to the first phase after completing service there proceed to the second and so on and finally emerge from the system after having service at the *kth* channel. Suppose that phase i behaves as an $M/M/c_i$ queue and that there are *ample* holding spaces in front of each queue. In steady state the output process of each phase is Poisson with rate equal to the initial input rate λ. From Burke's theorem we find that phase-by-phase decomposition is possible in such a case. The probability $P(n_1, \ldots, n_k)$ that there are n_i units in the system in the ith phase (which is an $M/M/c_i$ queue, $i = 1, 2, \ldots, k$) is therefore given by

$$P(n_1, \ldots, n_k) = P_1(n_1) \ldots P_k(n_k)$$
$$= \prod_{i}^{k} P_i(n_i)$$

where $P_i(n_i) = Pr\{\text{there are } n_i \text{ in the system in an } M/M/c_i \text{ queue in steady state}\}$.

The result is obtained in a *product form*. We shall see that such product-form results also hold good even in more general cases of network of queues.

So far studies have been confined to networks of *Markovian* queues only, such that each service center is an $M/M/c_i$ queueing system. Further it is assumed that *steady state* exists, a sufficient condition for which is that each traffic intensity $\rho_i = \lambda/\mu_i < 1$. In the usual terminology of *networks of queues*, each phase or service center is called a *node*; a node may be one with one or more than one identical servers. In the more general system, customers enter the system at various nodal points, join the queue (if any), and receive service. In a tandem queue, the customers proceed sequentially from one node to the next one and in a feedforward queue customers proceed from one node to the next node in the forward direction.

Again we may have an *open* network where a customer after completing his service at a node may leave the system; on the other hand, in a closed network there is a fixed and finite number of customers, say K, such that neither can they leave the system nor can they be joined from outside by others. *A closed network in series or tandem* in which the customers after completing service at the last node again go back and join the first node is called a *cyclic* network.

Jackson (1957) considered a more general network of queues, which is known as a *Jackson network*. There are various situations that can be modeled by some of these network patterns. The idea of modeling a job shop as a queueing network is old. More recently, similar ideas appear in models of computer systems, road-traffic systems, command and control systems, data-transmitting systems, teletraffic systems, and others. The applications of queueing networks also extend to biology (neural networks), disease (compartmental models), and polymerization (cluster models). Kelly (1979) gives many varied examples of queueing networks. All these applications make the study of queueing networks more significant.

Remark 1. We have assumed so far that the waiting space before each of the servers in series is large enough. However, this assumption may not always be valid. Use the notation $M/M/1/L_1 + 1$ to indicate that the server has before him a waiting space for L_1 customers besides the one being served, if any (i.e. in all $L_1 + 1$ places). We may have the following three types of disciplines.

(a) The customer at the server j who, upon completion of service, cannot gain immediate access to the server $(j + 1)$ overflows the sever $(j + 1)$ and goes immediately to server $j + 2$. This is called an *overflow discipline.*

(b) A customer who cannot gain immediate access to the server $(j + 1)$ leaves the system. This is called a *loss discipline.*

(c) A customer who cannot gain immediate access to the server $(j + 1)$ stays idle at server j until a space before the server $(j + 1)$ becomes available. This is called a *blocking discipline.*

Consider the system $M/M_1/1 \to ./M_2/1/L_2 + 1$ $(L_2 < \infty)$ with loss discipline (finite waiting space before the second server). The service-time distribution of server i is exponential with mean $1/\mu$ and $\rho_i = \lambda/\mu_i$, λ being the arrival rate to server 1. Here the queue-length processes are independent and

$$P(n_1, n_2) = [(1 = \rho_1)\rho_1^{n_1}]. \left[\frac{(1 - \rho_2)\rho_2^{n_2}}{(1 - \rho_2)^{L_2+2}}\right],$$

$$n_1 = 0, 1, \ldots, n_2 = 0, 1, \ldots, L_2 + 1.$$

The system has a product-form solution. The system

$$M/M_1/1/L_1 \to ./M_2/1/L_2$$

with $L_i < \infty$ has a blocking discipline that imposes dependence between queue-length processes before server 1 and server 2. The system does not have a product-form solution. There are a large number of variations for the series network of queues and many papers have been devoted to them.

Remark 2. Chen (1989) considers two single-server tandem queues, the first being an $M/D/1$ queue and the second having exponential service times, i.e., the tandem queue $M/D/1 \to ./M/1$. He shows that the steady-state distribution $p(n_1, n_2)$ does not have a product form and hence that the two queues are not independent, and that the sojourn times of the two queues are not independent.

Remark 3. If the input is Poisson, and service times at all the channels are exponential, then the output is Poisson; further sojourn times at different channels are independent irrespective of the order of the channels. Weber (1979) shows that the final departure process is independent of the order of the channels for an arbitrary arrival process as long as the servers are exponential.

Remark 4. Boxma (1986) gives a review of results relating to (1) two queues in series, and (2) two parallel queues with a single server.

5.3. Jackson Network

In this model customers from one node i proceed to an arbitrary node and fresh customers may join a node from outside. Suppose that there are k nodes, where the ith node ($i = 1, \ldots, k$) consists of c_i exponential servers with parameter μ_i: customers after receiving service at the ith node proceed to the jth node with probability p_{ij}; suppose further that the ith node may receive customers from a Poisson stream with rate λ_i from outside the system as well. Customers at node i depart from the system with probability $q_i = 1 - \sum_{j=1}^{k} p_{ij}$.

This is Jackson's network model. The model is very general. For example, with $\lambda_i = 0$, and $p_{ij} = 1$, whenever $j = i + 1$, $1 \le i \le k - 1$, and $q_k = 1$, we get k queues in tandem. For a closed network with k members, $\lambda_i = 0$, $\sum_{j=1}^{k} p_{ij} = 1$ so that $q_i = 0$ for each i; for a cyclic network, $\lambda_i = 0$, $p_{ij} = 1$, $j = i + 1$, $1 \le i \le k - 1$ and $p_{k1} = 1$ (and $q_i = 0$ for all i).

Consider Jackson's general network model with k nodes. The input to ith node consists of outputs of the other nodes as well as the external input λ_i. The total average arrival rate of customers α_i to node i is the sum of the Poisson arrival rates λ_i from outside the system plus the arrival rate from arrivals to node i from (other) internal nodes $\sum_j p_{ji}\alpha_j$.

Figure 5.1 provides a diagram of node i.

It easily can be seen that the parameters α_i satisfy the equation

$$\alpha_i = \lambda_i + \sum_{j=1}^{k} p_{ji}\alpha_j, \qquad i = 1, 2, \ldots, k; \tag{3.1}$$

α_i gives the effective arrival rate to the node i or effective rate of flow through node i.

The preceding equations are known as *traffic equations*, *flow balance equations*, or *conservation equations* for the rate of flow through an arbitrary node. Note that $\alpha_i q_i$ is the rate of departure from the system from the node

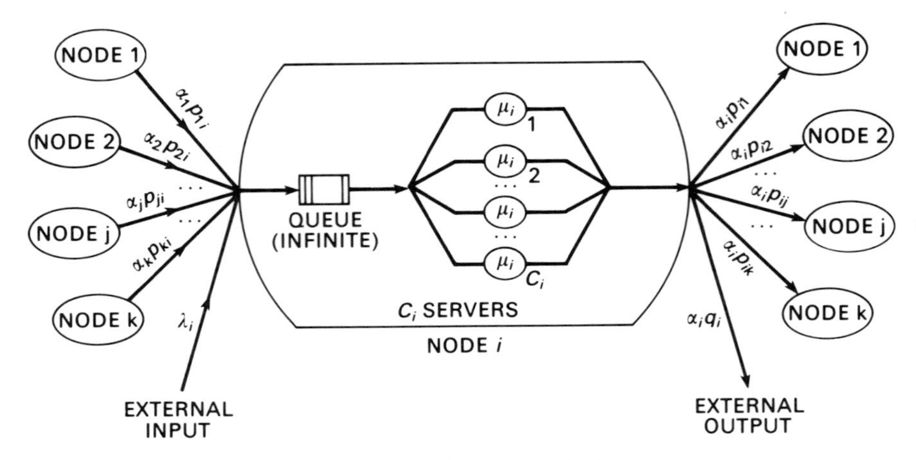

FIGURE 5.1 Diagram of node i in a Jackson network.

i. The existence of the solution of the preceding set of equations is a necessary condition for the existence of the steady-state distribution in a Jackson network.

In his fundamental paper Jackson (1957) shows that for a Jackson network of (Markovian) queues, the particular product-form result of marginal distribution holds in equilibrium, implying the independence of various nodes in the network. This is commonly referred to as Jackson's theorem, which we consider next. The proof given by Jackson is straightforward; it uses the usual probability arguments employed to derive balance equations. It is assumed that the network is completely open.

Theorem 5.1. (Jackson's Theorem). *Let $(n_1, n_2, \ldots, n_k)$ denote the state of the complete system in which there are n_i (in the queue and in service) at node i in a Jackson network of Markovian queues in equilibrium and let $p(n_1, \ldots, n_k)$ be the probability that the system is in the state $(n_1, \ldots, n_k)$.*

Assume that

$$\rho_i = \frac{\alpha_i}{\mu_i} < 1, \qquad i = 1, 2, \ldots, k,$$

where $\{\alpha_i\}$ are given by the balance equations

$$\alpha_i = \lambda_i + \sum_j \alpha_j p_{ji}, \qquad i = 1, 2, \ldots, k.$$

If $p_i(n_i)$ denotes the probability that there are n_i in the system (in queue plus *service) for the $M/M/c_i$ queue with input rate α_i, and service rate μ_i for each of the c_i servers,*

$$p_i(n) = p_i(0)\frac{\left(\frac{\alpha_i}{\mu_i}\right)^n}{n!}, \; n = 0, 1, 2, \ldots, c$$

$$= p_i(0)\frac{\left(\frac{\alpha_i}{\mu_i}\right)^n}{\left[n!c_i^{n-c_i}\right]}, \; n = c_i + 1, \ldots. \tag{3.2}$$

Then

$$p(n_1, \ldots, n_k) = p_1(n_1)p_2 \cdots p_k(n_k). \tag{3.3}$$

Proof. Let $p_t(n_1, \ldots, n_k)$ be the probability that the complete system is in state $(n_1, \ldots, n_k)$ at time t; then considering in the usual manner the infinitesimal interval $(t, t+h)$ following the interval $(0, t)$ we can write the differential equation satisfied by p_t. Denote

$$q_i = 1 - \sum_j p_{ij}, \qquad a_i(n) = \min\{n, c_i\} = n, \quad \text{if } n < c_i$$

$$= c_i \quad \text{if } n \geq c_i$$

$$\delta_i = \min\{n_i, 1\} = 1, \qquad n_i \geq 1$$

$$= 0, \qquad n_i = 0.$$

The state at $t + h$ can be reached from $(n_1, \ldots, n_k)$, the state at t, in one of the following four mutually exclusive ways.

(A) State at t is $(n_1, \ldots, n_k)$ and no arrival occurs to any node from external source nor does any departure occur from any node in the interval $(t, t+h)$ of length h. We get

$$Pr(A) = p_t(n_1, \ldots, n_k)[1 - (\sum \lambda_i)h - \sum a_i(n_i)\mu_i h] + o(h). \tag{3.4}$$

(B) State at t is $(n_1, \ldots, n_i + 1, \ldots, n_k)$ for some i $(i = 1, \ldots, k)$, there is one service completion at that node i in $(t, t+h)$ and this completion departs from the system (with probability q_i). We get

$$Pr(B) = \sum_{i=1}^{k} p_t(n_1, \ldots, n_i + 1, \ldots, n_k)[a_i(n_i + 1)h\mu_i]q_i + o(h). \tag{3.5}$$

(C) State at t is $(n_1, \ldots, n_i - 1, \ldots, n_k)$ for some i $(i = 1, 2, \ldots, k)$ and there is one arrival from the external source to node i in the interval $(t, t + h)$. We get

$$Pr(C) = \sum_{i=1}^{k} p_t(n_1, \ldots, n_i - 1, \ldots, n_k)\{\lambda_i h \delta_i\} + o(h). \tag{3.6}$$

(Here $\delta_i = 1$, if $n_i \geq 1$, and $\delta_i = 0$ if $n_i = 0$.)

(D) State at t is $(n_1, \ldots, n_i + 1, \ldots, n_j - 1, \ldots, n_k)$, and there is one service completion at node i in $(t, t + h)$ and the one whose service is completed moves to node j with probability p_{ij}. Thus,

$$\begin{aligned} Pr(D) = \sum_i \sum_j & \, p_t(n_1, \ldots, n_i + 1, \ldots, n_j - 1, \ldots, n_k) \\ & \times [a_i(n_i + 1)\mu_i h p_{ij}] + o(h). \end{aligned} \tag{3.7}$$

Thus,

$$p_{t+h}(n_1, \ldots, n_k) = Pr(A) + Pr(B) + Pr(C) + Pr(D). \tag{3.8}$$

Following the usual procedure of transferring the term $p_t(n_1, \ldots, n_k)$ (here the term $Pr(A)$) to the left-hand side, then dividing by h, and finally taking limits as $h \to 0$, we get the differential-difference equation for $p'(t)$.

Putting $p_t(n_1, \ldots, n_k) = p(n_1, \ldots, n_k)$ and $p'(t) = 0$, we get the equations satisfied by the steady-state probabilities as follows:

$$\begin{aligned} \left[\sum_i \lambda_i + \sum_i a_i(n_i)\mu_i\right] & p(n_1, \ldots, n_k) \\ = & \sum_i a_i(n_i + 1)\mu_i q_i p(n_1, \ldots, n_i + 1, \ldots, n_k) \\ & + \sum_i \lambda_i \delta_i p(n_1, \ldots, n_i - 1, \ldots, n_k) \\ & + \sum_i \sum_j a_i(n_i + 1)\mu_i p_{ij} p(n_1, \ldots, n_i + 1, \ldots, n_j - 1, \ldots, n_k). \end{aligned} \tag{3.9}$$

We shall now show that $p(n_1, \ldots, n_k)$ given by (3.3) is a solution of the preceding equation. From the form of $p(n_1, \ldots, n_k) = p_1(n_1) \cdots p_k(n_k)$, (given in (3.3)) and the form of $p_i(n_i)$ (given in (3.2)), we get

$$\begin{aligned} \frac{p(n_1, \ldots, n_i + 1, \ldots, n_k)}{p(n_1, \ldots, n_i, \ldots, n_k)} &= \frac{p_i(n_i + 1)}{p_i(n_i)} \\ &= \frac{\alpha_i}{\mu_i a_i(n_i + 1)} \end{aligned} \tag{3.10}$$

$$\frac{p(n_1, \ldots, n_i - 1, \ldots, n_k)}{p(n_1, \ldots, n_i, \ldots, n_k)} = \frac{\mu_i a_i(n_i)}{\alpha_i} \quad \text{and} \tag{3.11}$$

$$\frac{p(n_1, \ldots, n_i + 1, \ldots, n_j - 1, \ldots, n_k)}{p(n_1, \ldots, n_i, \ldots, n_j, \ldots, n_k)} = \frac{\alpha_i}{\mu_i a_i(n_i + 1)} \frac{\mu_j a_j(n_j)}{\alpha_j}. \tag{3.12}$$

Dividing both sides of (3.9) by $p(n_1, \ldots, n_i, \ldots, n_k)$ and using (3.10), (3.11), and (3.12), we get

$$\begin{aligned}\left[\sum_i \lambda_i + \sum_i a_i(n_i)\mu_i\right] &= \sum_i a_i(n_i + 1)\mu_i q_i\left[\frac{\alpha_i}{\mu_i a_i(n_i + 1)}\right] \\ &\quad + \sum_i \frac{\lambda_i \delta_i \mu_i a_i(n_i)}{\alpha_i} \\ &\quad + \sum_i \sum_j a_i(n_i + 1)\mu_i p_{ij} \frac{\alpha_i \mu_j a_j(n_j)}{\alpha_j \mu_i a_i(n_i + 1)}.\end{aligned} \tag{3.13}$$

We have

$$\begin{aligned}\sum_i \alpha_i q_i &= \sum_i \alpha_i\left[1 - \sum_{j=1}^{k} p_{ij}\right] \\ &= \sum \alpha_i - \sum_j \left(\sum_i \alpha_i p_{ij}\right) \\ &= \sum_i \alpha_i - \sum_j (\alpha_j - \lambda_j) = \sum_j \lambda_j = \sum_i \lambda_i\end{aligned} \tag{3.14}$$

$$\begin{aligned}\sum_i \sum_j \frac{\alpha_i \mu_j a_j(n_j) p_{ij}}{\alpha_j} &= \sum_j \frac{a_j(n_j)}{\alpha_j} \mu_j \left\{\sum_i \alpha_i p_{ij}\right\} \\ &= \sum_j \frac{a_j(n_j)}{\alpha_j} \{\mu_j(\alpha_j - \lambda_j)\} \\ &= \sum_j \mu_j a_j(n_j) - \sum_j \frac{\lambda_j \mu_j a_j(n_j)}{\alpha_j} \\ &= \sum_i \mu_i a_i(n_i) - \sum_i \frac{\lambda_i \mu_i a_i(n_i)}{\alpha_i}.\end{aligned} \tag{3.15}$$

Noting that $\delta_i a_i(n_i) = a_i(n_i)$, we get that the RHS of (3.13) is equal to

$$\sum \lambda_i + \sum \frac{\lambda_i \mu_i a_i(n_i)}{\alpha_i} + \sum \mu_i a_i(n_i) - \sum \frac{\lambda_i \mu_i a_i(n_i)}{\alpha_i}$$

$$= \sum \lambda_i + \sum \mu_i a_i(n_i)$$

$$= \text{LHS of Eq. (3.13).}$$

This shows that (3.3) (with 3.2)) satisfies (3.9), in other words, (3.3) is a solution of (3.9). It can be verified by direct summation that (3.3) also satisfies the normalizing condition $\sum_{n_i} p(n_1, \ldots, n_k) = 1$, $(n_i \geq 0)$.

The justification that (3.3) gives the steady-state distribution follows from a result of Markov process theory that can be stated as follows:

If there is a positive solution of the balance equations of an irreducible Markov process and if such a solution satisfies the normalizing condition, then the steady-state distribution of the process exists and is given by that solution. This establishes the uniqueness of the solution. This completes the proof.

Note 1. An alternative proof of Jackson's theorem (using the concept of partial balance) has been given by Lemoine (1977).

Note 2. The equilibrium-queue-length distribution in a Jackson network is of the product form.

Note 3. Jackson's theorem contains the remarkable result that whenever the equilibrium condition exists, each node in the network behaves as if it were an independent $M/M/c_i$ queue with Poisson input. It is implied that the states n_i of individual nodes i $(i = 1, 2, \ldots, k)$ in steady state are independent random variables.

Note 4. In equilibrium, the external departure streams or outputs from individual nodes are independent Poisson processes with rate $q_i \alpha_i$ for node i, $i = 1, 2, \ldots, k$. The relation (3.14) shows that $\sum \lambda_i = \sum \alpha_i q_i$. That is, under equilibrium the total *external* input flow rate is equal to the total external output flow rate. However, the arrival and departure processes at an $M/M/1$ node are not always equivalent (Walrand, 1982a).

Note 5. In general, the total *input* process into an individual node is not necessarily Poisson even under equilibrium conditions. For a demonstration of this, see Gelenbe and Mitrani (1980, pp. 85–86). Here non-Poisson total arrivals or inputs see time averages and this accounts for the simple state probability results.

Note 6. The traffic equations

$$\alpha_i = \lambda_i + \sum_j \alpha_j p_{ji}, \qquad i = 1, \ldots, k,$$

have a unique solution. Writing $\boldsymbol{\alpha} = (\alpha_1, \ldots, \alpha_k)$, $\boldsymbol{\lambda} = (\lambda_1, \ldots, \lambda_k)$, $\mathbf{P} = (p_{ij})$, we get

$$\begin{aligned} \boldsymbol{\alpha} &= \boldsymbol{\lambda} + \boldsymbol{\alpha}\mathbf{P} \quad \text{or} \\ \boldsymbol{\alpha}(\mathbf{I} - \mathbf{P}) &= \boldsymbol{\lambda}. \end{aligned} \tag{3.16}$$

Now since in an open network, any customer in any queue eventually leaves the network, it follows that each element of $\mathbf{P}^n$ converges to 0 as $n \to \infty$. Thus, $(\mathbf{I} - \mathbf{P})^{-1}$ converges, that is $(\mathbf{I} - \mathbf{P})$ has an inverse, so that the rank of $(\mathbf{I} - \mathbf{P})$ is k. This demonstrates that $\boldsymbol{\alpha} = \boldsymbol{\lambda} + \boldsymbol{\alpha}\mathbf{P}$, for given $\boldsymbol{\lambda}$ and $\mathbf{P}$, has a unique solution in $\boldsymbol{\alpha}$. $\mathbf{P}$ is called the *transfer*, *switching*, or *routing* matrix. The square matrix $\mathbf{P}$ from an open network is not stochastic.

Note 7. Consider a state $\mathbf{N} = (n_1, \ldots, n_i, \ldots, n_k)$ of the network in equilibrium. For each node i, the rate of flow out of state $\mathbf{N}$ due to a departure of a customer from node i is equal to the rate of flow into the state $\mathbf{N}$ due to the arrival of a customer into node i due to either external input or internal transfer. This gives *local-balance* relations; Eq. (3.9) involving the rates of flow of the complete network constitutes the *global-balance* equation.

The local-balance equation can be obtained by equating the rate of flow out of state $(n_1, \ldots, n_i, \ldots, n_k)$ due to a customer leaving node i with the rate of flow into state $(n_1, \ldots, n_i, \ldots, n_k)$ due to arrival of a customer to node i, either from outside or from some of the other internal nodes. We have

$$\text{rate of flow out of } i = a_i(n_i)\mu_i p(n_1, \ldots, n_i, \ldots, n_k)(1 - p_{ii}) \quad \text{and} \tag{A}$$

$$\text{rate of flow into } i = \lambda_i \delta_i p(n_1, \ldots, n_i - 1, \ldots, n_k) \text{ (from outside)} \tag{B}$$

$$= \sum_{j \neq i} a_j(n_j + 1)\mu_j p_{ji} p(n_1, \ldots, n_i - 1, \ldots, n_j + 1, \ldots, n_k). \tag{C}$$

Assuming that p is of the form (3.3), and using (3.10)–(3.12), we get the following form of local-balance equation:

$$a_i(n_i)\mu_i(1 - p_{ii}) = \frac{\lambda_i \delta_i \mu_i a_i(n_i)}{\alpha_i} + \sum_{j \neq i} \frac{\alpha_j p_{ji} a_i(n_i)\mu_i}{\alpha_i}.$$

Note 8. In notes (3) and (5) we observe that though the equilibrium distribution has the product form as if the service facilities were independent and all arrival processes were Poisson, but the arrival process within the network is, in general, not Poisson. Further, the facilities and the associated queue-length processes (time-dependent) are not independent (Melamed, 1979).

This shows that equilibrium distribution does not capture the transient or time-dependent behavior. Similar behavior is also noted in the case of a $G/M/1$ queue. This points to the limitations of the steady-state result. (See also Section 2.5.)

Note 9. Goodman and Massey (1984) have generalized Jackson's theorem to the nonergodic case; their results completely characterize the large-time behavior ($t \to \infty$) of Jackson networks.

5.4. Closed Markovian Network (Gordon and Newell Network)

In contrast to Jackson's open network, Gordon and Newell (1967a) considered a *closed* network of Markovian queues, in which a fixed and finite number of customers, say, K, circulate through the network, there being no external input or departure from the network. This corresponds to Jackson's model in which $\lambda_i = 0$ and $q_i = 0$ for each i and $n_1 + n_2 + \cdots + n_k = K$ (fixed).

With $\lambda_i = 0$, the traffic equation reduces to $\boldsymbol{\alpha} = \boldsymbol{\alpha}\mathbf{P}$ and $q_i = 0$ for each i. This implies that the switching matrix $\mathbf{P}$ is stochastic. Assuming that $\mathbf{P}$ is irreducible, one can therefore regard $\boldsymbol{\alpha}$ as the unique stationary distribution of a discrete parameter Markov chain with transition probability matrix $\mathbf{P}$. For this closed network $\mathbf{N}(t) = (n_1(t), \ldots, n_k(t))$ is an irreducible continuous-time Markov chain with finite state space and therefore possesses an equilibrium distribution.

Theorem 5.2. (*Gordon and Newell*, 1967*a*). *Consider a closed network with k nodes within which a total of K customers (jobs) circulate according to the* *routing matrix* $\mathbf{P}$ *(of order k). The node i has* C_i *identical exponential servers with rate* μ_i, $i = 1, 2, \ldots, k$.

Let $\boldsymbol{\alpha}'$ *be any nonzero solution of* $\boldsymbol{\alpha}' = \boldsymbol{\alpha}'\mathbf{P}$ *and*

$$\rho_i = \frac{\alpha_i}{\mu_i}, \qquad d_i(n_i) = \begin{cases} n_i!, & n_i \le c_i \\ (c_i!)(c_i)^{n_i - c_i}, & n_i > c_i \end{cases}. \tag{4.1}$$

Then the steady-state probability that the state of the system is $(n_1, \ldots, n_k)$ *is given by*

$$p(n, \ldots, n_k) = \frac{1}{A(K)} \prod_{i=1}^{k} \frac{\rho_i^{n_i}}{d_i(n_i)} \tag{4.2}$$

where

$$A(K) = \sum_{\substack{n_i \ge 0 \\ \Sigma n_i = K}} \prod_{i=1}^{k} \frac{\rho_i^{n_i}}{d_i(n_i)} \tag{4.3}$$

(A^{-1} being the normalizing constant).

Proof. The proof of the theorem is also by direct verification, that is, by showing that (4.2) is a solution of the corresponding balance equation (obtained by putting $\lambda_i = 0$ and $q_i = 0$ in Eq. (3.9).)

Particular case. For $c_i = 1$ for all i (each node having only one server) $d_i(n_i) = 1$.

Notes.

(1) The equilibrium distribution is still of the product form. The product, however, is not the product of distributions of states of the individual nodes. The states of the nodes are not independent as is evident also from the fact that $\sum n_i = K$ (fixed).

Compare this with the corresponding result for a Jackson (open) network.

(2) The routing matrix $\mathbf{P}$ is a stochastic matrix.

(3) Kiessler (1989) gives a simple proof of the equivalence of input and output intervals (interinput and interoutput distributions) at each node in a Jackson network of single-server nodes; he also shows that the proof is valid for a closed (Gordon–Newell) network of single-server nodes.

Example 5.2. Two-node system with feedback

Consider a closed network having two nodes and K jobs circulating among these two nodes (node 1 corresponding to CPU and node 2 to an I/O device of a computer system having a fixed number of K circulating programs). Suppose that the lengths of successive service times at node i (CPU execution bursts and I/O bursts) are IID exponential with mean $1/\mu_i$, $i = 1, 2$. At the end of service at node 1 a customer goes to node 2 with probability p or is fed back to node 1 with probability $q = 1 - p$. (At the end of a CPU burst a program requests an I/O operation with probability p.) Let n be the number of customers at node 1 (programs in the CPU queue) including the one being served, so that $K - n$ is the number at node 2 (in the I/O queue), and $p(n, K - n)$ be the steady-state probability. Here

$$\mathbf{P} = \begin{pmatrix} q & p \\ 1 & 0 \end{pmatrix}$$

and $\boldsymbol{\alpha}' = \boldsymbol{\alpha}'\mathbf{P}$ leads to $\boldsymbol{\alpha}' = (a, ap)$ where a is an arbitrary constant. We have $\rho_1 = a/\mu_1$ and $\rho_2 = ap/\mu_2$. We have

$$\begin{aligned} p(n, K - n) &= \frac{1}{A(K)} \left(\frac{a}{\mu_1}\right)^n \left(\frac{ap}{\mu_2}\right)^{K-n} \\ &= \frac{1}{A(K)} \left(\frac{1}{r}\right)^K r^n \end{aligned}$$

where

$$r = \frac{\mu_2}{p\mu_1} \quad \text{and}$$

$$\begin{aligned} A(K) = \sum_{n=0}^{K} \left(\frac{1}{r}\right)^K r^n &= \left(\frac{1}{r}\right)^K \cdot \frac{1 - r^{K+1}}{1 - r}, \qquad r \neq 1 \\ &= \left(\frac{1}{r}\right)^K \cdot \frac{1}{K + 1}, \qquad r = 1. \end{aligned}$$

Thus, for $n = 0, 1, \ldots, K$,

$$\begin{aligned} p(n, K - n) &= \frac{(1 - r)r^n}{1 - r^{K+1}}, \qquad r \neq 1, \\ &= \frac{1}{K + 1}, \qquad r = 1. \end{aligned} \tag{4.4}$$

In other words,

$$p(K-m, m) = \frac{\left[1-\left(\frac{1}{r}\right)\right]\left(\frac{1}{r}\right)^m}{1-\left(\frac{1}{r}\right)^{K+1}}, \qquad r \neq 1,$$

$$= \frac{1}{K+1}, \qquad r = 1. \tag{4.5}$$

The server-utilization factors u_1 and u_2 at node 1 (CPU) and at node 2 (I/O), respectively, are given by

$$\begin{aligned} u_1 = \sum_n p(n, K-n) = 1 - p(0, K) &= \frac{r - r^{K+1}}{1 - r^{K+1}} = \frac{r(1-r^K)}{1-r^{K+1}}, \qquad r \neq 1 \\ &= \frac{K}{K+1}, \qquad r = 1, \quad \text{and} \end{aligned} \tag{4.6}$$

$$u_2 = \sum_n p(n, K-n) = 1 - p(K, 0) = \frac{\frac{1}{r} - \left(\frac{1}{r}\right)^{K+1}}{1 - \left(\frac{1}{r}\right)^{K+1}} \tag{4.7}$$

$$\begin{aligned} &= \frac{1-r^K}{1-r^{K+1}}, \qquad r \neq 1 \\ &= \frac{K}{K+1}, \qquad r = 1. \end{aligned} \tag{4.8}$$

Notes.

(1) We have

$$\begin{aligned} &(1) \quad \frac{u_1}{u_2} = r = \frac{\mu_2}{p\mu_1} = \frac{\rho_1}{\rho_2} \qquad \text{and} \\ &(2) \quad u_1 = \frac{A(K-1)}{A(K)}. \end{aligned} \tag{4.9}$$

(2) Queueing networks serve as models for multiprogrammed computer systems and communication networks and certain parts-manufacturing systems. The number K of circulating programs in a computer network is known as the *level* (or *degree*) of *multiprogramming*.

5.5. Cyclic Queue

Consider a closed network of K nodes such that the output of the node i goes to the next node $i+1$ $(1 \le i \le k-1)$ whereas the output of the last node k feeds back to node 1 and so on. Such a queue is called *cyclic queue.* A cyclic queue is a special kind of closed-network queue having routing matrix

$$\mathbf{P} = (p_{ij}),$$

where

$$\begin{aligned} p_{ij} &= 1, \quad j = i+1, \quad 1 \le i \le k-1 \\ &= 1, \quad i = k, \quad j = 1 \\ &= 0, \quad \text{otherwise.} \end{aligned}$$

The results of a closed-queueing network will apply. The solution of $\boldsymbol{\alpha}' = \boldsymbol{\alpha}'\mathbf{P}$ leads to

$$\begin{aligned} &\rho_1\mu_1 = \rho_k\mu_k \\ &\rho_i\mu_i = \rho_{i-1}\mu_{i-1}, \qquad i = 1, 2, \ldots, k \\ &\left(\rho_i = \frac{\alpha_i}{\mu_i}\right) \end{aligned}$$

so that

$$\begin{aligned} \rho_i &= \frac{\mu_k}{\mu_i}\rho_k, \qquad i = 1 \\ &= \frac{\mu_{i-1}}{\mu_i}\rho_{i-1}, \qquad i = 2, 3, \ldots, k. \end{aligned}$$

Thus,

$$\rho_2 = \left(\frac{\mu_1}{\mu_2}\right)\rho_1, \qquad \rho_3 = \left(\frac{\mu_2}{\mu_3}\right)\rho_2 = \left(\frac{\mu_1}{\mu_3}\right)\rho_1$$

$$\rho_k = \left(\frac{\mu_1}{\mu_k}\right)\rho_1.$$

Using Theorem 5.2 (with $c_i = 1$) we get

$$\begin{aligned} p(n_1, \ldots, n_k) &= \frac{1}{A(K)} \prod_{i=1}^{k} \rho_i^{n_i} \\ &= \frac{1}{A(K)} \rho_1^{n_1} \frac{(\mu_1)^{K-n_1}}{(\mu_2)^{n_2} \cdots (\mu_k)^{n_k}}. \end{aligned}$$

Thus, we get

$$p(n_1, \ldots, n_k) = \frac{1}{A_1(K)} \frac{\mu_1^{K-n_1}}{(\mu_2)^{n_2} \cdots (\mu_k)^{n_k}}$$

where $[A_1(K)]^{-1}$ is the normalizing constant; $[A_1(K)]^{-1}$ equals the sum of the second factor on the RHS over n_is such that $\sum_{i=1}^{k} n_i = K$.

Notes.

(1) The corresponding result for multiple servers at the nodes can be written down easily.
(2) Efficient and stable computational algorithms are put forward for calculation of the normalization constant, for example (Buzen, 1973), [see also Harrison (1985)].

Example 5.3. Consider a cyclic queue with two nodes and K circulating jobs. The first node (node 1) has an exponential server with rate μ, and the second node (node 2) has an exponential server with rate λ. (This can be treated as a special case of Example 5.2 with $\mu_1 = \mu$, $\mu_2 = \lambda$, and $p = 1$.) We have

$$r = \frac{\mu_2}{p\mu_1} = \frac{\lambda}{\mu} = \rho \text{ (say)} \quad \text{and}$$

$$\begin{aligned} p(n, K-n) &= \frac{(1-\rho)\rho^n}{1-\rho^{K+1}}, \qquad \rho \neq 1 \\ &= \frac{1}{K+1}, \qquad \rho = 1. \end{aligned} \tag{5.1}$$

Now for an $M/M/1/K$ queue, the steady probability that there are n $(\leq K)$ in the system is also given by (5.1). Thus, a two-node cyclic queue as described previously may be considered as equivalent to a limited-space $M/M/1/K$ queue. (See Problem 5.6 for a more general result.) The result was also discussed in Section 3.3.3.

Remarks.

(1) Distributions of sojourn time and cycle time in a cyclic exponential network have been considered, as noted in concluding remarks. See Problems 5.9 and 5.10 for some specific cases where distributions are of product form.

(2) In 1963, Jackson introduced a more general Markovian model that includes the Jackson networks (1957) and the Gordon and Newell networks

(1967a, b) as particular cases. Here he considers state-dependent external-arrival rate (dependent upon the total number of customers in the system) and state-dependent exponential-service rate (dependent on the number of customers at a node). Jackson gives sufficient conditions for the existence of equilibrium distribution at the nodes and shows that the probability distribution of the state of the network is of product form.

(3) Posner and Bernholtz (1968a, b) generalize Gordon and Newell's model by allowing, among other things, travel time between pairs of nodes to have arbitrary distribution and by providing for different service rates.

5.6. BCMP Networks

We now examine a more general network: the network considered by Baskett, Chandy, Muntz, and Palacios. It is called the BCMP network. Here new service disciplines as well as a number $R(\geq 1)$ of classes of jobs (customers) are introduced.

Let k be the number of nodes in a *closed* network and let there be $R(\geq 1)$ classes of customers (jobs). Customers circulate in the network and may change class as they move from one node to another. A job of class r and node i, after completing its service at node i, moves to node j as a job of class s with probability $p_{ir,js}$. Thus,

$$\mathbf{P} = (p_{ir,js}), \qquad 1 \leq i \leq k,$$
$$1 \leq j \leq k, \qquad 1 \leq r, \qquad s \leq R,$$

is the TPM of a Markov chain.

Let μ_{ir} be the service rate of a customer of class r at node i.

Nodes can be divided into four types according to the service discipline.

Type 1: The node has a single server, with exponential service time; the rates of service are the same for all R types of jobs and the service discipline is FIFO. Here $\mu_{ir} = \mu_i$ for all r.

Type 2: The node has a single server with processor-sharing service discipline. Each class of job may have a distinct service-time distribution that is given by a differentiable function.

Type 3: The node has sufficient number of servers so that there is no queue at the node (and so that a job starts receiving service immediately on arrival).

Service-time distribution, which is to be differentiable, can be distinct for a distinct class of customers.

Type 4: The node has a single-server, preemptive LIFO service discipline such that a new arrival interrupts the customer being served, if any, and the displaced customer returns to the head of the queue and starts receiving service as soon as the customer who caused the interruption completes his service. Service-time distribution, which is to be differentiable, can be distinct for a distinct class of customers.

Service-time distribution at nodes of Types 2, 3, and 4 can be more general Coxian type. (See Gelenbe and Pujolle, 1987, for a description of Coxian distribution and its properties.) Further, distributions can be different for different classes.

We examine the system-state probability distribution in steady state.

First, solve the traffic equations for each distinct class. Let the solution be designated by α_{ir} where α_{ir} satisfies

$$\alpha_{ir} = \sum_{r'=1}^{R} \left(\sum_{j=1}^{k} \alpha_{jr} p_{jr', ir} \right), \qquad i = 1, 2, \ldots, k, \ r = 1, 2, \ldots, R.$$

Thus, α_{ir} gives the relative frequency of the number of visits of a customer of class r to the node i.

Let n_r be the number of jobs of class r in the network and n_{ir} be the number of jobs of class r at node i. Then

$$n_r = \sum_{i=1}^{k} n_{ir}, \qquad r = 1, 2, \ldots, R,$$

and $m_i = \sum_{r=1}^{R} n_{ir}$ is the total number of jobs at node i, $i = 1, 2, \ldots, k$. Denote vector $\mathbf{N}_i$ by

$$\mathbf{N}_i = (n_{i1}, n_{i2}, \ldots, n_{iR}), \qquad i = 1, \ldots, k.$$

Then a state of the system can be denoted by the vector

$$\mathbf{N} = (\mathbf{N}_1, \mathbf{N}_2, \ldots, \mathbf{N}_k).$$

We now state without proof the BCMP theorem.

Theorem 5.3. BCMP Theorem (Baskett *et al.* 1975). *The steady-state probability that the system is at state* $\mathbf{N}$ *is given by*

$$P(\mathbf{N}) = \frac{1}{A(n_1, \ldots, n_k)} \prod_{i=1}^{k} f_i(\mathbf{N}_i)$$

where

$$f_i(\mathbf{N}_i) = (m_i)! \prod_{r=1}^{R} \frac{1}{(n_{ir})!} (\alpha_{ir})^{n_{ir}} \left(\frac{1}{\mu_i}\right)^{m_i},$$

if node i is of Type 1

$$= (m_i)! \prod_{r=1}^{R} \frac{1}{(n_{ir})!} \left(\frac{\alpha_{ir}}{\mu_{ir}}\right)^{n_{ir}},$$

if node i is of Type 2 or 4;

$$= \prod_{r=1}^{R} \frac{1}{(n_{ir})!} \left(\frac{\alpha_{ir}}{\mu_{ir}}\right)^{n_{ir}}$$

if node i is of Type 3;

and A^{-1} is the normalizing constant.

Notes.

(1) The result is of product form.

(2) Only the mean service times enter in the result (even for more general types of service distribution considered).

(3) The case of the open network can also be covered by considering two fictitious nodes: node O (origin) and node $k + 1$ (sink). Then we shall have the TPM.

$$\mathbf{P} = (p_{ir,jr'}), \qquad i = 0, 1, \ldots, k,$$
$$j = 1, 2, \ldots, k + 1,$$
$$r, r' = 1, 2, \ldots, R.$$

(4) The BCMP network's essential advantage is in its introduction of different classes of jobs and service disciplines other than FIFO. These considerations lead to a wider field of applications.

(5) The main difficulty in using the result of the BCMP network is the computation of the normalization constant. For computational techniques see, for example, Reiser (1977, 1982), and Sauer and Chandy (1981).

(6) When only FIFO service discipline is considered, then the case of only Type 1 node in BCMP network will arise. Here service time is exponential and all classes have the same service-time distribution in a node. Thus, the case can be covered by the Jackson network as well.

(7) Kelly (1975) gives a generalization of the BCMP theorem by allowing jobs to follow arbitrary paths in the network and not Bernoulli branches.

5.7. Concluding Remarks

Networks have been found to be very useful in the formulation and modeling of computer, communication, and other such systems. Several studies have been directed toward networks. Williams and Bhandiwad (1976) consider a model for multiprogrammed computers. For a general review and survey of the Jackson network, refer to Lemoine (1977, 1978), who points out the limitations of classical work and lists some open questions. For an exhaustive survey of random processes in queueing networks, see Disney and König (1985) (which contains a list of 314 references). For a review of closed network and cyclic queues, see Koenigsberg (1982). For a review of queueing-network models in computer-system design, see Kobayashi (1978), Gelenbe and Mitrani (1980), Gelenbe and Pujolle (1987), Sauer and Chandy (1981), and Geist and Trivedi (1982). For a survey of performance evaluation of data-communication systems, see Reiser (1982) (containing a list of 142 references) and Kobayashi (1978). See also Walrand (1988).

Closed Markovian queueing networks have emerged as an important tool for modeling computer systems, communication systems, on-line computer networks, and other real-time computer-based systems. This has been possible because of the discovery of an important class of networks, the so-called product-form networks, which are analytically tractable (Sauer and Chandy 1981, Kelly, 1979). In addition to networks, discrete-time queueing processes provide suitable models for analysis of data traffic in computer and communication systems. (See Kobayashi, 1983, for discrete-time queueing systems.)

Progress in the theory of Markovian networks has resulted in several algorithms for efficient computation of performance measures such as utilization, throughput, average response time, and marginal distributions. The convolution method has been treated by Buzen (1973) and Reiser and Kobayashi (1975); Kobayashi (1978, 1983) proposes Polyatheoretic algorithms.

The design and optimal operation in routing queueing networks have also been of considerable importance; such questions arise in management of computer systems and in other areas. The problem relates to finding appropriate parameter values for optimization of a specified objective function. (See, for example, Lazar, 1982, 1983); and Schwartz, 1976). Algorithmic results for determination of optimal routing strategies in a network have been given, for example, by Agnew (1976), Gallanger (1977), and Towsley (1980).

Sojourn time in queueing networks has also been engaging considerable attention. Refer to papers, for example, by Boxma and Donk (1982), Boxma *et al.* (1984), Chow (1980), Daduna (1986a, b), Kelly and Pollett (1983), Lemoine (1979, 1987), Melamed (1982), Schassberger and Daduna (1983, 1987), Walrand and Varaiya (1980) and Balsamo and Donatiello (1989).

The literature on queueing networks has been growing at a very rapid pace. For some details and for references, see Gelenbe and Pujolle (1987) and Walrand (1988).

Problems and Complements

5.1. Consider an open network with two nodes having a single exponential server at each of the two nodes with service rates μ_i, $i = 1, 2$. Suppose that arrivals to node 1 occur in accordance with a Poisson process having rate λ. After being served at node 1, the customer (job) goes to node 2 with probability p or leaves the system with probability $(1 - p) = q$. From node 2 it again goes to node 1. Assume that the system is at steady state. Show that the queues at nodes 1 and 2 behave like independent $M/M/1$ queues with intensities

$$\rho_1 = \frac{\lambda}{(q\mu_1)} \quad \text{and}$$

$$\rho_2 = \frac{\lambda p}{(q\mu_2)},$$

and that the system state (n_1, n_2) with n_i at node i, $i = 1, 2$, has the probability

$$p(n_1, n_2) = (1 - \rho_1)\rho_1^{n_1}(1 - \rho_2)\rho_2^{n_2}.$$

Show that the average response time at the two nodes is given by

$$E(R) = \frac{1}{\lambda}\left[\frac{\rho_1}{1 - \rho_1} + \frac{\rho_2}{1 - \rho_2}\right].$$

Let $E(B_i)$ be the average service-time requirements at node i, $i=1, 2$. Then show that

$$E(R) = \sum_{i=1}^{2} \frac{E(B_i)}{1 - \lambda E(B_i)}.$$

5.2. Consider an open network such that arrivals to node 0 occur from outside in accordance with a Poisson process with rate λ. After receiving service at node 0, the job (customer) may leave the system with probability q or may go one of the k nodes, the probability that it goes to node i being p_i, $i = 1, \ldots, k$, $\sum_{i=1}^{k} p_i = 1 - q$. From node i, $i = 1, 2, \ldots, k$, it goes back to node 0. Each of the $(k + 1)$ nodes has an exponential server, the rate at node i being μ_i, $i = 0, 1, 2, \ldots, k$.

Represent the network by a suitable diagram. Write down the routing matrix $\mathbf{P}$ and find $\boldsymbol{\alpha}' = (\alpha_0, \alpha_1, \ldots, \alpha_k)$. Assume that the system is at steady state. Show that the queues at the $(k + 1)$ nodes behave as independent $M/M/1$ queues. Denoting

$$\rho_i = \frac{\alpha_i}{\mu_i}, \qquad i = 0, 1, 2, \ldots, k,$$

show that the system state $(n_0, n_1, \ldots, n_k)$ has the probability

$$p(n_0, n_1, \ldots, n_k) = \prod_{i=0}^{k} (1 - \rho_i)\rho_i^{n_i}.$$

Show that the average total response time is given by

$$\begin{aligned} E(R) &= \frac{1}{\lambda} \sum_{i=0}^{k} \frac{\rho_i}{1 - \rho_i} \\ &= \sum_{i=0}^{k} \frac{E(B_i)}{1 - \lambda E(B_i)} \end{aligned}$$

where $E(B_i)$ is the average service-time requirement at node i (Trivedi, 1982).

5.3. *M/M/1* Queue with Bernoulli feedback

Consider an $M/M/1$ FCFS queue with Bernoulli feedback such that after completion of service, the job may leave the system with probability q or may be fed back into the system with probability p.

(a) Show that the effective average arrival rate to the queue is λ/q.

(b) Show that the number of jobs N in the system has a geometric distribution

$$Pr\{N = n\} = \left(1 - \frac{\lambda}{q\mu}\right)\left(\frac{\lambda}{q\mu}\right)^n, \qquad n = 0, 1, 2, \ldots.$$

(c) Show that the time Y from the last input (to the server) to the next feedback has the distribution

$$R_Y(t) = Pr\{Y \geq t\} = \frac{p\mu}{\mu - \lambda} e^{-(\mu-\lambda)t} + \frac{q\mu - \lambda}{\mu - \lambda},$$

and that the interinput time I has hyperexponential distribution.

(d) Show that the number of jobs N_d left behind a customer departing from the system has the same distribution as N, that is,

$$Pr\{N_d = n\} = Pr\{N = n\}, \qquad n = 0, 1, 2, \ldots.$$

(e) Further, show that the departure process is Poisson with rate λ (Burke, 1976).

5.4. Consider Jackson's open-network model with a single server at each of the k nodes. Let T and S denote respectively the total time spent in the system and the total service time received by a unit. Assume that the system is in equilibrium and that the network is such that a unit can never visit any node more than once. If μ_i is the rate of service for the server (exponential) at node i and $\rho_i = \alpha_i/\mu_i$, $i = 1, 2, \ldots, k$, then show that

$$E(T) = A^{-1} \sum_{i=1}^{k} \frac{\rho_i}{1 - \rho_i}$$

$$E(S) = A^{-1} \sum_{i=1}^{k} (1 + \rho_i)$$

and

$$E(W) = E(T - S) = A^{-1} \sum_{i=1}^{k} \frac{\rho_i^2}{1 - \rho_i^2}$$

where $A = \sum_{i=1}^{k} \lambda_i$, λ_i being the rate of arrival from outside to node i.

Show that the LST of T is the transform of a mixture of exponential distributions and find the same (Lemoine, 1977).

5.5. Consider a closed network with three nodes and K circulating jobs; suppose that the service times at the nodes are independent exponential RVs with rates μ_1, μ_2, and μ_3, respectively. Suppose that after receipt of service at node 1, the job is fed back into node 1 with probability p_1 or goes to nodes 2 (or 3) with probability p_2 (or p_3), with $(p_1 + p_2 + p_3 = 1)$. Draw a diagram and find $\boldsymbol{\alpha} = (\alpha_1, \alpha_2, \alpha_3)$. Show that the probability that the system is at state (n_1, n_2, n_3), $\sum_{i=1}^{3} n_i = K$ is given by

$$p(n_1, n_2, n_3) = \frac{1}{A(K)} \left(\frac{p_2 \mu_1}{\mu_2}\right)^{n_2} \left(\frac{p_3 \mu_1}{\mu_3}\right)^{n_3}$$

where

$$A(K) = \sum_{\substack{n_i \geq 0 \\ \Sigma n_i = K}} \left(\frac{p_2 \mu_1}{\mu_2}\right)^{n_2} \left(\frac{p_3 \mu_1}{\mu_3}\right)^{n_3}.$$

Show that the server utilization at node 1 equals

$$u_1 = \sum_{\substack{n_1 > 0 \\ \Sigma n_i = K}} p(n_1, n_2, n_3) = \frac{A(K-1)}{A(K)}.$$

5.6. $M/G/1/K$ queue as Cyclic Network (Refer to the next chapter for an analysis of $M/G/1$ queues.)

Consider a cyclic network with K circulating jobs and two nodes, with a single server at each of them. The server at node 1 has a general service-time distribution (G) with rate μ and the server at node 2 has an exponential service-time distribution (M) with rate λ. Suppose that the system is in steady state. Let $p(n, K-n) = Pr\{n$ units at node 1 and $K-n$ units at node 2 in the cyclic network$\}$, $p_K(n) = Pr\{n$ units in the system in an $M/G/1/K$ limited–waiting-space queue), and $p_n = Pr\{n$ units in an unrestricted $M/G/1/\infty$ queue$\}$.

(a) Show that $p_K(n) = p(n, K-n)$, $n = 0, 1, \ldots, K$ (that is, the cyclic queue described previously is equivalent to an $M/G/1/K$ queue).

(b) Show further that, for $\rho = \lambda/\mu$,

$$\begin{aligned} p_K(n) &= C(K)p(n), \qquad n = 0, 1, \ldots, K-1 \\ &= 1 - \frac{\{1 - C(K)(1-\rho)\}}{\rho}, \; n = K \end{aligned}$$

where

$$C(K) = \left\{1 - \rho\left[1 - \sum_{n=0}^{K-1} p(n)\right]\right\}^{-1}.$$

Gelenbe and Pujolle (1987).

(c) Show that $u_1 \equiv 1 - p(0, K)$, $u_2 = 1 - p(K, 0)$ are the utilization factors at nodes 1 and 2 and that $u_1/u_2 = \rho$.

Verify the results for the particular case $G \equiv M$.

Note. Carroll *et al.* (1982) consider such a two-node loop system, by taking a Coxian server at one node. The results are extended by considering two Coxian servers at the two nodes by Van de Liefvoort (1986).

5.7. $M/G/1$–PS model

Consider an $M/G/1$ queue where the queue discipline is processor-sharing or time-sharing. This discipline implies that if there are already $(n-1)$ customers in the system, then the arriving customer as well as the other (waiting) customers in the system all start receiving service immediately at the average rate of μ/n. There is no queue as such and the rate at which units receive service changes each time a new arrival joins the system and each time a unit whose service requirement is fully met departs from the system.

We denote the system by $M/G/1$–PS.

(a) Show that the steady-state distribution of the number in the system N has the same geometric distribution as that in a standard $M/M/1$ queue (with FIFO discipline), that is, for $\rho = \lambda/\mu < 1$,

$$Pr\{N = n\} = (1-\rho)\rho^n, \qquad n = 0, 1, 2, \ldots .$$

(b) If B is DF of the service-time distribution S with mean $E(S)$ and W is the response time (waiting time in the system), then show that

$$E\{W|S = t\} = \frac{t}{1-\rho}$$

and that

$$E\{W\} = \int_0^\infty \frac{t}{1-\rho}\, dB(t) = \frac{E(S)}{1-\rho}.$$

(c) Show that the average conditional delay (see Note (1) below) experienced by a customer is given by

$$E[D|S = t] = E[W|S = t] - t$$

$$= \frac{\rho t}{1-\rho}$$

and that

$$E[D] = \frac{\rho E(S)}{1 - \rho}.$$

(d) Show further that the output process is Poisson (Kleinrock, 1967).

Notes.

(1) A customer experiences a delay D as the full service time required by the customer cannot be had in a single installment because the discipline is processor-sharing (though there is no queue as such and an arrival starts receiving service immediately on arrival).

(2) For an $M/G/1$–PS queue, $E[W]$ is independent of the distribution of the service time but depends only on its expected value $E[S]$, whereas in case of a standard $M/G/1$–FIFO service, both the first and second moments enter in the expression for $E[W]$ (as is given by Pollaczek–Khinchine formula).

(3) $M/G/1$–PS can be used as model of some time-shared computer systems.

(4) The distribution of D is not known in the general case. See the next problem for the special case $G \equiv M$.

5.8. $M/M/1$–PS: conditional delay

Let $F(x; t) = Pr\{D \le x | S = t\}$ be the DF of the conditional delay given that total service requirement is of length t, and let $F^*(s; t)$ be its LST. Then, when $\rho < 1$,

$$F^*(s; t) = \frac{(1 - \rho)(1 - \rho r^2) \exp\{-\lambda(1 - r)t\}}{\{(1 - \rho r)\}^2 - \rho(1 - r)^2 \exp\left\{\dfrac{-\mu(1 - \rho r^2)t}{r}\right\}}$$

where r is the (smaller) root of

$$\lambda x^2 - (\lambda + \mu + s)x + \mu = 0.$$

Deduce that

$$E\{D|S = t\} = \frac{\rho t}{1 - \rho} \quad \text{and}$$

$$\operatorname{var}\{D|S = t\} = \left[\frac{2\rho t}{\mu(1 - \rho)^3} - \frac{2\rho}{\mu(1 - \rho)^4}\right] \times [1 - \exp\{-(1 - \rho)\mu t\}].$$

Compare $E\{D|S = t\}$ for $M/M/1$–PS with average queueing time $E\{W_q\}$ for $M/M/1$–FIFO discipline.

Show that for $t < 1/\mu$,

$$E\{D|S = t\} < E\{W_q\},$$

that is, for arrivals whose service-time requirement t is less than the average service time $1/\mu$, the mean delay under the PS discipline is less than the mean delay (queueing time) under the FIFO discipline.

Show that, for $\rho < 1$,

$$E\{D|S = t, N = n\} = \frac{\rho t}{1-\rho} + \frac{[n(1-\rho) - \rho][1 - \exp\{-(1-\rho)\rho t\}]}{\mu(1-\rho)^2}$$

where N is the number in the system in an $M/M/1$–PS system.

Verify that

$$\sum_{n=0}^{\infty} E\{D|S = t, N = n\}Pr\{N = n\} = E\{D|S = t\}.$$

Show further that

$$\lim_{\rho \to 1} E\{D|S = t, N = n\} = nt + \frac{\mu t^2}{2}$$

(Coffman *et al.*, 1970).

5.9. Sojourn-time distribution in cyclic exponential network.

Consider a cyclic network with K circulating jobs among two nodes: 1 and 2. The service time at node i is exponential with parameter μ_i, $i = 1, 2$, and the service processes at the two nodes are independent. Let W_i be the sojourn time or response time (queueing plus service time) at node i, $i = 1, 2$. Show that the joint distribution of the consecutive sojourn times W_1, W_2 has the LST

$$E[\exp(-s_1 W_1 - s_2 W_2)] = \sum_{n=0}^{K-1} a(n)\left(\frac{1}{1 + \frac{s_1}{\mu_1}}\right)^{n+1}\left(\frac{1}{1 + \frac{s_2}{\mu_2}}\right)^{K-n}, \qquad \text{Re } s_1, s_2 \geq 0$$

where

$$a(n) = \frac{1 - \left(\frac{\mu_2}{\mu_1}\right)}{1 - \left(\frac{\mu_2}{\mu_1}\right)^K} \left(\frac{\mu_2}{\mu_1}\right)^n, \qquad \mu_1 \neq \mu_2$$

$$= \frac{1}{K}, \qquad \mu_1 = \mu_2.$$

Show that the correlation between W_1 and W_2 is nonpositive (Boxma and Donk, 1982).

Notes.

(1) Boxma *et al.* (1984) extend (by induction) the result to $M(\geq 1)$ node exponential cyclic network.

(2) The distribution has a product form.

5.10. Cycle-time distribution in cyclic exponential network

Consider an M-node cyclic network with K circulating jobs, the nodes having independent exponential servers with parameters μ_i, $i = 1, 2, \ldots, M$. Let W_i be the sojourn time at node i and let

$$T = W_1 + \cdots + W_M$$

be the cycle time (sum of consecutive sojourn times). Show that the LST of T is given by

$$E[\exp(-sT)] = E\{\exp[-s(W_1 + \cdots + W_M)]\}$$

$$= \sum_{n_i \in C} a(n_1, \ldots, n_M) \prod_{i=1}^{M} \left(\frac{1}{1 + s/\mu_i}\right)^{n_i + 1} \qquad \operatorname{Re} s \geq 0,$$

where

$$C \equiv C(M, K-1) = \left\{(n_1, \ldots, n_M)\colon n_i \geq 0, \sum_{i=1}^{M} n_i = K - 1\right\} \quad \text{and}$$

$$a(n_1, \ldots, n_M) = \prod_{i=1}^{M} \frac{\left(\frac{1}{\mu_i}\right)^{n_i}}{\sum_{n_i \in c} \prod_{i=1}^{M} \left\{\frac{1}{\mu_i}\right\}^{n_i}}.$$

Find $E(T)$ for $M = 2$ (Schassberger and Daduna, 1983). Note that the distribution has a product form.

References

Agnew, C. E. (1976). On quadratic adaptive routing algorithms. *Com. Ass. Comp. Mach*, **19**, 18–22.

Albin, S. L. (1982). On Poisson approximation for superposition arrival processes in queues. *Mgmt. Sci.* **28**, 126–137.

Albin, S. L. (1984). Approximating a point process by a renewal process II: Superposition arrival processes to queues. *Opns. Res.* **32**, 1133–1162.

Balsamo, S., and Donatiello, L. (1989). On the cycle time distribution in a two-stage cyclic network with blocking. *IEEE Trans. Software Eng.* **15**, 1206–1216.

Barbour, A. D. (1976). Networks of queues and the method of stages. *Adv., Appl. Prob.* **8**, 584–591.

Baskett, F., Chandy, K. M., Muntz, R. R., and Palacios, F. G. (1975). Open, closed and mixed networks of queues with different classes of customers. *J. Ass. Comp. Mach.* **22**, 248–260.

Boxma, O. J. (1983). The cyclic queue with one general and one exponential server. *Adv. Appl. Prob.* **15**, 857–873.

Boxma, O. J. (1986). Models of two queues: a few new views in *Teletraffic Analysis and Computer Performance Evaluation*, 75–98 (Boxma, O. J., Cohen, J. W., and Tijms, H. C., Eds.), Elsevier Science Publishers B. V., North Holland, Amsterdam. (The paper contains a list of 94 references.)

Boxma, O. J. and Donk, P. (1982). On response time and cycle time distributions in a two-stage cyclic queue. *Perf. Ev.* **2**, 181–194.

Boxma, O. J., Kelly, F. P., and Konheim, A. G. (1984). The product form for sojourn time distributions in cyclic exponential queues. *J. Ass. Comp. Mach.* **31**, 128–133.

Brandwajn, A., and Jow, Y. L. (1988). An approximation method for tandem queues with blocking. *Opns. Res.* **36**, 73–88.

Burke, P. J. (1956). The output of a queueing system. *Opns. Res.* **4**, 699–704.

Burke, P. J. (1976). Proof of a conjecture on the interarrival time distribution in an $M/M/1$ queue with feedback. *IEEE Trans. On Comm.* **24**, 175–178.

Buzacott, J. A., and Yao, D. D. (1986). On queueing network models of flexible manufacturing systems. *Queueing Systems* **1**, 5–27.

Buzen, J. P. (1973). Computational algorithms for closed queueing networks with exponential servers. *Comm. Ass. Comp. Mach.* **16**, 527–531.

Carroll, J. K., Van de Liefvoort, A., and Lipsky, L. (1982). Solutions of $M/G/1//N$ type loops with extensions to $M/G/1$ and $GI/M/1$ queues. *Opns. Res.* **30**, 490–514.

Chandy, K., Herzog, U. and Woo, L. (1975a). Parametric analysis of queueing networks. *IBM J. Res. Dev.* **19**, 36–42.

Chandy, K. M., Herzog, U., and Woo, L. (1975b). Approximate analysis of general queueing networks. *IBM J. Res. Dev.* **19**, 43–49.

Chandy, K. M., Howard, J. H. and Towsley. D. F. (1977). Product form and local balance in queueing networks. *J. Ass. Comp. Mach.* **24**, 250–263.

Chen, T. M. (1989). On the independence of sojourn times in tandem queues. *Adv. Appl. Prob.* **21**, 488–489.

Chow, W. M. (1980). The cycle time distribution of exponential queues. *J. Ass. Comp. Mach.* **27**, 281–286.

Coffman, E. G., Jr., Muntz, R. R., and Trotter, H. (1970). Waiting time distributions for processor sharing systems. *J. Ass. Comp. Mach*, **17**, 123–130.

Daduna, H. (1986a). Cycle times in two-stage closed queueing networks: applications to multi-programmed computer systems with virtual memory. *Opns. Res.* **34**, 281–288

Daduna, H. (1986b). Two-stage cyclic queues with nonexponential servers: steady state and cyclic time. *Opns. Res.* **34**, 455–459.

Disney, R. L. (1981). Queueing Networks. *American Math. Soc. Proceedings of Symposia in Applied Mathematics* **25**, 53–83.

Disney, R. L. and König, D. (1985). Queueing Networks. A survey of their random processes. *SIAM Review* **27**, 335–403. (Contains a list of 314 references.)

Disney, R. L. and Kiessler, P. C. (1987). *Traffic Processes in Queueing Networks: A Markov Renewal Approach.* The Johns Hopkins University Press, Baltimore, MD.

Dukhovny, I. M., and Koenigsberg, E. (1981). Invariance properties of queueing networks and their application to computer communication systems. *INFOR* **19**, 185–204.

Gallanger, R. R. (1977). A minimum delay routing algorithm using distributed component. *IEEE Trans. on Comm.* **25**, 73–84.

Geist, R., and Trivedi, K. (1982). Queueing network models in computer system design. *Mathematics Magazine* **55**, 2, 67–80.

Gelenbe, E. (1975). On approximate computer system models *J. Ass. Comp. Mach.* **22**, 261–269.

Gelenbe, E., and Mitrani, I. (1980). *Analysis and Synthesis of Computer Systems*, Academic Press, New York.

Gelenbe, E., and Pujolle, G. (1987). *Introduction to Queueing Networks*, John Wiley, New York.

Gnedenko, B. V., and Kovalenko, I. N. (1968). *Introduction to Queueing Theory*, (2nd ed. 1980). Israel program for Sci. Tran, Jerusalem.

Goodman, J. B., and Massey, W. A. (1984). The non-ergodic Jackson network. *J. Appl. Prob.* **21**, 860–869.

Gordon, W. J., and Newell, G. P. (1967a). Closed queueing systems with exponential servers. *Opns. Res.* **15**, 254–265.

Gordon, W. J., and Newell, G. P. (1967b). Cyclic queueing systems with restricted length queues. *Opns. Res.* **15**, 266–277.

Gross, D., and Harris, C. M. (1985). *Fundamentals of Queueing Theory*, 2nd ed., Wiley, New York.

Heidelberger, P., and Lavenberg, S. (1984). Computer performance evaluation methodology. *IEEE Trans. on Computers* **C-33**, 1195–1220.

Jackson, J. R. (1957). Networks of waiting lines. *Opns. Res.* **5**, 518–522.

Jackson, J. R. (1963). Jobshop like queueing systems. *Mgmt. Sci.* **10**, 131–142.

Jackson, R. R. P. (1954). Queueing systems with phase type service. *Opl. Res. Q.* **5**, 109–120.

Jansen, V., and König, D. (1980). Insensitivity and steady state probabilities in product form for queueing networks. *J. Inf. Proc. & Cyber (EIK)* **16**, 385–397.

Kelly, F. P. (1975). Networks of queues with customers of different types. *J. Appl. Prob.* **12**, 542–554.

Kelly, F. P. (1976). Networks of queues. *Adv. Appl. Prob.* **8**, 416–432.

Kelly, F. P. (1979). *Reversibility and Stochastic Networks.* Wiley, New York.

Kelly, F. P. (1985). Stochastic models of computer communication systems. *J.R.S.S.* **B47**, 379–395.

Kelly, F. P. (1989). On a class of approximation for closed queueing networks. *Queueing Systems* **4**, 69–76.

Kelly, F. P. and Pollett, P. K. (1983). Sojourn times in closed queueing networks. *Adv. Appl. Prob.* **15**, 638–656.

Kiessler, P. C. (1989). A simple proof of the equivalence of input and output intervals in Jackson networks. *Opns. Res.* **37**, 645–647.

Kleinrock, L. (1967). Time shared systems—a theoretical treatment. *J. Ass. Comp. Mach.* **14**, 242–261.

Kobayashi, H. (1978). *Modeling and Analysis: An Introduction to System Performance Evaluation Methodology*, Addison-Wesley, Reading, Mass.

Kobayashi, H. (1983). Stochastic Modeling: Queueing Models in: *Probability Theory and Computer Science* (Louchard, G. and Latouche, G. eds.), Academic Press, London.

Kobayashi, H., and Reiser, M. (1975). On generalization of job routing behavior in a queuing network model. *IBM Research Report*, **RC** 5252.

Kobayashi, H., and Konheim, A. G. (1977). Queueing models for computer communication system analysis. *IEEE Trans. Comm.* **COM–25**, 2–29.

Koenigsberg, E. (1985). Cyclic queues. *Opns. Res. Qrly*, **9**, 22–35.

Koenigsberg, E. (1982). Twenty-five years of cycle queues and closed queue networks: a review. *J. Opl. Res. Soc.* **33**, 605–619.

Koenigsberg, E. (1983). Dominance, invariance, conservation, decomposition, and equivalence applied to queue networks. Preprint.

Lavenberg, S. S. (Ed.) (1983). *Computer Performance Modeling Handbook*. Academic Press, New York.

Lazar, A. A. (1982). The throughput time delay function of an $M/M/1$ queue. *IEEE Trans. Inf. Th.* **29**, 914–918.

Lazar, A. A. (1983). Optimal flow control of a class of queueing networks in equilibrium. *IEEE Trans. Aut. Cont.*, **AC-28**, 1001–1007.

Lemoine, A. J. (1977). Network of queues—a survey of equilibrium analysis. *Mgmt. Sci.*, **24**, 464–481.

Lemoine, A. J. (1978). Networks of queues—a survey of weak convergence results. *Mgmt. Sci.* **24**, 1175–1193.

Lemoine, A. J. (1979). On total sojourn time in networks of queues. *Mgmt. Sci.* **25**, 1034–1035.

Lemoine, A. J. (1987). On sojourn time in Jackson networks of queues. *J. Appl. Prob.* **24**, 495–510.

Massey, W. (1984a). Open networks of queues: their algebraic structure and estimating their transient behavior. *Adv. Appl. Prob.* **16**, 176–201.

Massey, W. (1984b). An operator-analytic approach to Jackson network. *J. Appl. Prob.* **21**, 379–393.

McKenna, J. (1989). A generalization of Little's law to moments of queue length and waiting times in a closed product form queueing network. *J. Appl. Prob.* **26**, 121–133.

Melamed, B. (1979). Characterization of Poisson traffic streams in Jackson queueing networks. *Adv. Appl. Prob.* **11**, 422–438.

Melamed, B. (1982). Sojourn times in queueing networks. *Math. Opns. Res.* **7**, 233–244.

Newell, G. F. (1984). Approximations for superposition arrival processes in queues. *Mgmt. Sci.* **30**, 623–630.

Perros, H. G. (1984). Queuing networks with blocking: a bibliography. *Acta Sigmetrics* **10**, 139–148.

Perros, H. G., and Altiok, T. (eds.) (1989). *Queueing Networks with Blocking.* North Holland, New York.

Posner, M., and Bernholtz, B. (1968a). Closed finite queueing networks with time lags. *Opns. Res.* **16**, 962–976.

Posner, M., and Bernholtz, M. (1968b). Closed finite queueing networks with time lags and several classes of units. *Opns. Res.* **16**, 977–985.

Reiser, M. (1977). Numerical methods in separable queueing networks, in: *Algorithmic Methods in Probability* (Neuts, M. F., ed) *TIMS Studies in Management Sciences.* Vol. 7, 113–142. North-Holland, Amsterdam.

Reiser, M. (1982). Performance evaluation of data communication systems. *Proc. of the IEEE* **70**, 171–196. (Contains a list of 142 references.)

Reiser, M., and Kobayashi, H. (1975). Queueing networks with multiple closed chains: theory and computational algorithms. *IBM J. Res. Develop.* **19**, 282–294.

Reiser, M., and Lavenberg, S. S. (1980). Mean value analysis of closed multichain queueing networks. *J. Ass. Comp. Mach.* **27**, 313–322.

Sauer, C. H., and Chandy, K. (1981). *Computer System and Performance Modeling.* Prentice-Hall, Englewood Cliffs, NJ.

Schassberger, R., and Daduna, H. (1983). The time for a round trip in a cycle of exponential queues. *J. Ass. Comp. Mach.* **30**, 146–150.

Schassberger, R., and Daduna, H. (1987). Sojourn times in queueing networks with multiserver nodes. *J. Appl. Prob.* **24**, 511–521.

Schwartz, M. (1977). *Computer Communication Network Designs and Analysis.* Prentice-Hall, Englewood-Cliffs, NJ.

Schwartz, M. (1987). *Telecommunication Networks: Protocols, Modeling and Analysis.* Addison-Wesley, Reading, Mass.

Shalmon, M., and Kaplan, M. (1984). A tandem network of queues with deterministic service and intermediate arrivals. *Opns. Res.* **32**, 753–773.

Shanthikumar, J. G., and Yao, D. D. (1988a). Stochastic monotonicity of the queue length in closed queueing networks. *Opns. Res.* **35**, 583–588.

Shanthikumar, J. G., and Yao, D. D. (1988b). Second order properties of the throughput of a closed queueing network. *Maths. Opns. Res.* **13**, 524–534.

Shanthikumar, J. G., and Yao, D. D. (1989). Stochastic monotonicity in general queueing networks. *J. Appl. Prob.* **26**, 413–417.

Sigman, K. (1989). Notes on stability of closed queueing networks. *J. Appl. Prob.* **26**, 678–682.

Towsley, D. (1980). Queueing network models with state-dependent routing. *J. Ass. Comp. Mach.* **27**, 323–337.

Trivedi, K. S. (1982). *Probability and Statistics with Reliability, Queueing and Computer Science Applications.* Prentice-Hall, Englewood Cliffs, NJ.

Van de Liefvoort, A. (1986). A matrix-algebraic solution to two K_m servers in a loop. *J. Ass. Comp. Mach.* **33**, 207–223.

Walrand, J. (1982a). On the equivalence of flows in a network of queues. *J. Appl. Prob.* **19**, 195–203.

Walrand, J. (1982b). Poisson flows in single class open networks of quasi reversible queues, *Stoch. Proc. & Appl.* **13**, 293–303.

Walrand, J. (1988). *Introduction to Queueing Networks.* Prentice-Hall, Englewood-Cliffs, NJ.

Walrand, J. and Varaiya, P. (1980). Sojourn times and the overtaking condition in Jacksonian networks, *Adv. Appl. Prob.* **12**, 1000–1018.

Walrand, J., and Varaiya, P. (1981). Flows in queueing networks: a martingale approach. *Math. of Opns. Res.* **6**, 387–404.

Weber, R. R. (1979). The interchangeability of tandem queues. *J. Appl. Prob.* **16**, 690–695.

Whitt, W. (1983). The queueing network analyzer. *Bell. Sys. Tech. J.* **62**, 2779–2815.

Whitt, W. (1984). Open and closed models of network of queues. *Bell. Sys. Tech. J.* **63**, 1911–1979.

Williams, A., and Bhandiwad, R. (1976). A generating function approach to queueing network analysis of multiprogrammed computers. *Network* **6**, 1–22.

Yao, D. D., and Kim, S. C. (1987). Some order relations in closed network of queues with multiserver stations. *Nav. Res. Log. Qrly*, **34**, 53–66.

6 Non-Markovian Queueing Systems

6.1. Introduction

We have so far been discussing queueing processes that are either birth–death or non–birth–death processes, in either case the processes being Markovian. The theory of Markov processes could be applied in studying them. The models in which both the interarrival-time and the service-time distributions are exponential are birth–death Markovian; models in which the distributions of either or both are Erlangian are non–birth–death but nevertheless can be treated as Markovian. We shall now consider where the distributions of the interarrival and service times are not necessarily exponential or Erlangian.

We shall first examine a model of the type $M/G/1$ where G denotes that the distribution is general. The process $\{N(t), t \geq 0\}$ where $N(t)$ gives the state of the system or the system size at time t is then non-Markovian. However, the analysis of such a process could be based on a Markovian process that can be extracted out of it. There are a number of techniques or approaches that are used for this purpose.

(1) *Imbedded–Markov-chain technique.* Kendall (1951) uses the concept of regeneration point (due to Palm, 1943) by suitable choice of regeneration points which involves extraction from the process $\{N(t), t \geq 0\}$ Markov chains in discrete time at those points. The technique is known as the imbedded–Markov-chain technique.

(2) *The supplementary-variable technique.* This involves inclusion of such variables, for example, the service time $X(t)$ already received by the customer in service, if any. The method is based on the treatment of the couplet $\{N(t), X(t)\}$. Cox (1955), Kendall (1953), Keilson and Kooharian (1960) have indicated this technique. Henderson (1972) considers the couplet $\{N(T), Y(t)\}$, where $Y(t)$ is the remaining or residual service time of the customer in service, if any. These methods involve inclusion of a supplementary variable $X(t)$ or $Y(t)$.

(3) *Lindley's integral-equation method.* This method, which is suitable for a $G/G/1$ system, takes the customer-arrival times as regeneration points and considers the waiting time of the nth customer (which is a Markov process) as the object of study.

(4) Other methods. Besides the preceding there are other methods of dealing with non-Markovian systems such as the random-walk and combinatorial approaches (Takács, 1967) and the method of Green's function (Keilson, 1965).

Next we shall discuss the imbedded–Markov-chain technique.

6.2. Imbedded–Markov-Chain Technique for the System with Poisson Input

We are concerned at any instant t with a pair of RVs $N(t)$, the number in the system at time t, and $X(t)$, the service time already received by the customer in service, if any; while $\{N(t), t \geq 0\}$ is non-Markovian, the vector $\{N(t), X(t), t \geq 0\}$ is a Markov process. Whereas in the case of an $M/M/1$ system (because of the memoryless property of service-time distribution), attention can be confined to $N(t)$ alone, for the system $M/G/1$ we have to consider $X(t)$ also along with $N(t)$. Now by observing the number in the system at a select set of points rather than at all points of time t, it is possible to simplify matters to a great extent. These special sets of points or instants should be such that by considering the number in the system at any such point and other inputs, it should be possible to calculate the number in the system at the next such point or instant. There are several such sets of points. A very suitable set of points is the set of *departure instants* (from service channel) at which successive customers leave the system on completion of service. Let the departure instants of the customers $C_1, C_2, \ldots, C_n, \ldots$ be $t_1, t_2, \ldots, t_n \ldots$, respectively. At such a point of time, say, the departure instant t_n of C_n, the time spent in service by the next customer C_{n+1} is zero and, thus, given $N(t_n)$ at any departure instant (that is, the number of customers left behind by the departing customer C_n) and given the additional input to the system (arrivals during the time of service of the next customer C_{n+1}) it is possible to calculate $N(t_{n+1})$, the number left behind by the next departing customer C_{n+1}. Thus, we get $N(t_{n+1})$ given $N(t_n)$, and the number of arrivals during the service time of customer C_{n+1}; and so $\{N(t_n), n \geq 1\}$ defines a Markov chain, the instants $t_1, t_2, \ldots, t_n$ being imbedded Markovian points. Thus, we can get $N(t_n)$ and its distribution, that is, the distribution of the number in the system at departure epochs t_n, $n \geq 1$.

For a queueing system (in steady state) with Poisson arrivals we have the following properties:

(1) The probability a_n of the number n found by an arriving customer is equal to the probability d_n of the number n left behind by a departing customer. Again, Poisson arrivals see time averages. When equilibrium is reached in a queueing system with Poisson arrivals, we have $a_n = p_n$, where p_n is the probability that the number in the system at any time (in steady state) is n. Thus,

$$a_n = d_n = p_n.$$

Thus, the probability distribution of the number in the system at the imbedded Markov points is the same as the probability distribution of the number in the system at all points of time. Thus, it suffices to consider the process $\{N(t_n), n \geq 0\}$ at the departure instants or, to be more specific, the process $\{N(t_n + 0), n \geq 0\}$ where $N(t_n + 0)$ is the number immediately following the nth departure. (See also Remark, Section 6.3.3.)

(2) The transitions of the process occur at departure instants t_n. The numbers in the system immediately following these instants form a Markov chain such that the transitions occur at the departure instants. The interval between two transitions (that is, between two departures) is equal to the service time when the departure leaves at least one in the system and is equal to the convolution of the interarrival time (which is exponential) and the service time when the departure leaves the system empty. If $Y(t)$ denotes the

number of customers left behind by the most recent departing customer, that is, $Y(t) = N(t_n)$, $t_n \le t \le t_{n+1}$, then $Y(t)$ will be a semi-Markov process having $\{N(t_n + 0), n = 0, 1, 2, \ldots\}$ for its imbedded Markov chain. The sequence of intervals $(t_{n+1} - t_n)$, $n = 0, 1, 2, \ldots$, being the interdeparture time of successive units forms a renewal process.

Assume that the input process is Poisson with rate λ, and the service times are IID RVs having a general distribution with DF $B(t)$ and mean $(1/\mu)$. Let $B^*(s) = \int_0^\infty e^{-st}\, dB(t)$ be its LST, then $-B^{*(1)}(0) = 1/\mu$.

Let A be the number of arrivals during the service time of a unit. Conditioning on the duration of the service time of a unit, we get

$$k_r = Pr\{A = r\} = \int_0^\infty \frac{e^{-\lambda t}(\lambda t)^r}{r!}\, dB(t), \qquad r = 0, 1, 2, \ldots. \tag{2.1}$$

The PGF $K(s)$ of $\{k_r\}$ is given by

$$\begin{aligned} K(s) &= \sum_{r=0}^{\infty} k_r s^r = \sum_{r=0}^{\infty} s^r \left\{ \int_0^\infty \frac{e^{-\lambda t}(\lambda t)^r}{r!}\, dB(t) \right\} \\ &= \int_0^\infty e^{-\lambda t}\, dB(t) \left\{ \sum_{r=0}^{\infty} \frac{(\lambda t s)^r}{r!} \right\} = \int_0^\infty e^{-(\lambda - \lambda s)t}\, dB(t). \\ &= B^*(\lambda - \lambda s). \end{aligned} \tag{2.2}$$

We have

$$E\{A\} = K'(1) = -\lambda B^{*(1)}(0) = \frac{\lambda}{\mu} = \rho.$$

Suppose that the arrivals occur in bulk with distribution $a_j = Pr(X = j)$, the arrival instants being its accordance with a Poisson process with rate λ. Then the probability distribution of the total number of arrivals A in an interval of time t is given by

$$Pr(X = j) = \sum_{k=0}^{j} \frac{e^{-\lambda t}(\lambda t)^k}{k!}\, a_j^{(k)*}, \qquad j = 0, 1, 2, 3, \ldots \tag{2.3}$$

where $a_j^{(k)*}$ is the k-fold convolution of a_j with itself. The distribution of the total number of arrivals A during the service period of a unit is given by

$$Pr\{A = j\} = \int_0^\infty \sum_{k=0}^{j} \frac{e^{-\lambda t}(\lambda t)^k}{k!}\, a_j^{(k)*}\, dB(t). \tag{2.4}$$

Let $A(s) = \sum_j a_j s^j$ be the PGF of the bulk size X; then $\sum_j a_j^{(k)*} s^j = [A(s)]^k$.

The PGF of A is then given by

$$\begin{aligned}
K(s) &= \sum_{j=0}^{\infty} Pr(A = j)s^j = \sum_{j=0}^{\infty} s^j \left\{ \sum_{k=0}^{j} \int_0^{\infty} \frac{e^{-\lambda t}(\lambda t)^k}{k!} a_j^{(k)*} \, dB(t) \right\} \\
&= \int_0^{\infty} \sum_{k=0}^{\infty} \frac{e^{-\lambda t}(\lambda t)^n}{n!} \left\{ \sum_j s^j a_j^{(k)*} \, dB(t) \right\} \\
&= \int_0^{\infty} \sum_{k=0}^{\infty} \frac{e^{-\lambda t}(\lambda t)^k}{k!} [A(s)]^k \, dB(t) \\
&= \int_0^{\infty} e^{-\lambda t} e^{\lambda t A(s)} \, dB(t) \\
&= \int_0^{\infty} e^{-[\lambda t - \lambda A(s) t]} \, dB(t).
\end{aligned}$$

Thus, for a bulk arrival system

$$K(s) = B^*[\lambda - \lambda A(s)], \quad \text{and} \tag{2.5a}$$

$$\rho = \frac{\lambda E(X)}{\mu}. \tag{2.5b}$$

6.3. The *M/G/1* Model: Pollaczek–Khinchin Formula

6.3.1. *Steady-State Distribution of Departure Epoch System Size*

Let t_n, $n = 1, 2, \ldots$, $t_0 = 0$ be the instant at which the nth unit, C_n leaves the system immediately on completion of his service. Then $\{t_n, n \geq 0\}$ is a renewal process. Denote

$$\begin{aligned}
N(t) &= \text{number of units at time } t, \\
X_n &= \text{number of units left behind by the } n\text{th departing unit} \\
&= N(t_n + 0), \qquad n = 0, 1, 2, \ldots.
\end{aligned}$$

and

$$A_n = \text{number of units that arrive during the service time of } C_n.$$

We have

$$\begin{aligned}
X_{n+1} &= X_n - 1 + A_{n+1}, \; X_n \geq 1 \\
&= A_{n+1}, \qquad X_n = 0.
\end{aligned} \tag{3.1}$$

Since A_n is the same for all n, we may write $A_n = A$. It can be seen that $\{X_n, n \geq 0\}$ is a Markov chain—it is the imbedded Markov chain of the process $\{N(t), t \geq 0\}$, and is the Markov chain extracted from the process at the regeneration points t_n. Denote

$$k_i = Pr\{A = i\}, \qquad i = 0, 1, 2, \ldots .$$

Let us find the transition probabilities of the denumerable Markov chain $\{X_n, n \geq 0\}$. Define

$$p_{ij} = Pr\{X_{n+1} = j | X_n = i\}. \tag{3.2}$$

Then p_{ij} are given by

$$\begin{aligned} p_{ij} &= k_{j-i+1}, \qquad i \geq 1, \qquad j \geq i-1 \\ &= 0, \qquad i \geq 1, \qquad j < i-1 \end{aligned}$$

and

$$p_{0j} = p_{ij} = k_j, \qquad j \geq 0. \tag{3.3}$$

Thus, the TPM of the chain can be put as:

$$\mathbf{P} = (p_{ij}) = \begin{bmatrix} k_0 & k_1 & k_2 & \cdots \\ k_0 & k_1 & k_2 & \cdots \\ 0 & k_0 & k_1 & \cdots \\ \cdots & \cdots & \cdots & \cdots \\ \cdots & \cdots & \cdots & \cdots \end{bmatrix}. \tag{3.4}$$

This chain has been discussed in Example 1.3., Chapter 1. It can be easily seen that the chain is irreducible and aperiodic. It can also be shown that when $\rho < 1$, the chain is persistent nonnull and hence ergodic. We can then apply the ergodic theorem of Markov chains. The limiting probabilities (that a departing customer leaves behind j customers in the system)

$$v_j = \lim_{n \to \infty} p_{ij}^{(n)}, \quad j = 0, 1, 2, \ldots$$

exist and are independent of the initial state i. Then the PGF

$$V(s) = \sum_j v_j s^j, \quad \text{of } \{v_j\} \tag{3.5}$$

is given by

$$V(s) = \frac{(1 - K'(1))(1 - s)K(s)}{K(s) - s} \tag{3.6}$$

(see Example 1.3., Chapter 1).

Putting $K'(1) = \rho$, we get, for $0 < \rho < 1$,

$$\begin{aligned} V(s) &= \frac{(1-\rho)(1-s)K(s)}{K(s)-s} \\ &= \frac{(1-\rho)(1-s)B^*(\lambda-\lambda s)}{B^*(\lambda-\lambda s)-s}. \end{aligned} \tag{3.7}$$

This is known as the *Pollaczek–Khinchine (P–K) formula.*

Remark. What we have obtained is the distribution of the *departure epoch* system size. We shall consider *general time* system size later.

Example 6.1. The $M/D/1$ Model.
Consider that the service time is Erlang-k (with mean $1/\mu$) having PDF

$$b(t) = \frac{(\mu k)^k t^{k-1} e^{-k\mu t}}{\Gamma(k)}, \qquad 0 < t < \infty$$

and LST

$$B^*(s) = \left(\frac{\mu k}{s+\mu k}\right)^k.$$

We get

$$V(s) = \frac{(1-\rho)(1-s)}{1 - s\left\{1 + \dfrac{\rho(1-s)}{k}\right\}^k}.$$

(See also Section 4.1.1.) The Model $M/D/1$ with constant service time can be considered a limiting case of the above. As $k \to \infty$, the whole mass of the distribution E_k is concentrated at the mean $(1/\mu)$, so that E_k can be considered deterministic (constant $= 1/\mu$). Now, as $k \to \infty$, $B^*(s) \to e^{-s/\mu}$ and

$$V(s) \to \frac{(1-\rho)(1-s)}{1 - se^{\rho(1-s)}}.$$

It can be seen that

$$\begin{aligned} v_0 &= 1 - \rho \\ v_1 &= (1-\rho)(e^\rho - 1) \\ v_2 &= (1-\rho)(e^{2\rho} - 2(\rho+1)\rho) \end{aligned}$$

and so on.

6.3.2. Waiting Time Distribution

Assume that the steady state exists ($\rho < 1$). Let W and W_q denote the waiting times in the system and in the queue, respectively, and let $W(t)$ and $W_q(t)$ denote their distribution functions. We have, for a Poisson input process,

$$p_n = v_n = Pr\{\text{departing unit leaves } n \text{ in the system}\}.$$

Conditioning on the waiting time of the unit, we get

$$p_n = \int_0^\infty Pr\{\text{departing unit leaves } n \text{ in the system} | t \le W < t + dt\}$$
$$\times Pr\{t \le W < t + dt\}.$$

Again, the event that the departing unit leaves n when his waiting time is t is the event than n arrivals occur during t. Thus,

$$p_n = \int_0^\infty \frac{e^{-\lambda t}(\lambda t)^n}{n!} dW(t), \qquad n=0, 1, 2, \ldots. \tag{3.8}$$

We have

$$\begin{aligned} V(s) = \sum_{n=0}^\infty v_n s^n &= \int_0^\infty \{e^{-\lambda t} dW(t)\} \left\{ \sum_{n=0}^\infty \frac{(\lambda t)^n}{n!} s^n \right\} \\ &= \int_0^\infty e^{-\lambda t(1-s)} dW(t) \\ &= W^*(\lambda - \lambda s) \end{aligned} \tag{3.9}$$

where

$$W^*(s) = \int_0^\infty e^{-st} dW(t)$$

is the LST of W. Since $p_n = v_n$, we have

$$P(s) = W^*(\lambda - \lambda s).$$

From (3.7) and (3.9), we get

$$W^*(\lambda - \lambda s) = \frac{(1-\rho)(1-s)B^*(\lambda - \lambda s)}{B^*(\lambda - \lambda s) - s}.$$

Writing α for $\lambda - \lambda s$, we get

$$W^*(\alpha) = \frac{\alpha(1-\rho)B^*(\alpha)}{\alpha - \lambda[1 - B^*(\alpha)]}$$

so that

$$W^*(s) = \frac{s(1-\rho)B^*(s)}{s - \lambda[1 - B^*(s)]}. \tag{3.10}$$

This relation connects the LSTs of the distribution of the service time v and the waiting time W in the system. To find the expression for LST $W_q^*(s)$, we note that $W = W_q + v$ (with v being independent of W_q) and

$$W_q^*(s) = \frac{s(1-\rho)}{s - \lambda[1 - B^*(s)]}. \tag{3.11a}$$

The formulas (3.10) and (3.11) are known as *Pollaczek–Khinchin transform formulas.*

6.3.2.1. Waiting Time in the Queue: Alternative Approach. From renewal theory, we find that the residual service time X of the unit receiving service at the time the test unit arrives, has, for large t, the DF

$$G(x) = Pr\{X \le x\} = \frac{1}{E(v)} \int_0^x \{1 - B(y)\} dy.$$

Its LST equals

$$G^*(s) = \frac{1 - B^*(s)}{sE(v)} = \frac{\mu\{1 - B^*(s)\}}{s}.$$

Writing (3.11) as

$$W_q^*(s) = \frac{1-\rho}{1 - \rho\left[\frac{\mu}{s}(1 - B^*(s))\right]}$$

$$= \frac{1-\rho}{1 - \rho G^*(s)}, \tag{3.11b}$$

we find that the waiting time, when the service time is general, involves residual service time X, as is to be expected.

Expanding (3.11b), we get

$$\begin{aligned} W_q^*(s) &= (1-\rho)\left\{1 + \sum_{k=1}^{\infty} \rho^k [G^*(s)]^k\right\} \\ &= (1-\rho) + (1-\rho)\sum_{k=1}^{\infty} \rho^k [G^*(s)]^k. \end{aligned} \tag{3.11c}$$

Inversion of (3.11b) gives the DF $W_q(x)$ of W_q.

If g denotes the PDF of the residual service times X, then inversion of $[G^*(s)]^k$ gives $g^{(k)*}(x)$, the k-fold convolution of g with itself. Writing $\delta(x)$ as the Dirac delta function and inverting (3.11c) we can get the PDF $w_q(x)[=dW_q(x)/dx]$ as follows:

$$w_q(x) = (1-\rho)\delta(x) + (1-\rho)\sum_{k=1}^{\infty} \rho^k g^{(k)*}(x), \qquad x \geq 0. \tag{3.11d}$$

As observed by Kleinrock (1975 p. 201), no satisfactory intuitive explanation has been found for the preceding form of waiting-time PDF in terms of weighted sum of convolved residual service time PDFs.

Particular case. For $M/M/1$, X is again exponential with mean $(1/\mu)$ and

$$g^{k*}(x) = \frac{(\mu^k x^{k-1} e^{-\mu x})}{\Gamma(x)}.$$

Thus, we get

$$\begin{aligned} w_q(x) &= (1-\rho)\delta(x) + (1-\rho)\sum_{k=1}^{\infty} \frac{[(\mu\rho)^k x^{k-1} e^{-\mu x}]}{\Gamma(k)} \\ &= (1-\rho)\delta(x) + \mu\rho(1-\rho)\exp\{-\mu(1-\rho)x\}. \end{aligned} \tag{3.11e}$$

Note. Shanthikumar (1988) has shown that waiting-time distribution has DFR (decreasing failure rate) for service time having IMRL (increasing mean residual life). He also shows that waiting time distribution in a $G/G/1$ queue with DFR service time is DFR.

6.3.2.2. *Expected Waiting Time or Expected Number in the System.*

The moments of $E\{N\}$ and $E\{W_q\}$ and $E\{W\}$ can be easily obtained as follows: We have

$$B^{*(k)}(0) = \frac{d^k}{ds^k} B^*(s)\bigg|_{s=0} = (-1)^k E(v^k), \qquad k = 1, 2, \ldots.$$

($B^{*(k)}(s)$ denotes the kth derivative of $B^*(s)$). From (3.11), we get

$$\frac{d}{ds}\frac{W_q^*(s)}{1-\rho} = \frac{s-\lambda+\lambda B^*(s) - s(1+\lambda B^{*(1)}(s))}{[(s-\lambda+\lambda B^*(s)]^2},$$

which takes the form 0/0 as $s \to 0$. Using L'Hôpital's rule we get

$$\lim_{s\to 0} \frac{[1 + \lambda B^{*(1)}(s)] - (1 + \lambda B^{*(1)}(s)) - s(\lambda B^{*(2)}(s))}{2[s - \lambda + \lambda B^*(s)][1 + \lambda B^{*(1)}(s)]}$$

$$= \frac{-\lambda B^{*(2)}(0)}{2[1 + \lambda B^{*(1)}(0)]^2} = \frac{-\lambda E(v^2)}{2\left(1 - \dfrac{\lambda}{\mu}\right)^2}.$$

Thus,

$$E(W_q) = -W_q^{*(1)}(0) = \frac{\lambda}{2(1-\rho)} E(v^2)$$

$$= \frac{\lambda}{2(1-\rho)}\left(\sigma_v^2 + \frac{1}{\mu^2}\right), \tag{3.12}$$

$$= \frac{\rho}{2\mu(1-\rho)}(1 + c_v^2)$$

where σ_v^2 is the variance of the service time v, and c_v is the coefficient of variation of v, i.e., $c_v = \mu\sigma_v$. By differentiating $W_q^*(s)$ k times WRT s and putting $s = 0$, we can obtain the kth moment of W_q. It can be seen that

$$\text{var}(W_q) = \frac{\lambda}{12(1-\rho)^2}[4(1-\rho)E(v^3) + 3\lambda\{E(v^2)\}^2] \tag{3.13}$$

The mean waiting time in the system is given by

$$E\{W\} = E\{W_q + v\} = E(v) + E\{W_q\}$$

$$= \frac{1}{\mu} + \frac{\lambda}{2(1-\rho)} E(v^2). \tag{3.14}$$

Using Little's formula we can obtain $E\{N\}$ from (3.14) and $E\{L_q\}$ from (3.12). We have

$$E\{N\} = \lambda E\{W\} = \rho + \frac{\lambda^2}{2(1-\rho)} E(v^2)$$

$$= \rho + \frac{\rho^2}{1-\rho}\{1 + c_v^2\} \tag{3.15}$$

$$E\{L_q\} = \lambda E\{W_q\} = \frac{\lambda^2}{2(1-\rho)} E(v^2).$$

The moments $E\{N\}$, $E\{N^2\}, \ldots$ can be obtained directly from (3.7).

From (3.12) and (3.14) it follows that for an $M/G/1$ queue (with parameters λ, μ) the expected waiting time (whether in the queue or in the system) is the least when $\sigma_v = 0$, i.e., service-time distribution is constant $(=1/\mu)$. The model then becomes $M/D/1$. Thus, determinism in the service time distribution minimizes mean waiting time in a single server queue with Poisson input. A similar result holds also for a $G/M/1$ queue. This is discussed in Section 6.7.3.

Notes.

(1) The first $(k + 1)$ moments of the service-time distribution determine the first k moments of the waiting-time distribution and vice-versa.

(2) Lemoine (1976) obtains the moments of the waiting time without using the PK formula by utilizing a relationship between the single-server queues and random walks.

(3) For an alternative derivation of the PK formula, see Fakinos (1982).

Example 6.2. (Sphicas and Shimshak, 1978). Denote

$$c = \frac{\mathrm{SD}\{W_q\}}{E\{W_q\}} = \text{coefficient of variation of } W_q \quad \text{and}$$

$$m_i = E(v^i), \text{ } i\text{th moment of the service time.}$$

We get

$$E(W_q) = \frac{\lambda/\mu}{2(1-\rho)}\,\mu E(v^2) = \frac{\rho m_2}{2m_1(1-\rho)} \quad \text{and}$$

$$E(W_q^2) = \frac{\rho m_3}{3m_1(1-\rho)} + \frac{\rho^2 m_2^2}{2m_1^2(1-\rho)^2},$$

so that

$$c^2 = \frac{4}{3}\left(\frac{1}{\rho} - 1\right)\left(\frac{m_1 m_3}{m_2^2}\right) + 1 \geq 1.$$

Now $B(\cdot)$ being the DF of service time v, we find that $x\,dB(x)/m_1$ is a PDF with mean m_2/m_1 and variance $[m_1 m_3 - m_2^2]/m_1^2$, so that $m_1 m_3 \geq m_2^2$ and hence

$$c^2 \geq \frac{4}{3}\left(\frac{1}{\rho} - 1\right) + 1 = \frac{\left(\frac{4}{\rho} - 1\right)}{3} \geq 1.$$

We now examine a particular case:

$M/E_k/1$

Here, $$m_1 = \frac{1}{\mu}, \qquad m_2 = \frac{k+1}{k\mu^2}, \qquad m_3 = \frac{(k+1)(k+2)}{k^2\mu^3}$$

so that

$$\frac{\left(\frac{4}{\rho} - 1\right)}{3} \leq c^2 \leq \frac{2}{\rho} - 1.$$

6.3.3. *General Time System Size Distribution of an M/G/1 Queue Supplementary Variable Technique (Cox, 1955)*

We first consider representation of the PDF of an RV through its hazard function.

Suppose that a general distribution G (of an RV v) has the hazard function $r(x)$, so that

$$r(x)dx = \frac{Pr\{x \leq v < x + dx\}}{Pr\{v \geq x\}}$$

$$= \frac{B(x + dx) - B(x)}{1 - B(x)}, \qquad B(x) = Pr\{v \leq x\}.$$

In other words,

$$r(x) = \frac{B'(x)}{1 - B(x)} = \frac{-R'(x)}{R(x)}, \qquad R(x) = Pr\{v \geq x\}$$

so that

$$R(x) = \exp\left\{ - \int_0^x r(y)dy \right\}$$

and the PDF $f(x)$ of v is given by

$$f(x) = -R'(x) = r(x)\exp\{-N(x)\}$$

where

$$N(x) = \int_0^x r(y)dy$$

$$\left(N(0) = 0 \quad \text{and} \quad \frac{d}{dx} N(x) = r(x) \right).$$

If v is the service time, then $r(x)dx = Pr$ {service will be completed in $(x, x + dx)$ given that service time exceeds x}.

We shall now outline the approach through supplementary variable technique (due to Cox, 1955), which is an important technique for obtaining a transient solution of a non-Markovian system. Inclusion of a supplementary variable enables one to write down the differential equations, as in the case of a Markovian system. The supplementary variable $X(t)$ considered here is defined below.

Let

$$N(t) = \text{system size at time } t$$

$$X(t) = \text{time already spent in service by time } t \text{ of a unit receiving service (spent service time by time } t)$$

$$r(x) = \text{hazard rate function of the service time}$$

$$B(x) = Pr\{v \leq x\}$$

$$B^*(s) = \text{LST of } v$$

$$p_n(t) = Pr\{N(t) = n\} \text{ (with } p_0(0) = 1)$$

$$p_n(t, x) = Pr\{N(t) = n, x \leq X(t) < x + dx\}, \qquad n \geq 1$$

$$Q(t, z) = \sum_{n=0}^{\infty} p_n(t)z^n$$

and

$$Q(t, x, z) = \sum_{n=1}^{\infty} p_n(t, x)z^n.$$

We have

$$p_n(t) = \int_0^{\infty} p_n(t, x)dx$$

and

$$p_0(t + \delta t) = \{1 - \lambda\delta t + o(\delta t)\}p_0(t) + \int_0^{\infty} p_1(t, x)r(x)dx\, \delta t \tag{3.16}$$

(the second term is obtained by conditioning the amount of service already received by the unit in service at time t).

The above equation yields (as $\delta t \to 0$)

$$\frac{\partial}{\partial t} p_0(t) = -\lambda p_0(t) + \int_0^\infty p_1(t, x) r(x) dx. \tag{3.16a}$$

Again, for $\delta x > 0$

$$p_1(t + \delta t, x + \delta x) \tag{3.17}$$
$$= [1 - \lambda \delta t + o(\delta t)][1 - r(x)\delta x + o(\delta x)] p_1(t, x).$$

Subtracting and adding a term $p_1(t, x + \delta x)$ to the LHS, then dividing by $\delta t(\delta x)$ and taking limits as $\delta t \to 0(\delta x \to 0)$, we get

$$\frac{\partial}{\partial t} p_1(t, x) + \frac{\partial}{\partial x} p_1(t, x) = -(\lambda + r(x)) p_1(t, x). \tag{3.17a}$$

For $n \geq 2$, we shall get an additional term on the RHS of (3.17): $p_{n-1}(t, x)(\lambda \delta t)$ [equal to the probability of arrival of a unit in $(t, t + \delta t)$ when there are already $(n - 1)$ units in the system]. Thus we shall have, for $n \geq 2$

$$\frac{\partial}{\partial t} p_n(t, x) + \frac{\partial}{\partial x} p_n(t, x) = -[\lambda + r(x)] p_n(t, x)$$
$$+ \lambda p_{n-1}(t, x). \tag{3.18}$$

We shall have the following boundary conditions corresponding to the case where $x = 0+$, with a new service commencing before time t:

$$p_1(t, 0) = \int_0^\infty p_2(t, x) r(x) dx + \lambda p_0(t) \tag{3.19}$$

and

$$p_n(t, 0) = \int_0^\infty p_{n+1}(t, x) r(x) dx, \qquad n \geq 2. \tag{3.20}$$

Multiplying (3.18) by z^n, $n = 2, 3, \ldots$ and (3.17a) by z, then adding all the terms, we get

$$\frac{\partial}{\partial t} \left\{ \sum_{n=1}^\infty p_n(t, x) z^n \right\} + \frac{\partial}{\partial x} \left\{ \sum_{n=1}^\infty p_n(t, x) z^n \right\}$$
$$= -(\lambda + r(x)) \sum_{n=1}^\infty p_n(t, x) z^n$$
$$+ \lambda \sum_{n=2}^\infty p_{n-1}(t, x) z^n$$

whence

$$\frac{\partial}{\partial t} Q(t, x, z) + \frac{\partial}{\partial x} Q(t, x, z)$$

$$= -(\lambda - \lambda z + r(x))Q(t, x, z). \tag{3.21}$$

This is a partial differential equation of the Lagrangian type.

Again multiplying (3.20) by z^n, $n = 2, 3, \ldots$ and (3.19) by z and adding the terms, one gets

$$Q(t, 0, z) = \int_0^\infty \left\{ \sum_{n=1}^\infty p_{n+1}(t, x)z^n \right\} r(x)dx + \lambda z p_0(t). \tag{3.22}$$

Now

$$\int_0^\infty \left\{ \sum_{n=1}^\infty p_{n+1}(t, x)z^n \right\} r(x)dx$$

$$= \int_0^\infty \left(\frac{1}{z}\right) \left\{ \sum_{n=1}^\infty p_{n+1}(t, x)z^{n+1} \right\} r(x)dx$$

$$= \left(\frac{1}{z}\right) \int_0^\infty \{Q(t, x, z) - p_1(t, x)z\} r(x)dx$$

$$= \left(\frac{1}{z}\right) \left[\int_0^\infty Q(t, x, z)r(x)dx - z \int_0^\infty p_1(t, x)r(x)dx \right]$$

$$= \left(\frac{1}{z}\right) \left[\int_0^\infty Q(t, x, z)r(x)dx - z\{p_0'(t) + \lambda p_0(t)\} \right]$$

from (3.16a). Thus, (3.22) reduces to

$$zQ(t, 0, z) = \int_0^\infty Q(t, x, z)r(x)dx - zp_0'(t) + \lambda z(z - 1)p_0(t). \tag{3.22a}$$

The partial differential equation (3.21) can be solved using the boundary condition (3.22a) and the normalizing condition $\sum_{n=0}^\infty p_n(t) = 1$.

6.3.3.1. *Steady-State Distribution of the General Time System Size.*

Suppose that

$$\lim_{t \to \infty} p_n(t) = p_n, \qquad n \geq 0$$

and

$$\lim_{t \to \infty} p_n(t, x) = p_n(x), \qquad x > 0, \qquad n \geq 1$$

$$= p_0(x) = 0, \qquad x > 0,$$

exist. Then $\{p_n, n \geq 0\}$ gives the distribution of the general time system size. Let

$$\begin{aligned} Q(x, z) &= \sum_{n=1}^{\infty} p_n(x) z^n \\ &= \sum_{n=1}^{\infty} \left\{ \lim_{t \to \infty} p_n(t, x) \right\} z^n \\ &= \lim_{t \to \infty} Q(t, x, z) \end{aligned}$$

and

$$Q(z) = \int_0^{\infty} Q(x, z) dx.$$

Then equations (3.16a), (3.17a)–(3.18), (3.19), and (3.20) reduce, respectively, to

$$\lambda p_0 = \int_0^{\infty} p_1(x) r(x) dx \tag{3.23}$$

$$\frac{\partial}{\partial x} p_n(x) = -(\lambda + r(x)) p_n(x) + \lambda p_{n-1}(x), \qquad n \geq 1 \\ (\text{with } p_0(x) = 0, \qquad x > 0) \tag{3.24}$$

$$p_1(0) = \int_0^{\infty} p_2(x) r(x) dx + \lambda p_0 \tag{3.25}$$

and

$$p_n(0) = \int_0^{\infty} p_{n+1}(x) r(x) dx, \qquad n \geq 2. \tag{3.26}$$

The partial differential equation (3.21) and the boundary condition (3.22a) reduce, respectively, to

$$\frac{d}{dx} Q(x, z) = -[\lambda - \lambda z + r(x)] Q(x, z) \tag{3.27}$$

and

$$z Q(0, z) = \int_0^{\infty} Q(x, z) r(x) dx + \lambda z(z - 1) p_0; \tag{3.28}$$

the normalizing condition is

$$p_0 + Q(1) = 1. \tag{3.29}$$

The equation (3.27) is a linear first-order differential equation whose solution is

$$Q(x, z) = Q(0, z) \exp\{-\lambda(1-z)x - N(x)\}. \tag{3.30}$$

Substituting the above expression for $Q(x, z)$ in (3.28), we get

$$\begin{aligned} zQ(0, z) &= Q(0, z) \int_0^\infty \exp\{-\lambda(1-z)x - N(x)\} r(x) dx + \lambda z(z-1)p_0 \\ &= Q(0, z)\left[\int_0^\infty e^{-\lambda(1-z)x}\{e^{-N(x)} r(x)\} dx\right] + \lambda z(z-1)p_0 \\ &= Q(0, z) B^*(\lambda - \lambda z) + \lambda z(z-1)p_0 \end{aligned}$$

so that

$$Q(0, z) = \frac{\lambda z(z-1)p_0}{z - B^*(\lambda - \lambda z)}. \tag{3.31}$$

We then get from (3.30)

$$\begin{aligned} Q(z) &= \int_0^\infty Q(x, z) dx \\ &= Q(0, z) \int_0^\infty \exp\{-\lambda(1-z)x - N(x)\} dx \\ &= Q(0, z) \int_0^\infty e^{-N(x)} e^{-\lambda(1-z)x} dx \\ &= Q(0, z)\left[e^{-N(x)} \frac{e^{-\lambda(1-z)x}}{-\lambda(1-z)} \bigg|_{x=0}^{\infty} \right. \\ &\quad \left. - \int_0^\infty e^{-N(x)} \left\{ -\frac{d}{dx} N(x) \right\} \frac{e^{-\lambda(1-z)x}}{-\lambda(1-z)} dx \right] \\ &= \frac{Q(0, z)}{\lambda(1-z)} \left[1 - \int_0^\infty e^{-\lambda(1-z)x}\{e^{-N(x)} r(x)\} dx \right] \\ &= \frac{Q(0, z)}{\lambda(1-z)} [1 - B^*(\lambda - \lambda z)]. \end{aligned} \tag{3.32}$$

Thus, from (3.31) and (3.32), we get

$$Q(z) = \frac{z[B^*(\lambda - \lambda z) - 1]p_0}{z - B^*(\lambda - \lambda z)}. \tag{3.33}$$

Using L'Hôpital's rule, we get

$$\begin{aligned}
Q(1) &= \lim_{z \to 1} Q(z) \\
&= p_0 \lim_{z \to 1} \frac{z[B^*(\lambda - \lambda z) - 1]}{z - B^*(\lambda - \lambda z)} \\
&= p_0 \lim_{z \to 1} \frac{[B^*(\lambda - \lambda z) - 1] + z[-\lambda B^{*(1)}(\lambda - \lambda z)]}{1 + \lambda B^{*(1)}(\lambda - \lambda z)} \\
&= p_0 \frac{\rho}{1 - \rho}, \quad \text{since } -B^{*(1)}(0) = E(v) = 1/\mu.
\end{aligned}$$

The normalizing condition

$$p_0 + Q(1) = 1$$

yields

$$p_0 = 1 - \rho$$

(a result that holds for any single server queueing system in steady state.) Finally, we get from (3.33)

$$Q(z) = \frac{(1 - \rho)z[B^*(\lambda - \lambda z) - 1]}{z - B^*(\lambda - \lambda z)}. \tag{3.34}$$

Now the PGF of the general time system size in steady state is given by

$$\begin{aligned}
P(z) &= \sum_{n=0}^{\infty} p_n z^n = p_0 + \sum_{n=1}^{\infty} p_n z^n = p_0 + Q(z) \\
&= (1 - \rho) \frac{B^*(\lambda - \lambda z)(1 - z)}{B^*(\lambda - \lambda z) - z}.
\end{aligned}$$

Thus, $P(z) = V(z)$ where

$$V(z) = \sum_{n=0}^{\infty} v_n z^n$$

is the Pollaczek–Khinchin formula (for departure-epoch system-size distribution). We find that in steady state the general time system size has the same distribution as the departure epoch system size.

Remark. Incidentally, we have established the equality $P(z) = V(z)$, that is, $v_n = p_n$ for all $n \geq 0$. This also follows from the result that in a system with

Poisson input PASTA holds and so $v_n = p_n$ (also noted in (1) Section 6.2). Further, d_n and v_n denote the same limiting probability.

Notes.

(1) See Cox (1955), also Keilson and Kooharian (1960), and Cohen (1982) for details of the supplementary variable technique applied to $M/G/1$ queue. Borthakur and Medhi (1974) apply the technique to a more general model $M^X/G(a, b)/1$. The technique is applied in another situation by Borthakur, Medhi, and Gohain (1987).

(2) Characterization problems of equilibrium system size distribution in $M/G/1$ queues have been studied. Rego (1988) shows that of all $M/G/1$ queues, the $M/M/1$ queue is the only one in which both $\{p_j\}$ and $\{k_j\}$ have geometric distributions. Disney *et al.* (1973) studied characterization through renewal departure processes; they showed that the departure process from an $M/G/1$ queue with infinite capacity is a renewal process *iff* it is in steady state and service times are exponential (i.e., the queue is an $M/M/1$ queue).

6.3.4. Semi-Markov Process Approach

The system-size $N(t)$ is neither semi-Markovian (nor Markovian). Consider that a transition occurs with a service completion (departure of a unit), i.e., t_n, $n = 0, 1, 2, \ldots$ are the nth departure epochs. Using the notation of Section 1.9,

$$X_n = N(t_n + 0) = \text{system size at the } n\text{th departure}$$
$$\text{(i.e., number left behind by the } n\text{th departure)}$$

$$Y(t) = X_n, \qquad t_n \le t < t_{n+1}$$

we see that $\{Y(t), t \ge 0\}$ is a semi-Markov process that has $\{X_n, n \ge 0\}$ for its embedded Markov chain. We get

$$\begin{aligned} Q_{i,j}(t) &= 0, \qquad i \ge 1, \qquad j < i - 1 \\ &= \int_0^t \frac{e^{-\lambda y}(\lambda y)^{j-i+1}}{(j-i+1)!}\, dB(y), \qquad i \ge 1, j \ge i - 1 \\ &= \int_0^t \{1 - e^{-\lambda(t-y)}\} \frac{e^{-\lambda y}(\lambda y)^j}{j!}\, dB(y), \qquad i = 0, j \ge 0. \end{aligned}$$

One can proceed as in Section 1.9 to find

$$v_j = \lim_{n \to \infty} p_{ij}^{(n)}$$

and

$$p_j = \lim_{t \to \infty} Pr\{Y(t) = j\}.$$

See Fabens (1961) for a relationship between v_j and π_j where

$$\pi_j = \lim_{t \to \infty} Pr\{N(t) = j\}.$$

See also Neuts (1967) for semi-Markov analysis of the more general model $M/G(a, b)/1$.

6.3.5. Approach via Martingale

Approach via martingale is also being considered in the literature; such an approach leads to interesting results. Baccelli and Makowski (1989) consider an exponential martingale associated with the (embedded) Markov chain $\{X_n, n \geq 0\}$, where X_n is the number left behind by the nth departure in an $M/G/1$ queue in steady state. Using the basic regularity properties of martingales (Baccelli and Makowski, 1986) and standard renewal theoretic arguments, they show that this martingale provides a unified probabilistic framework for deriving several interesting well-known results, such as the Pollaczek–Khinchine formula, the transient generating function of the departure epoch system size, and the generating function of the number served during a busy period. Rosenkrantz (1983) obtains the busy period distribution of an $M/G/1$ queue via a martingale. (See Remark (5) Section 6.4.2.)

6.4. Busy Period

6.4.1. Introduction

Assume that a busy period is initiated by a single customer: it commences at the instant of the arrival of a customer to an empty system and terminates at the instant the server becomes free for the first time. The initiating customer will be called the "ancestor" and the customers who arrive during the service time of the initiating customer will be called its "descendants"—the ith customer who arrives during the service time of the initiating customer will be called the ith descendant. The descendants of the

ith descendant will be called the "progenies" of the ith descendant. As usual, let

$$
\begin{aligned}
T &= \text{duration of the busy period} \\
G(t) &= Pr\{T \le t\} \\
B(t) &= Pr\{v \le t\}, \quad \text{where } v \text{ is the service time} \\
G^*(s) &= LST \text{ of } T(=E\{e^{-sT}\}) \\
B^*(s) &= LST \text{ of } v(=E\{e^{-sv}\}) \\
G^{(n)*}(t) &= n\text{-fold consolution of } G(t) \text{ with itself} \\
A(t) &= \text{number of arrivals during } (0, t] \\
& \quad (\equiv \text{a Poisson random variable with mean } \lambda t) \\
g_r &= E\{T^r\}
\end{aligned}
$$

and

$$T_i = \text{total service time of the } i\text{th descendant and all its progeny.}$$

It can be seen that T_i is a busy period initiated by a single customer (the ith customer) and as such T_i has the same distribution as the busy period T; T_i is called a *sub–busy* (or *pseudo-busy*) period. The T_i's are IID random variables distributed as T.

6.4.2. Busy Period Distribution: Takács Integral Equation

The distribution of a busy period was obtained by Takács (1962). The result is discussed below.

Theorem 6.1. *The LST of the busy period can be expressed as a functional equation*

$$G^*(s) = B^*[s + \lambda - \lambda G^*(s)]. \tag{4.1}$$

Proof. To obtain the busy period distribution we condition on two events, namely the duration of the service time v of the initiating customer (ancestor) and on the number A of arrivals during the service time of the ancestor. Given that $v = x$ and $A \equiv A(x) = n$, then n sub–busy periods $T_1, \ldots, T_n$ are generated by the n descendants, and

$$T = x + T_1 + \cdots + T_n. \tag{4.2}$$

Since the T_i's are IID as T and are also independent of x, we have

$$\begin{aligned}
E\{e^{-sT}|v = x, A = n\}& \\
&= E\{e^{-s(x+T_1+\cdots+T_n)}\} \\
&= [E\{e^{-sx}\}][E\{e^{-s(T_1+\cdots+T_n)}\}] \\
&= E\{e^{-sx}\}[E\{e^{-sT}\}]^n \qquad (\text{since } T_i \equiv T) \\
&= e^{-sx}[G^*(s)]^n, \qquad (4.3)
\end{aligned}$$

(since x is a fixed quantity). To get the LST of T we have to remove the conditions on v and A. We have

$$\begin{aligned}
E\{e^{-sT}|v = x\} &= \sum_{n=0}^{\infty} E\{e^{-sT}|v = x, A = n\}Pr\{A = n\} \\
&= \sum_{n=0}^{\infty} e^{-sx}[G^*(s)]^n \frac{e^{-\lambda x}(\lambda x)^n}{n!} \\
&= e^{-sx}e^{-\lambda x}e^{xG^*(s)\lambda} \\
&= e^{-(s+\lambda-\lambda G^*(s))x}. \qquad (4.4)
\end{aligned}$$

Finally,

$$\begin{aligned}
E\{e^{-sT}\} &= \int_0^{\infty} E\{e^{-sT}|v = x\}dB(x) \\
&= \int_0^{\infty} e^{-(s+\lambda-\lambda G^*(s))x}\, dB(x)
\end{aligned}$$

Thus,

$$G^*(s) = B^*(s + \lambda - \lambda G^*(s)).$$

Particular Case. For an $M/M/1$ system, $B^*(s) = \mu/(s + \mu)$ and then (4.1) reduces to a quadratic in $G^*(s)$; its unique root (such that $G^*(0) = 1$) is z_2/ρ. (See Eqs. (9.5) and (9.26), Chapter 3.)

Remarks.

(1) It was Good who first noted that the possibility of analyzing a busy period as a branching process.

(2) The functional equation has a unique root $G^*(s) \leq 1$. Further, $G^*(s)$ is

the LST of a proper probability distribution *iff* $\rho \le 1$. [For a proof, see Example, III, (4.a) p. 417, Feller, Vol. II (1966).]

(3) *Takács integral equation*: We have

$$\begin{aligned} Pr\{T \le t \mid v = x, \quad A = n\} \\ = Pr\{T_1 + \cdots + T_n \le t - x\} \\ = G^{(n)*}(t - x), \end{aligned}$$

since T_i's are independent. Thus

$$G(t) = Pr\{T \le t\} = \int_0^\infty \sum_{n=0}^{\infty} \frac{e^{-\lambda x}(\lambda x)^n G^{(n)*}(t-x)}{n!} \, dB(x), \tag{4.5}$$

which is known as Takács integral equation for the busy period. Taking LST of the above, and interchanging the order of integration, on simplification one gets the functional equation (4.1).

(4) Conway *et al.* (1967) show how one can use the analysis of a busy period to obtain the distribution of the waiting time in the system.

(5) Rosenkrantz (1983) gives a new formula for $G^*(s)$ through a martingale approach. The formula that expresses $G^*(s)$ in terms of $B^*(s)$ is of independent interest in itself. Writing

$$f(s) = \lambda[B^*(s) - 1] + s$$

and

$$f(f^{-1}(s)) = f^{-1}(f(s)) = s,$$

($f^{-1}(s)$, the inverse of $f(s)$ is assumed to exist in the neighborhood of 0), he shows that

$$G^*(f(s)) = B^*(s); \tag{4.6}$$

or, taking inverses

$$G^*(s) = B^*(f^{-1}(s)). \tag{4.7}$$

The formula immediately yields $E(T)$ and higher moments of T.

(6) The expected busy period $E(T)$ can be obtained even without finding the distribution of T by applying a result from renewal theory (see Remark 4, Section 3.9.2): the same result holds since the input is Poisson (and thus $E(I) = 1/\lambda$).

6.4.3. Further Discussion of the Busy Period

6.4.3.1. Moments of the Busy Period. The functional equation (4.1) is difficult to invert. However, one can easily find the moments of T. We have

$$\begin{aligned} E(T) &= -\frac{d}{ds} G^*(s)\bigg|_{s=0} \\ &= E(v)\{1 + \lambda E(T)\} \end{aligned}$$

so that

$$E(T) = \frac{E(v)}{(1-\rho)} = \frac{1}{\mu(1-\rho)}. \tag{4.8}$$

The expected duration of the busy period of an $M/G/1$ queue is the same as that of an $M/M/1$ queue; it is independent of the form of the service time distribution (that occurs only through its mean).

$$\begin{aligned} E(T^2) &= \frac{d^2}{ds^2} G^*(s)\bigg|_{s=0} \\ &= G^{*(2)}(s)|_{s=0} \\ &= B^{*(2)}(0)\left[\frac{d}{ds}(s + \lambda - \lambda G^*(s))\bigg|_{s=0}\right]^2 \\ &\quad + B^{*(1)}(0)\frac{d^2}{ds^2}(s + \lambda - \lambda G^*(s))\bigg|_{s=0} \\ &= B^{*(2)}(0)[1 - \lambda G^{*(1)}(0)]^2 + B^{*(1)}(0)[-\lambda G^{*(2)}(0)] \\ &= E(v^2)[1 + \lambda E(T)]^2 + \lambda E(v)E(T^2). \end{aligned}$$

Thus,

$$\begin{aligned} E(T^2) &= \frac{E(v^2)[1 + \lambda E(T)]^2}{1 - \lambda E(v)} \\ &= \frac{E(v^2)}{(1-\rho)^3}. \end{aligned} \tag{4.9}$$

We have

$$\operatorname{var}(T) = \frac{E(v^2)}{(1-\rho)^3} - \left[\frac{1}{\mu(1-\rho)}\right]^2 = \frac{1}{(1-\rho)^3}[\sigma_v^2 + \rho(E(v))^2]. \tag{4.10}$$

6.4.3.2. Number Served during a Busy Period. Let

$N(T) \equiv N$ = number served during a busy period (T)

N_i = number served during the sub-busy period initiated by the ith descendant

= number served during the sub-busy period T_i (i.e., total service time of the ith descendant and all its progeny)

$P(z) = \sum_{k=1}^{\infty} Pr(N = k)z^k$ (PGF of N or $N(T)$), and

$h_r = E(N^r), \qquad r = 1, 2, \ldots.$

It is clear that all the N_i's are independently distributed as N, and

$$P_i(z) = \sum_{k=1}^{\infty} Pr(N_i = k)z^k = P(z)$$

for all i. Given that the number of arrivals A during the service time of the ancestor is n, we have

$$N = 1 + N_1 + \cdots + N_n. \tag{4.11}$$

Because of independence of the N_i's, the conditional generating function of N can be written as

$$\begin{aligned} E(z^N \mid A = n) &= E(z^{1 + N_1 + \cdots + N_n}) \\ &= E(z) \prod_{i=1}^{n} E(z^{N_i}) \\ &= z[P(z)]^n. \end{aligned} \tag{4.12}$$

Removing the condition on A, we get

$$\begin{aligned} P(z) = E(z^N) &= \sum_{n=0}^{\infty} E(z^N \mid A = n) Pr\{A = n\} \\ &= z \sum_{n=0}^{\infty} [P(z)]^n Pr\{A = n\}. \end{aligned} \tag{4.13}$$

We have

$$\sum_{n=0}^{\infty} Pr(A = n)z^n$$

= PGF of the number of arrivals during the service time of a unit

$$= B^*(\lambda - \lambda z) \text{ for a Poisson input system} \tag{4.14}$$

so that

$$\sum_{n=0}^{\infty} Pr\{A = n\}[P(z)]^n = B^*(\lambda - \lambda P(z)). \tag{4.15}$$

Finally, we get from (4.13) and (4.15)

$$P(z) = zB^*(\lambda - \lambda P(z)). \tag{4.16}$$

The PGF $P(z)$ of the number served during a busy period satisfies the above functional equation.

Notes.

(1) The joint bivariate distribution of T and N of an $M/G/1$ system has been obtained by Prabhu (1960, 1965). Enns (1969) and Scott and Ulmer (1972) consider a joint trivariate distribution of T, N, and M (the maximum number served during a busy period).

(2) Busy period of an $M/G/1/K$ queue has been considered by Harris (1971) and Miller (1975) (see Problems and Complements 6.7).

(3) Shanthikumar (1988) shows that the number served during a busy period is DFR for IFR service times.

(4) The number N served during a busy period can also be expressed as a first passage time of a certain Markov chain with an absorbing state 0.

6.4.3.3. Moments of the Number Served during a Busy Period. The functional equation (4.16) is difficult to solve. However, moments of N can be obtained easily from it. Note that

$$\begin{aligned} \frac{d^k}{dz^k} B^*(z)\bigg|_{z=0} &= B^{*(k)}(0) \\ &= (-1)^k \mu_k \end{aligned}$$

where $\mu_k = E(v^k) = k$th moment of the service time distribution (with $\mu_1 = 1/\mu$). We get from (4.16)

$$\begin{aligned} h_1 &= P'(1) \\ &= B^*(0) - \lambda B^{*(1)}(0)P'(1) \\ &= B^*(0) + \frac{\lambda}{\mu} h_1 = 1 + \rho h_1 \end{aligned}$$

whence

$$h_1 = \frac{1}{1-\rho}. \tag{4.17}$$

Further,

$$\begin{aligned} P''(1) &= B^{*(1)}(0)[-\lambda P'(1)] + B^{*(1)}(0)[-\lambda P'(1)] \\ &\qquad + B^{*(2)}(0)[-\lambda P'(1)]^2 + B^{*(1)}(0)[-\lambda P''(1)] \\ &= 2\frac{\lambda}{\mu} h_1 + \mu_2(-\lambda h_1)^2 + \frac{\lambda}{\mu} P''(1), \end{aligned}$$

whence

$$\begin{aligned} P''(1) &= \frac{\left[\dfrac{2\rho}{(1-\rho)} + \dfrac{\mu_2\lambda^2}{(1-\rho)^2}\right]}{(1-\rho)} \\ &= \frac{2\rho(1-\rho) + \lambda^2\mu_2}{(1-\rho)^3}. \end{aligned}$$

Since $P''(1) = h_2 - h_1$, we get

$$h_2 = \frac{2\rho(1-\rho) + \lambda^2\mu_2}{(1-\rho)^3} + \frac{1}{1-\rho}. \tag{4.18}$$

Thus, the variance of the number served during a busy period is given by

$$\sigma_h^2 = \frac{\rho(1-\rho) + \lambda^2\mu_2}{(1-\rho)^3}. \tag{4.19}$$

6.4.3.4. Particular Cases

(a) *The model $M/M/1$.* Here

$$\mu_1 = \frac{1}{\mu}, \qquad \mu_2 = \frac{2}{\mu^2}, \qquad B^*(s) = \frac{\mu}{s+\mu}.$$

Then the functional equation (4.16) reduces to

$$P(z) = \frac{z\mu}{\mu + \lambda - \lambda P(z)} \quad \text{or}$$

$$\lambda P^2(z) - (\lambda + \mu)P(z) + \mu z = 0. \tag{4.20}$$

Solving and considering the root for which $P(1) = 1$, we get,

$$P(z) = \frac{1+\rho}{2\rho}\left\{1 - \left[1 - \frac{4\rho z}{(1+\rho)^2}\right]^{1/2}\right\}. \tag{4.21}$$

Expanding $P(z)$ in a power series in z and comparing the coefficients of z^n we get

$$Pr[N(T) = n] = \frac{\frac{1}{n}\binom{2n-2}{n-1}\rho^{n-1}}{(1+\rho)^{2n-1}}, \qquad n \geq 1 \tag{4.22}$$

as the distribution of the number $N(T)$ served during a busy period T.

This distribution has been described by Haight (1961) as "analogous to the Borel–Tanner." The mean and variance of the number served during a busy period can be obtained from (4.17) and (4.19) or directly from (4.21). These are given by

$$E\{N(T)\} = \frac{1}{1-\rho} \quad \text{and} \tag{4.23}$$

$$\text{var}\{N\{T)\} = \frac{\rho(1+\rho)}{(1-\rho)^3}. \tag{4.24}$$

(b) *The Model M/D/1*. We have $\mu_1 = 1/\mu$, $\mu_2 = 0$, $B^*(s) = e^{-s/\mu}$ so that the functional equation becomes

$$\begin{aligned} P(z) &= ze^{-[\lambda - \lambda P(z)]/\mu} \\ &= ze^{-\rho}e^{\rho P(z)}. \end{aligned} \tag{4.25}$$

This is a particular case of what is known as a Borel–Tanner distribution. It can be shown that

$$Pr[N(T) = n] = \frac{(n\rho)^{n-1}e^{-n\rho}}{n!}, \qquad n = 1, 2, \ldots. \tag{4.26}$$

The mean and variance can be easily obtained from (4.25). These are given by

$$E\{N(T)\} = \frac{1}{(1-\rho)} \tag{4.27}$$

$$\text{var}\{N(T)\} = \frac{\rho}{(1-\rho)^2}. \tag{4.28}$$

In this particular case we can easily obtain the DF $G(t)$ of the busy period by using (4.26). For, if the number served during the busy period T is n, then the duration of the busy period is n/μ, since the service time of each of the customers is of constant duration $1/\mu$. Thus,

$$G(t) = Pr\{T \le t\} = Pr\left\{\frac{n}{\mu} \le t\right\} = Pr\{n \le \mu t\}$$
$$= \sum_{n=1}^{[\mu t]} \frac{(n\rho)^{n-1} e^{-n\rho}}{n!}$$

[using (4.26)] where $[\mu t]$ is the greatest integer not exceeding μt.

6.4.4. Delay Busy Period

Often it is useful to consider a more general kind of busy period that is initiated by the performance of some initial task before starting service (Conway *et al.* (1967).) The initiating task is called the delay and the time spent is called start-up time T_0. The time T_b spent in servicing units (subsequent to T_0) until none is left is the ordinary busy period. When there is no initial task, the ordinary busy period is denoted by T. Let $G_0^*(s)$, $G_b^*(s)$, $G_c^*(s)$ be the LST of T_0, T_b and $T_c (= T_0 + T_b)$, respectively; and $G^*(s)$ be the LST of ordinary busy period T (with no initial task).

Consider that the expected number of arrivals during T_0 is $\lambda E(T_0)$ and that the delay busy period is initiated by the number of arrivals during T_0. It can be seen that

$$E(T_b) = \lambda E(T_0)[E(T)]$$
$$= \frac{\lambda E(T_0) E(v)}{1-\rho} = \frac{\rho}{1-\rho} E(T_0). \tag{4.29}$$

Further, if we denote the duration of the busy cycle by T_c, so that

$$T_c = T_0 + T_b,$$

then

$$E(T_c) = E(T_0) + E(T_b)$$
$$= \frac{E(T_0)}{1-\rho}. \tag{4.30}$$

It is clear that $T_b = \sum_{i=1}^{N} T$, where N is the RV denoting the number of arrivals during T_0 and is independent of T. Now N has the PGF $G_0^*(\lambda - \lambda s)$, since arrivals are Poisson; T_b is the sum of a random number of RVs each equal to T. Hence the LST of T_b is given by

$$G_b^*(s) = G_0^*[\lambda - \lambda G^*(s)]. \tag{4.31}$$

Further, since the first task (the initiating task) of the delay busy period has LST $G_0^*(s)$ we can at once obtain the LST of T_c in the same manner as that obtained for T. We can thus write down the LST of T_c by writing G_0 in place of B on the RHS expression for $G^*(s)$ in (4.1). Thus,

$$G_c^*(s) = G_0^*[s + \lambda - \lambda G^*(s)]. \tag{4.32}$$

The mean is given by

$$E(T_c) = -\frac{d}{ds}[G_c^*(s)]\bigg|_{s=0};$$

on simplification, one gets

$$E(T_c) = \frac{E(T_0)}{1 - \rho}$$

as is given by (4.30).

6.4.5. Delay Busy Period under N-Policy

It is generally assumed that when the server completes his service on a customer and finds no one waiting to serve, he remains idle until the next customer arrives. However, instead of the server commencing service as soon as one customer arrives, the server may wait until the queue length reaches a desired level $N(\geq 1)$, i.e., each time the system becomes empty, the server waits (or gets busy with other work) until the queue length becomes exactly N, then commences serving customers one by one, and continues until the system becomes completely empty. This is called *N-policy* with *exhaustive service*. The period the server waits until the arrival of the Nth customer is also called a *server's vacation*. (See Section 8.3.) The length of a vacation depends on the arrival process. The model was studied by Yadin and Naor (1963) and Heyman (1968); the latter showed that this model possesses certain optimal properties. (See section 8.4.)

It may happen that the server returning from vacation when the queue length builds up to exactly N may not be immediately available for servicing the waiting customers; he may be engaged in some preservice work or in gearing up the service mechanism for service operation. (This kind of situation may also arise in an inventory problem). The time that the server remains occupied in such preservice work may be called start-up time (SUT). This model has been studied by Baker (1973) (see Problem 3.3 in Chapter 3), and by Borthakur *et al.* (1987) as a more generalized model. By considering that SUTs are identically zero, one may get the models earlier studied. While Baker analyzes the steady-state behavior of an $M/M/1$ queue under N-policy and exponential SUT, Borthakur *et al.* consider an $M/M/1$ queue with N-policy and general SUT.

We consider here an extension of the Takács integral equation for busy period of an $M/G/1$ model with N-policy and general SUT (Medhi and Templeton, 1990). Here the busy period is defined as the length of the interval B from the instant that the Nth customer arrives in a (server) idle system to the instant the server completes servicing of all the units, leaving the system empty for the first time. The RV B is the sum of two RVs: (a) the SUT denoted by U and (b) the period T during which the server is occupied with actual servicing of units consisting of (i) N units, (ii) those that arrive during the start-up time, and (iii) those that arrive during the service times of customers (i) and (ii).

Thus, $B = U + T$, where U and T are not independent. Hence B is the busy period of an $M/G/1$ queue initiated by N customers plus those customers who arrive during the SUT U.

Let λ and μ be the arrival and service rates $G_0^*(s) =$ LST of U, with $E(U) = u$, $G^*(s) =$ LST of a busy period V initiated by one customer in a standard $M/G/1$ queue, with DF $G(x)$, and $G_B^*(s) =$ LST of the busy period B with DF $G_B(x)$. By conditioning on the duration of the SUT, we get,

$$G_B(x) = Pr(B \leq x) = \int_0^x Pr\{B_1 \mid U = t\} dU(t)$$

where $(B_1 \mid U = t)$ is the event that the busy period generated by N customers plus those arriving during SUT of length t is less than or equal to $(x - t)$. Thus,

$$G_B(x) = \int_0^x \sum_{n=0}^{\infty} \frac{e^{-\lambda t}(\lambda t)^n}{n!} G^{(N+n)*}(x - t) dU(t) \tag{4.33}$$

where $G^{(N+n)*}$ is the $(N + n)$-fold convolution of G with itself.

Taking the LST of (4.33), one gets

$$G_B^*(s) = \int_0^{\infty} \int_0^x \sum_{n=0}^{\infty} \frac{e^{-\lambda t}(\lambda t)^n}{n!} e^{-sx} G^{(N+n)*}(x - t) dU(t) dx. \tag{4.34}$$

Changing the order of integration, one gets,

$$G_B^*(s) = \int_0^\infty \sum_{n=0}^\infty \frac{e^{-\lambda t}(\lambda t)^n}{n!} dU(t) \times \int_t^\infty e^{-sx} G^{(N+n)*}(x-t) dx$$

$$= \int_0^\infty \sum_{n=0}^\infty \frac{e^{-\lambda t}(\lambda t)^n}{n!} e^{-st}[G^*(s)]^{N+n} dU(t)$$

$$= [G^*(s)]^N \int_0^\infty e^{-\lambda t} e^{\lambda t G^*(s)} e^{-st} dU(t)$$

$$= [G^*(s)]^N \int_0^\infty e^{-[s+\lambda-\lambda G^*(s)]t} dU(t).$$

Thus,

$$G_B^*(s) = [G^*(s)]^N G_0^*[s + \lambda - \lambda G^*(s)], \qquad N \geq 1, \tag{4.35}$$

where $G^*(s)$ is given by the Takács integral equation

$$G^*(s) = B^*(s + \lambda - \lambda G^*(s)),$$

B^* being the LST of the service-time distribution. We can at once obtain the moments of B from the preceding.

From (4.35), we get

$$E(B) = -\frac{d}{ds} G_B^*(s)\bigg|_{s=0} = NE(V) + E(U)[1 + \lambda E(V)]$$
$$= \frac{N}{\mu(1-\rho)} + \frac{u}{1-\rho} \tag{4.36}$$

since

$$E(V) = \frac{1}{\mu(1-\rho)}, \qquad \rho = \frac{\lambda}{\mu}.$$

Remarks.

(1) By considering that on an average, the number of arrivals during U is λu, it can be seen that

$$E(T) = (N + \lambda u)E(V)$$

so that

$$
\begin{aligned}
E(B) &= E(U) + E(T) \\
&= u + \frac{N + \lambda u}{\mu(1 - \rho)} \\
&= \frac{N}{\mu(1 - \rho)} + \frac{u}{1 - \rho}.
\end{aligned}
$$

(2) If I is the idle period, then $E(I) = N/\lambda$. Denote by p_0 the steady-state probability that the server is idle with number in the queue less than N. Then using

$$p_0 = \frac{E(I)}{E(I) + E(B)},$$

we get

$$p_0 = \frac{N(1 - \rho)}{N + \lambda u} \tag{4.37}$$

as the fraction of time the server remains idle; the fraction of time the server remains busy (either with preservice work or with servicing of units) equals

$$1 - p_0 = \frac{N\rho + \lambda u}{N + \lambda u} = \rho + \frac{\lambda u(1 - \rho)}{N + \lambda u},$$

Another interesting point is that when $u = 0$ (i.e., start-up time is zero), then $1 - p_0 = \rho$, independently of the control parameter N.

(3) Equation (4.32) gives the LST of the delay busy period of the system where the server becomes engaged in start-up work as soon as the system becomes empty. Equation (4.35) gives the LST of the delay busy period of the system where the server becomes engaged in start-up work as the queue size, after an empty queue, builds up to $N(\geq 1)$. It may be seen that Eq. (4.35) also holds for $N = 0$, so that Eq. (4.32) can be obtained from (4.35) by putting $N = 0$

(4) When SUT is zero, then $G_0^*(s) = 1$, so that

$$G_B^*(s) = [G^*(s)]^N$$

as can be expected.

(5) In particular, for the $M/M/1$ queue under N-policy and with exponential SUT with mean u, the LST of the busy period is given by

$$G_B^*(s) = \left[\frac{\beta(s)}{\rho}\right]^N \frac{1}{1 + u\{s + \lambda - \mu\beta(s)\}}$$

where

$$\beta(s) = \frac{(s + \lambda + \mu) \pm \sqrt{[(s + \lambda + \mu)^2 - 4\lambda\mu]}}{2\mu}.$$

6.5. The $M^X/G/1$ Model with Bulk Arrival

6.5.1. *The Number in the System at Departure Epochs in Steady-State (Pollaczek–Khinchin Formula)*

Assume that the arrival epochs occur in accordance with a Poisson process with rate λ and the number of arrivals at each epoch is given by a RV X having distribution $a_j = Pr(X = j)$, and PGF

$$A(s) = \sum_j a_j s^j \quad \text{and } E(X) = \sum ja_j = A'(1) = a. \tag{5.1}$$

The total arrivals A constitute a compound Poisson process having PGF $\exp\{-\lambda[1 - A(s)]\}$.

Suppose that N is the total number of arrivals during the service time of a customer. Then the PGF of N is given by

$$E[s^N] = K(s) = B^*[\lambda - \lambda A(s)] \tag{5.2}$$

where $B^*(s) = \int_0^\infty e^{-st}\, dB(t)$. (See Eq. (2.5a).)

The traffic intensity is $\rho = \lambda E(X)/\mu = \lambda a/\mu$. Assume that $\rho < 1$ so that the steady state is reached. The Pollaczek–Khinchin formula can now be extended for $M^X/G/1$. Writing the preceding expression of $K(s)$ in the Pollaczek–Khinchine formula given by (3.7) for $M/G/1$ we get the expression for the PGF $V(s)$ of the number in the system at departure epochs in steady state. $V(s)$ is then given by

$$V(s) = \frac{(1 - \rho)(1 - s)B^*([\lambda - \lambda A(s))]}{B^*([\lambda - \lambda A(s)]) - s}. \tag{5.3}$$

which is the Pollaczek–Khinchin formula for $M^X/G/1$.

In the particular case when $a_1 = 1$, $a_j = 0$, $j > 1$, we get $A(s) = s$ and $K(s) = B^*(\lambda - \lambda s)$, and we have an $M/G/1$ queue.

6.5.2. *Waiting-Time Distribution*

Burke (1975) obtained the waiting-time distribution in an $M^X/G/1$ queueing system, thereby refining the results obtained by some authors earlier. We discuss next Burke's approach.

Consider a test unit and let D be the total waiting time of the unit in queue, that is, D is the queueing time of an arbitrary test unit. The delay D is seen by the test unit to consist of two independent delays, D_1 and D_2, where D_1 is the delay (or waiting time) of the first member to be served of the batch in which the test unit arrives and D_2 is the delay caused by the service times of the members of his batch who are served prior to the test unit, in other words $D = D_1 + D_2$. Let W, W_i be the DF of D, D_i, $i = 1, 2$, respectively, and let $W^*(s)$, $W_i^*(s)$ be the LST of W, W_i, respectively. Let $B(t)$ be the service-time distribution and $B^*(s)$ be its LST. Denote $\beta^*(s) =$ LST of the DF of the total service time of all customers belonging to the same arrival group. Then

$$\begin{aligned}\beta^*(s) &= \sum_{k=1}^{\infty} a_k[B^*(s)]^k \\ &= A[B^*(s)]. \end{aligned} \tag{5.4}$$

To find the delay D_1, consider a batch as a whole as a single *supercustomer*. Then the LST of the waiting time of the first member of the batch in which the test unit arrives can be obtained from the corresponding expression of an $M/G/1$ system with $B^*(s)$ replaced by $\beta^*(s)$. That is, if $\rho = \lambda a/\mu < 1$, then replacing $B^*(s)$ by $\beta^*(s)$ (given in (5.4)) in the Pollaczek–Khinchine formula (3.11), we get

$$\begin{aligned}W_1^*(s) &= \text{LST of the delay } D_1 \\ &= \frac{s(1-\rho)}{s - \lambda[1 - A(B^*(s))]}. \end{aligned} \tag{5.5}$$

Let p_i be the probability that the test customer arrives in a batch of size i. Let K be a significantly large number. Then in the first K batches of arrivals, the number of batches with i arrivals will be approximately $a_i K$, $i = 1, 2, \ldots,$ and the total number of customers arriving in batches of size i will be approximately $ia_i K$. Thus, the total number of arrivals in K batches is

$$\sum_{i=1}^{\infty} ia_i K$$

and the proportion of those arriving in batches of size i is

$$\frac{ia_i K}{\sum_i ia_i K} = \frac{ia_i}{\sum_i ia_i} = \frac{ia_i}{a}$$

where $a = E(X)$.

Thus, for large K,

$$p_i = \frac{ia_i}{a}. \tag{5.6}$$

Assume now that the test customer arrives in a batch of size i. Assume further that service within members of any batch is in random order. Then the probability that the test customer chosen is the jth in the batch of i is $1/i$, $j = 1, 2, \ldots, i$. Again, if he is the jth customer to be taken for service, his delay (or waiting time in the queue) will be equal to the service time of $(j - 1)$ customers of the batch (of size i) in which he arrives and who are served prior to him. Now conditioning on the size of the batch i on which the test customer arrives, we get

$$\begin{aligned} P(D_2 \le t) = W_2(t) &= \sum_{i=1}^{\infty} Pr\{\text{delay} \le t \,|\, \text{he arrives in a batch of size } i\} p_i \\ &= \sum_{i=1}^{\infty} \left[\sum_{j=1}^{i} B^{(j-1)*}(t) \frac{1}{i} \right] p_i \end{aligned} \tag{5.7}$$

where B^{k*} is the k-fold convolution of B with itself. Thus,

$$\begin{aligned} W_2^*(s) &= \text{LST of } W_2(t) \\ &= \sum_{i=1}^{\infty} \frac{p_i}{i} \left\{ \sum_{j=1}^{i} \text{LST of } B^{(j-1)*}(t) \right\} \\ &= \sum_{i=1}^{\infty} \frac{ia_i}{ia} \left\{ \sum_{j=1}^{i} [B^*(s)]^{j-1} \right\} \\ &= \sum_{i=1}^{\infty} \frac{a_i}{a} \frac{1 - [B^*(s)]^i}{1 - B^*(s)} \\ &= \frac{\sum_{i=1}^{\infty} a_i - \sum_{i=1}^{\infty} a_i [B^*(s)]^i}{a[1 - B^*(s)]} \\ &= \frac{1 - A[B^*(s)]}{a[1 - B^*(s)]}. \end{aligned} \tag{5.8}$$

Since $D = D_1 + D_2$, the LST of the total delay D or waiting time in the queue of the test customer has the LST given by

$$\begin{aligned} W^*(s) &= W_1^*(s)W_2^*(s) \\ &= \frac{s(1-\rho)}{s-\lambda+\lambda A[B^*(s)]} \frac{1-A[B^*(s)]}{a[1-B^*(s)]}. \end{aligned} \tag{5.9}$$

The waiting time in the system or response time of the test unit is given by $W_S = D + v$, where v is the service time. Thus, the LST $W_S^*(s)$ of W_S, the response time, is given by

$$W_S^*(s) = W^*(s)B^*(s).$$

Notes.

(1) Burke obtains result (5.9), by taking into account the difference between the distribution of the size of a batch when sampled over batches and when sampled over units.

(2) While the simple proof just given of the crucial result $p_i = ia_i/a$ is due to Ross (1980), a rigorous proof using Blackwell's renewal theorem has been given by Burke.

6.5.2.1. Particular Cases

(1) Single-arrival case: The Model $M/G/1$
Here

$$a_1 = 1, \qquad a_i = 0, \qquad i \neq 1, \qquad a = 1, \qquad A(s) = s, \quad \text{and}$$

$$B^*(s) = A[B^*(s)] = B^*(s)$$

so that

$$\begin{aligned} W_Q^*(s) = W^*(s) &= \frac{s(1-\rho)}{s-\lambda+\lambda B^*(s)} \frac{1-B^*(s)}{1-B^*(s)} \\ &= \frac{s(1-\rho)}{s-\lambda+\lambda B^*(s)}. \end{aligned}$$

If $G \equiv M$, i.e., for $M/M/1$, $B^*(s) = \mu/(s+\mu)$ and the waiting time in the queue has LST

$$W_Q^*(s) = \frac{(1-\rho)(s+\mu)}{s+\mu-\lambda};$$

and the waiting time in the system has LST

$$W_S^*(s) = W_Q^*(s)\,\frac{\mu}{s+\mu} = \frac{(1-\rho)\mu}{s+(\mu-\lambda)}$$

$$= \frac{\mu-\lambda}{s+(\mu-\lambda)}$$

and is exponential with mean $1/(\mu-\lambda)$.

(2) The system $M^X/G/1$ where X is geometric having distribution

$$P(X=k) = pq^{k-1}, \qquad k = 1, 2, \ldots, \qquad 0 < p, \qquad q < 1,$$

then

$$a = \left(\frac{1}{p}\right)$$

$$A[B^*(s)] = \sum_{k=1}^{\infty} pq^{k-1}[B^*(s)]^k$$

$$= \frac{pB^*(s)}{1-qB^*(s)}$$

and when $G \equiv M$, i.e., $M^X/M/1$ with geometric arrival distribution $B^*(s) = \mu/(s+\mu)$, we get

$$A[B^*(s)] = \frac{p\mu}{s+p\mu}.$$

Thus, the waiting time in the queue has LST

$$W^*(s) = \frac{p(1-\rho)(s+\mu)}{s+(p\mu-\lambda)}$$

and the waiting time in the system has LST

$$W_S^*(s) = W^*(s)B^*(s)$$

$$= \frac{p\mu-\lambda}{s+(p\mu-\lambda)};$$

the waiting time in the system is exponential with mean $1/(p\mu-\lambda)$. This is another example of memoryless distributions leading to another memoryless distribution.

6.5.2.2. Moments of $D = D_1 + D_2$ for $M^X/G/1$. We have

$$E(D_1) = E(D_1) + E(D_2)$$

$$E(D_1) = -\frac{d}{ds} W_1^*(s)\bigg|_{s=0} = -\frac{d}{ds}\frac{s(1-\rho)}{s-\lambda+\lambda A[B^*(s)]}$$

$$= -(1-\rho)\frac{s-\lambda+\lambda A(B^*(s)) - s\left\{1+\lambda A'(B^*(s))\dfrac{d}{ds}B^*(s)\right\}}{\{s-\lambda+\lambda A[B^*(s)]\}^2}\Bigg|_{s=0}$$

(which is of the form 0/0). Using L'Hôpital's rule and simplifying, we get

$$E(D_1) = \frac{\lambda(1-\rho)}{2(1-\rho)^2}\frac{d^2}{ds^2} A[B^*(s)]\bigg|_{s=0}$$

where $(d^2/ds^2)\, A[B^*(s)]$ is the second moment of the supercustomer's service time. On simplification, we get

$$E(D_1) = \frac{\lambda}{2(1-\rho)}\{A''[B^*(s)]\}\left[\frac{d}{ds}B^*(s)\right]^2 + A'[B^*(s)]\frac{d^2}{ds^2}B^*(s)\bigg|_{s=0}$$

$$= \frac{\lambda}{2(1-\rho)}\left[A''(1)\left(-\frac{1}{\mu}\right)^2 + A'(1)\mu_2\right] \tag{5.10}$$

where $\mu_i = E(v^i)$. Writing $m_2 = E(X^2)$, we get $A''(1) = E(X^2) - E(X) = m_2 - a$. Thus,

$$E(D_1) = \frac{\lambda}{2(1-\rho)}\left[\frac{m_2-a}{\mu^2} + a\mu_2\right]. \tag{5.11}$$

Again,

$$E(D_2) = -\frac{d}{ds} W_2^*(s)\bigg|_{s=0} \quad \text{or}$$

$$aE(D_2) = -\frac{d}{ds}\frac{1-A[B^*(s)]}{[1-B^*(s)]}\bigg|_{s=0}$$

$$= \frac{\left\{A'[B^*(s)]\dfrac{d}{ds}B^*(s)\right\}[1-B^*(s)] - \{1-A[B^*(s)]\}\dfrac{d}{ds}B^*(s)}{[1-B^*(s)]^2}\Bigg|_{s=0}$$

(which is of the form 0/0). Using L'Hôpital's rule and simplifying, one gets

$$E(D_2) = \frac{1}{2\mu}\left(\frac{m_2}{a} - 1\right). \tag{5.12}$$

Thus,

$$E(D) = \frac{\lambda}{2(1-\rho)}\left(\frac{m_2 - a}{\mu^2} + a\mu_2\right) + \left(\frac{m_2}{a} - 1\right)\frac{1}{2\mu}. \tag{5.13}$$

6.6. The $M/G(a, b)/1$ Model with General Bulk Service

Let A = number of arrivals during the service time of a unit and

$$q_r = Pr\{A = r\}, \qquad r = 0, 1, 2, \ldots .$$

Then A has the PGF

$$Q(s) = \sum_r q_r s^r = B^*(\lambda - \lambda s),$$

where $B^*(s)$ is the LST of $B(t)$. It can be easily seen that $\{X_n, n \geq 0\}$ (where $X_n = N(t_n + 0)$) is a Markov chain.

The transition probabilities

$$p_{ij} = Pr\{X_{n+1} = j | X_n = i\}$$

of the chain $\{X_n, n \geq 0\}$ can be written in terms of q_r as follows:

$$\begin{aligned} p_{ij} &= q_j, \qquad 0 \leq i \leq b, \qquad j \geq 0 \\ &= q_{j-i+b}, \qquad i \geq b, \qquad j \geq i - b \\ &= 0, \text{ in all other cases.} \end{aligned} \tag{6.1}$$

The TPM of the denumerable Markov chain can be written as follows:

$$\mathbf{P} = (p_{ij}) \equiv \begin{array}{c} \\ 0 \\ 1 \\ 2 \\ \vdots \\ b \\ b+1 \\ b+2 \\ \cdots \\ \cdots \\ \cdots \end{array} \begin{array}{c} \begin{array}{ccccc} 0 & 1 & 2 & 3 & \cdots \end{array} \\ \left[\begin{array}{ccccc} q_0 & q_1 & q_2 & q_3 & \cdots \\ q_0 & q_1 & q_2 & q_3 & \cdots \\ q_0 & q_1 & q_2 & q_3 & \cdots \\ \vdots & \vdots & \vdots & \vdots & \\ q_0 & q_1 & q_2 & q_3 & \cdots \\ 0 & q_0 & q_1 & q_2 & \cdots \\ 0 & 0 & q_0 & q_1 & \cdots \\ \cdots & \cdots & \cdots & \cdots & \\ \cdots & \cdots & \cdots & \cdots & \\ \cdots & \cdots & \cdots & \cdots & \end{array}\right] \end{array} \tag{6.2}$$

The chain is irreducible and aperiodic. It can be shown that it is persistent non-null when $\rho = (\lambda/b\mu) < 1$. It follows that probabilities

$$v_j = \lim_{n\to\infty} p_{ij}^{(n)}, \qquad j=0, 1, 2, \ldots$$

exist and $\mathbf{V} = (v_0, v_1, \ldots)$ is given as the unique solution of $\mathbf{V} = \mathbf{VP}$. We have

$$v_j = \left(\sum_{r=0}^{b-1} v_r\right)q_j + \left(\sum_{r=0}^{j} q_{j-r}v_{b+r}\right), \qquad j \geq 0.$$

The PGF of $\{v_j\}$ is given by

$$\begin{aligned}
V(s) &= \sum_{j=0}^{\infty} v_j s^j \\
&= \left(\sum_{r=0}^{b-1} v_r\right)Q(s) + \sum_{r=0}^{\infty}\left\{\sum_{j=r}^{\infty} q_{j-r}s^{j-r}\right\}v_{b+r}s^r \\
&= \left(\sum_{r=0}^{b-1} v_r\right)Q(s) + Q(s)\sum_{r=0}^{\infty} v_{b+r}s^r \\
&= \left(\sum_{r=0}^{b-1} v_r\right)Q(s) + \left(\frac{Q(s)}{s^b}\right)\left[V(s) - \sum_{r=0}^{b-1} v_r s^r\right].
\end{aligned}$$

On simplification, we get

$$V(s) = \frac{\sum_{r=0}^{b-1}(s^r - s^b)v_r}{Q(s) - s^b}. \tag{6.3}$$

Putting $B^*(\lambda - \lambda s)$ for $Q(s)$, we get

$$V(s) = \frac{\sum_{r=0}^{b-1}(s^r - s^b)v_r}{B^*(\lambda - \lambda s) - s^b}\, B^*(\lambda - \lambda s). \tag{6.4}$$

Notes.

(1) For the standard $M/G/1$ model, $a = b = 1$. Then $v_0 = p_0$ (as PASTA holds) and $p_0 = 1 - \rho$ (for a single server model), so that $v_0 = 1 - \rho$. We can get the Pollaczek–Khinchine formula (as given in (3.7a)) from the above formula (6.4). Thus, (6.4) can be considered as an extension of the Pollaczek–Khinchine formula.

(2) The distribution $\{v_j\}$ is independent of a and holds for $a = 1, 2, \ldots$ and even for $a = 0$.

(3) The transient distribution of $N(t)$ has been studied among others, by Neuts (1967), who also obtains π_j where

$$\pi_j = \lim_{t \to \infty} \{N(t) = j \mid N(0) = i\}.$$

See also Teghem *et al.* (1969) and Borthakur and Medhi (1974) for transient distribution.

6.7. The $G/M/1$ Model

6.7.1. Steady-State Arrival Epoch System Size

Let

$$u = \text{interarrival time}$$

$$A(t) = Pr\{u \le t\}$$

$$A^*(s) = \text{LST of } u$$

and t_n, $n = 1, 2, \ldots$ $(t_0 = 0)$ be the instant at which the nth arrival occurs (or nth unit arrives). Then $N(t_n - 0) = Y_n$, $n = 0, 1, 2, \ldots$ gives the number in the system immediately preceding the nth arrival. We have

$$Y_{n+1} = Y_n + 1 - B_{n+1}, \quad \text{if } Y_n \ge 0, \qquad B_{n+1} \le Y_n + 1, \tag{7.1}$$

where B_{n+1} is the number of units served during $(t_{n+1} - t_n)$, i.e., the interarrival time between the nth and $(n+1)$th arrivals. Clearly, $\{Y_n, n \ge 0\}$ is a denumerable Markov chain.

Now $B_n = B$ and

$$g_r = Pr\{B = r\} = \int_0^\infty \frac{e^{-\mu t}(\mu t)^r}{r!} \, dA(t)$$

$$r = 0, 1, 2, \ldots. \tag{7.2}$$

The PGF of A is given by

$$G(s) = \sum g_r s^r = \sum_r \int_0^\infty \frac{e^{-\mu t}(\mu t)^r}{r!} s^r \, dA(t) = \int_0^\infty e^{-\mu t} e^{s\mu t} \, dA(t)$$

$$= A^*(\mu - s\mu). \tag{7.3}$$

The transition probabilities of the chain, denoted by,

$$p_{ij} = Pr\{Y_{n+1} = j \mid Y_n = i\} \tag{7.4}$$

can be expressed in terms of the g_r's as follows:

$$\begin{aligned} p_{ij} &= g_{i+1-j}, \qquad i+1 \geq j \geq 1, \qquad i \geq 0 \\ &= 0, \qquad i+1 < j. \end{aligned} \tag{7.5}$$

For

$$j = 0,\ p_{i0} = 1 - \sum_{r=0}^{i} g_r = h_i \text{ (say)}. \tag{7.6}$$

The TPM of the chain can be put as:

$$\mathbf{P} = (p_{ij}) = \begin{bmatrix} h_0 & g_0 & 0 & 0 & 0 & \cdots \\ h_1 & g_1 & g_0 & 0 & 0 & \cdots \\ h_2 & g_2 & g_1 & g_0 & 0 & \cdots \\ \cdots & \cdots & \cdots & \cdots & \cdots & \cdots \\ \cdots & \cdots & \cdots & \cdots & \cdots & \cdots \end{bmatrix}. \tag{7.7}$$

The chain was considered in Example 1.4., Chapter 1. The chain is irreducible and aperiodic, and is persistent non-null when $\rho < 1$. Thus, when $\rho < 1$, the limiting arrival epoch system size probabilities

$$v_j = \lim_{n \to \infty} p_{ij}^{(n)} \tag{7.8}$$

exist and are given as the unique solution of

$$\begin{aligned} \mathbf{V} &= \mathbf{VP}, \\ \text{where } \mathbf{V} &= (v_0, v_1, \ldots), \textstyle\sum v_j = 1. \end{aligned} \tag{7.9}$$

Denote by r_0 the unique root inside $|z| = 1$ *of* $r(z) = z - A^*(\mu - \mu z) = 0$. We have

$$v_j = (1 - r_0) r_0^j, \qquad j \geq 0$$

as unique solution of (7.9).

***6.7.1.1. Alternative Method of Finding* V.** We consider an alternative method of finding the unique solution of (7.9). The equations

$$v_j = \sum_i v_i p_{ij}, \qquad j \geq 0, \quad \text{and}$$

$$\sum v_j = 1$$

reduce to

$$v_j = \sum_{i=j-1}^{\infty} v_i g_{i+1-j}$$

$$= \sum_{i=j-1}^{\infty} v_i \int_0^{\infty} \frac{e^{-\mu t}(\mu t)^{i+1-j}}{(i+1-j)!} \, dA(t), \qquad j \geq 1, \tag{7.10}$$

$$\sum v_j = 1.$$

As one equation is redundant, the equation for $j = 0$ is excluded.

Let us try a solution of the form $v_j = c\alpha^j$. Substituting in (7.9) we get

$$c\alpha^j = \sum_{i=j-1}^{\infty} c\alpha_i \int_0^{\infty} \frac{e^{-\mu t}(\mu t)^{i+1-j}}{(i+1-j)!} \, dA(t)$$

$$= c \int_0^{\infty} e^{-\mu t} \left[\sum_{i=j-1}^{\infty} \frac{(\mu t)^{i+1-j}}{(i+1-j)!} \alpha^i \right] dA(t)$$

(interchanging the summation and integration operators). Now

$$\sum_{i=j-1}^{\infty} \frac{(\mu t)^{i+1-j}}{(i+1-j)!} \alpha^i = \alpha^{j-1} \sum_{i=j-1}^{\infty} \frac{(\alpha \mu t)^{i+1-j}}{(i+1-j)!}$$

$$= \alpha^{j-1} \sum_{i=j-1}^{\infty} \frac{(\alpha \mu t)^k}{k!}$$

$$= \alpha^{j-1} e^{\alpha \mu t}.$$

Thus, from (7.10) we get

$$\alpha^j = \alpha^{j-1} \int_0^{\infty} e^{-\mu t} e^{\alpha \mu t} \, dA(t)$$

$$= \alpha^{j-1} \int_0^{\infty} e^{-\mu t(1-\alpha)} \, dA(t) \quad \text{or}$$

$$\alpha = A^*(\mu - \mu\alpha).$$

That is, α is a root of the equation

$$r(z) = z - A^*(\mu - \mu z) = 0.$$

From $\sum v_j = 1$, it follows that the root is of modulus less than 1 and that the constant is $c = 1 - \alpha$. As v_js are the unique solutions of (7.9) and $v_j = (1 - \alpha)\alpha^j$, $j \geq 0$, satisfy the same, we have

$$v_j = (1 - \alpha)\alpha^j, \qquad j \geq 0,$$

α being the root inside $|z| = 1$ of $r(z) = 0$. It can be shown that when $\rho < 1$, there is a unique root inside $|z| = 1$. Thus, we have $\alpha = r_0$, so that

$$v_j = (1 - r_0)r_0^j, \qquad j \geq 0. \tag{7.11}$$

Note. We have $v_j = (1 - r_0)r_0^j\, j = 0, 1, 2, \ldots$. The distribution is geometric having mean $r_0/(1 - r_0)$. An explanation for the occurrence of geometric distribution as steady state system size distribution of the $G/M/1$ queue has been put forward by Kingman (1963). An analytical proof of (7.11) via Wiener–Hopf technique has been given by Neuts (1966).

Example 6.3. The $E_k/M/1$ model. Here $A^*(s) = (\lambda k/(s + \lambda k))^k$ and the characteristic equation reduces to

$$\begin{aligned} r(z) &\equiv z - A^*(\mu - \mu z) \\ &= z - \left(\frac{\lambda k}{\mu - \mu z + \lambda k}\right)^k = 0. \end{aligned} \tag{7.12}$$

In particular, for the $M/M/1$ model, $k = 1$, so that (7.12) reduces to a quadratic $z^2 - (1 + \rho)z + \rho = 0$ with $r_0 = \rho$ as the unique root inside $|z| = 1$. We have thus

$$v_n = (1 - \rho)\rho^n, \qquad n \geq 0$$

and as PASTA holds

$$p_n = v_n = (1 - \rho)\rho^n, \qquad n \geq 0.$$

The $D/M/1$ model can be covered by taking the limit of $A^*(s)$ as $k \to \infty$. We have then $A^*(s) = e^{-s/\lambda}$, so that the characteristic equation reduces to

$$ze^{(1-z)/\rho} = 1.$$

Remark. The contrasting situation between $M/G/1$ and $G/M/1$ systems may be described as follows. Unlike the $G/M/1$ queue, the equilibrium queue length and delay distributions of the $M/G/1$ queue *completely* characterize the arrival and service processes, of course, to a scale factor. For an $M/G/1$ system, it is possible to reconstruct the entire model from the equilibrium distribution and thence also to calculate any desired fluctuations in the queue length process.

6.7.2. General Time System Size in Steady State

By considering the embedded Markov chain $\{Y_n, n \geq 0\}$ (where Y_n is the system size immediately preceding the nth arrival) we obtained the limiting probabilities

$$v_j = \lim_{n \to \infty} p_{ij}^{(n)}.$$

For a fixed j, v_j, is the probability that an arrival finds j in the system: we denote this probability by a_j, so that $v_j = a_j$.

Let $N(t)$ denote the system size at an arbitrary (general) time t and let

$$p_{ij}(t) = Pr\{N(t) = j \mid N(0) = i\}, \tag{7.13}$$

then

$$\lim_{t \to \infty} p_{ij}(t) = p_j$$

(when it exists) gives the probability that there are j in the system in steady state. Further, p_j equals the long-run proportion of times that there are j customers in the system, i.e., p_j are time averages. Since the arrival process is not Poisson we cannot conclude that $a_j = p_j$ holds (for an $M/G/1$ system, the equality holds). Thus, a_j and p_j are different for this $G/M/1$ system and we have to obtain the relation (if any) that exists between them. Tacács (1962) obtains a relation between them. We give below Ross's (1980) probabilistic derivation. To find p_j, we first observe that the rate at which the number in the system changes from $(j-1)$ to j must be equal to the rate at which it changes from j to $(j-1)$. The rate of change from state $(j-1)$ to state j is equal to λ multiplied by the proportion of arrivals who finds $(j-1)$ in the system; in other words, rate of change from state $(j-1)$ to state j equals $\lambda a_{j-1}, j \geq 1$. Now the rate at which the system changes from state j to $(j-1)$ is equal to the proportion of time that there are j in the system multiplied by the rate of service μ. In other words, rate of change from state j to state $j-1$ equals μp_j, $j \geq 1$. Since these two rates are equal, we have

$$\lambda a_{j-1} = \mu p_j, \qquad j \geq 1$$

so that

$$\begin{aligned} p_j &= \left(\frac{\lambda}{\mu}\right) a_{j-1}, j \geq 1 \\ &= \rho v_{j-1} \text{ (since } a_j = v_j). \end{aligned} \tag{7.14a}$$

Thus,

$$p_j = \left(\frac{\lambda}{\mu}\right)(1 - r_0)r_0^{j-1}, \qquad j \geq 1$$

$$= \rho(1 - r_0)r_0^{j-1}, \qquad j \geq 1.$$

Since $\sum_{j=0}^{\infty} p_j = 1$, we have $1 = p_0 + \rho(1 - r_0)/(1 - r_0)$ so that $p_0 = 1 - \rho$, and

$$p_j = \begin{cases} 1 - \rho, & j = 0 \\ \rho(1 - r_0)r_0^{j-1}, & j = 1, 2, \ldots. \end{cases} \tag{7.14b}$$

where r_0 is the unique root inside $|z| = 1$ of the characteristic equation $z - A^*(\mu - \mu z) = 0$. We have

$$E(N) = \frac{\rho}{1 - r_0} \tag{7.15}$$

and

$$\operatorname{var}(N) = \frac{\rho(1 + r_0 - \rho)}{(1 - r_0)^2}.$$

In particular, for an $M/M/1$ model, $r_0 = \rho$, so that $p_j = (1 - \rho)\rho^j, j \geq 0$. For a more general relation between p_j and a_j, see Fakinos (1982).

6.7.2.1. System Size at Most Recent Arrival. Besides the two stochastic processes $\{Y_n, n \geq 0\}$, and $\{N(t), t \geq 0\}$ another related stochastic process $\{Z(t), t \geq 0\}$, where $Z(t) = Y_n$, $t_n \leq t < t_{n+1}$, may be considered. $Z(t)$ denotes the system size at the most recent arrival. $\{Z(t), t \geq 0\}$ is a semi-Markov process having $\{Y_n, n \geq 0\}$ for its embedded Markov chain, the transitions occurring at the arrival epochs. Let

$$f_{ij}(t) = Pr\{Z(t) = j | Z(0) = i\}. \tag{7.16}$$

Then

$$\lim_{t \to \infty} f_{ij}(t) = f_j$$

(when it exists) gives the limiting probability that the system size at the most recent arrival is j. Now v_j is the limiting probability (associated with the embedded Markov chain Y_n) that an arrival finds j in the system, i.e.,

$$v_j = \lim_{n \to \infty} p_{ij}^{(n)}.$$

We have found a relationship between v_j and p_j (as given in (7.14a). Let us find a relationship, if any, existing between f_j and v_j (or p_j).

From the theory of semi-Markov process, we have

$$f_j = \frac{v_j m_j}{\sum_i v_i m_i} \tag{7.17}$$

where m_i is the expected time spent in the state i during each visit. Here $m_i = 1/\lambda$ for all i, as the expected sojourn time in any state i is the same as the expected interarrival time, transitions occurring only at the arrival points (which are the regeneration points). Thus,

$$f_j = \frac{v_j/\lambda}{\sum_i v_i/\lambda} = \frac{v_j}{\sum_i v_i} = v_j \quad \text{for all } j(\geq 0), \tag{7.18}$$

as is to be expected.

6.7.3. Waiting Time Distribution

As the distribution of $\{v_n\}$ is geometric and the service time is exponential, the waiting time in the queue W_q being random geometric sum of exponential variables, is modified exponential. We have $Pr\{W_q = 0\} = 1 - r_0$ and

$$Pr\{W_q \leq t\} = 1 - r_0 e^{-\mu(1-r_0)t}, \qquad t > 0. \tag{7.19}$$

We at once get $E\{W_q\}$ from (7.19). Alternatively,

$$\begin{aligned} E\{W_q\} &= \sum_{k=0}^{\infty} E\{\text{time in queue} \mid \text{an arrival finds } k \text{ in system}\} \\ &\quad \times Pr\{\text{an arrival finds } k \text{ in system}\} \qquad (7.20\text{a}) \\ &= \sum_{k=0}^{\infty} \frac{k}{\mu} (1 - r_0) r_0^k = \frac{r_0}{\mu(1 - r_0)} \end{aligned}$$

and using Little's formula, we get

$$\{L_Q\} = \lambda E\{W_q\} = \frac{\lambda r_0}{\mu(1 - r_0)}. \tag{7.20b}$$

The conditional distribution of waiting time in the queue W_q given that a unit has to wait, is again, exponential with mean $1/\mu(1 - r_0)$. For an $M/G/1$ system, we have from (3.12)

$$E\{W_q\} = \frac{\lambda}{2(1 - \rho)} \left\{ \sigma^2 + \frac{1}{\mu^2} \right\}$$

and it follows that for a Poisson input queue with specified arrival and service rates, $E\{W_q\}$ is the least when $G \equiv D$, i.e., when the service time is deterministic.

A similar result holds for the $G/M/1$ queue. This is stated as a folk theorem, which is as follows.

Folk Theorem. *Among all arrival processes for a single server queue having exponential service time and with given arrival and service rates, the arrival process for which the average waiting time is minimum (and all moments of waiting time as well as other related quantities) is the process with constant (or deterministic) interarrival times.*

For proof of the above theorem, see Hajek (1983) and Humblet (1982). They also show that the average waiting time in a $G/G/1$ queue is minimized by deterministic service time.

Note. The proof given by Hajek relies heavily on the convexity property of systems with exponential service time distribution. The property that is of independent interest in itself is as follows:

The expected queue length in a system with exponential service time at a given time is a convex function of the set of previous interarrival times.

The result holds for higher moments of queue length as well.

Another related result obtained by Fischer (1974) is as follows. The average waiting time in steady state of an arriving customer decreases monotonically in k for a fixed arrival and service rates for both $M/E_k/1$ and $E_k/M/1$ systems. As regards the average waiting time for an exponential server queue with specified arrival and service rates, the queues with Poisson input and with deterministic input have the greatest and the least average waiting times, among all Erlangian-type arrival processes. The folk Theorem implies that the average waiting time is the least for deterministic input among all arrival processes.

6.7.4. Expected Duration of Busy Period and Idle Period

Ross (1980) gives a simple and interesting method of finding the average duration of busy and idle periods of a $G/M/1$ system.

We have so far assumed that the queue discipline is FCFS (first-come/first-served). Even if the service discipline is not FCFS but LCFS (last-come/first-served) or service in random order, the distribution of the number of customers would be the same, and so also would be that of the lengths of the busy and idle periods. However, the waiting time in the queue will have a different distribution though the average waiting time will be the same irrespective of the type of the discipline.

We assume that the queue discipline is LCFS, and we denote by $W_q^{(L)}$ the waiting time under LCFS. An arriving test customer finds the server idle with probability $a_0 = v_0 = 1 - r_0$ or busy with probability r_0. If he finds the server busy, he will have to wait

(i) until the remaining service time of the customer under service (which has the same exponential distribution (as the full service time), and also

(ii) until the completion of services of each of the customers who arrives after him (service discipline being LCFS).

Now, the customer under service can be thought of initiating a server busy period B (its duration being the interval between the epoch of arrival of the test customer and the epoch of completion of services of all the customers who arrive after the test customer). Thus, if the test customer finds the server busy, he will have to wait in the queue for the same duration as the busy period B of the server. Conditioning on the state (idle or busy state of the server) in which the test customer finds the system on arrival, we get

$$\begin{aligned} E\{W_q^{(L)}\} &= E\{\text{wait}\,|\,\text{arrival finds server idle}\} \\ &\quad \times Pr\{\text{arrival finds none in system}\} \\ &\quad + E\{\text{wait}\,|\,\text{arrival finds server busy}\} \\ &\quad \times Pr\{\text{arrival finds server busy}\} \\ &= 0 \times a_0 + E\{B\}[(1 - a_0)] \\ &= E(B)r_0, \end{aligned}$$

but

$$E\{W_q^{(L)}\} = E\{W_q\} = \frac{r_0}{\mu(1 - r_0)} \quad \text{(from (7.20a))}.$$

Thus,

$$E(B) = \frac{1}{\mu(1 - r_0)}. \tag{7.21}$$

Again, we have

$$\frac{E(I)}{E(I) + E(B)} = p_0 = 1 - \rho, \tag{7.22}$$

where I is the idle period so that we get

$$E(I) = \frac{1-\rho}{\lambda(1-r_0)}.$$

Note. For Poisson input queue $r_0 = \rho$; we have then the results for an $M/M/1$ queue.

6.8. Multiserver Models

6.8.1. *The $M/G/\infty$ Model: Transient State Distribution*

Let

$N(t)$ = number in the system (or number of busy channels) at epoch t

$A(t)$ = number of arrivals by time t, i.e. in the interval $(0, t)$

$D(t)$ = number of departures by time t

$p_n(t) = Pr\{N(t) = n\}$

so that

$$A(t) = N(t) + D(t).$$

Assume that $p_0(0) = 1$, $p_n(0) = 0$, $n \neq 0$. We have, by conditioning on $A(t)$,

$$\begin{aligned} p_n(t) &= Pr\{N(t) = n\} \\ &= \sum_{k=n}^{\infty} Pr\{N(t) = n \mid A(t) = k\} Pr\{A(t) = k\} \\ &= \sum_{k=n}^{\infty} Pr\{N(t) = n \mid A(t) = k\} \frac{e^{-\lambda t}(\lambda t)^k}{k!}, \end{aligned} \tag{8.1}$$

since the arrival process is Poisson. Now,

$$\begin{aligned} &Pr\{N(t) = n \mid A(t) = k\} \\ &\quad = Pr\{\text{out of } k \text{ arrivals in } (0, t) \text{ the number still in service at epoch } t \text{ is } n\} \\ &\quad = \binom{k}{n}(r(t))^n[1 - r(t)]^{k-n}, \qquad k \geq n \end{aligned} \tag{8.2}$$

where

$$\begin{aligned} r(t) &= Pr\{\text{a unit arriving in } (0, t), \text{ is still in service at epoch } t\} \\ &= \int_0^t Pr\{\text{service time of a unit exceeds } (t - x), \\ &\qquad \text{given that the unit arrives at epoch } x(<t)\} \\ &\qquad \times Pr\{\text{the unit's arrival epoch is } x\}dx \end{aligned}$$

Now, given that an arrival (from a Poisson process) occurs in $(0, t)$, the interval x in $(0, t)$, is uniformly distributed over $(0, t)$ (see Section 1.5.1). Hence, $r(t)$ equals the probability that a unit's service time does not terminate by time t, given that his arrival is uniformly distributed over the interval $(0, t)$. Thus,

$$\begin{aligned} r(t) &= \frac{1}{t}\int_0^t Pr\{v \geq (t - x)\}dx \\ &= \frac{1}{t}\int_0^t \{1 - B(t - x)\}dx \\ &= \frac{1}{t}\left\{\int_0^t (1 - B(u))du\right\} \\ &= 1 - \frac{1}{t}\int_0^t B(u)du. \end{aligned} \tag{8.3}$$

We get

$$\begin{aligned} p_n(t) &= \sum_{k=n}^{\infty}\binom{k}{n}(r(t))^n[1 - r(t)]^{k-n}\frac{e^{-\lambda t}(\lambda t)^{k-n+n}}{k!} \\ &= \sum_{k=n}^{\infty}\frac{e^{-\lambda t}}{n!(k-n)!}\{\lambda t r(t)\}^n\{\lambda t(1 - r(t))\}^{k-n} \\ &= \frac{\{\lambda t r(t)\}^n}{n!}e^{-\lambda t}\sum_{k=n}^{\infty}\frac{\{\lambda t(1 - r(t)\}^{k-n}}{(k-n)!} \\ &= \frac{\{\lambda t r(t)\}^n}{n!}e^{-\lambda t}e^{\lambda t(1 - r(t))}. \end{aligned}$$

Finally,

$$p_n(t) = \frac{\{\lambda t r(t)\}^n}{n!}e^{-\{\lambda t r(t)\}}, \qquad n = 0, 1, 2, \ldots . \tag{8.4}$$

The distribution of $N(t)$ is Poisson with mean

$$\{\lambda t[r(t)]\} = \lambda \int_0^t \{1 - B(u)\} du. \tag{8.5}$$

Thus, $\{N(t), t \geq 0\}$ is a non-homogeneous Poisson process.

6.8.1.1. Steady-State Solution. As $t \to \infty$

$$\lambda t[r(t)] = \lambda \int_0^t \{1 - B(u)\} du \to \frac{\lambda}{\mu}$$

so that the steady-state distribution of the system size (or the number of busy channels) is Poisson with mean λ/μ, irrespective of the magnitude of λ/μ. Thus, when the number of channels is large (so that an arrival always finds an empty channel), the steady-state distribution of the system size is Poisson, irrespective of the form of the service time distribution and of the magnitude of (λ/μ). It may be recalled that the distribution is truncated Poisson for a c-server (exponential server) $M/M/c$ model.

The busy period distribution of the system $M/G/\infty$ earlier considered by Takács (1962) has been considered by Stadje (1985). Hall (1985) considers heavy traffic approximation.

Remark. Invariant or insensitivity property. The steady-state probabilities in the model considered, are said to be insensitive of the service time distribution (the same occurs only through its first moment $1/\mu$). This property is called the invariant or insensitivity property and the system is said to be invariant or insensitive. The property is quite a powerful one; it enables one to infer and to derive properties of systems with complex service time distribution from the properties of systems with more tractable exponential service time distribution. Further, approximations can be made in cases where the conditions for the validity of invariance are *almost* satisfied. Another important aspect is the connection between the invariant property and the product form of state probabilities. The discovery of such a property has contributed, to a large degree, toward advancement in applications to computer and communication systems. For a description of the class of queueing models that possess such a property, refer to Dukhovny and Koenigsberg (1981), and also to Disney and König (1985).

6.8.2. The Model $G/M/c$

We can now generalize the results of the Section 6.7.1 for a $G/M/1$ model to the c-server model $G/M/c$.

Define Y_n as the number in the system immediately before the arrival of the nth unit. Then $\{Y_n, n \geq 0\}$ is a Markov chain. We have

$$Y_{n+1} = Y_n + 1 - B_{n+1}$$

where B_{n+1} is the number of units served (or number of departures) during $[T_n = (t_{n+1} - t_n)]$, i.e., the interval between the nth and the $(n+1)$th arrivals. B_n is independent of n. However, the service rate during T_n would be state-dependent. We shall now find the transition probabilities

$$p_{ij} = Pr\{Y_n = j \mid Y_{n-1} = i\}.$$

Case (i): $i + 1 < j$. As j cannot exceed $i + 1$, $p_{ij} = 0$.

Case (ii): $i + 1 \geq j \geq c$. This case implies that all the servers are busy and the rate of service is $c\mu$ so that

$$p_{ij} = g_{i+1-j}$$

where now

$$g_r = \int_0^\infty \frac{e^{-c\mu t}(c\mu t)^r}{r!}\, dA(t), \qquad r = 0, 1, 2, \ldots. \tag{8.6}$$

Two more cases arise for $c > 1$.

Case (iii): $j \leq i + 1 \leq c$. If an arrival (say, nth arrival) finds i in the system, then as $i \leq c - 1 < c$, he will find an empty channel and will immediately enter service, so that $(i + 1)$ customers will be simultaneously receiving service (or so that $(i + 1)$ channels will be busy). In order that the next arrival (the $(n + 1)$th) finds exactly j on arrival, $(i + 1 - j)$ customers out of $(i + 1)$ receiving service will complete service during the interarrival interval T_n. Thus,

$p_{ij} = Pr\{$out of $(i + 1)$ customers being served, services of $(i + 1 - j)$ will be completed during an interarrival interval$\}$.

Conditioning on the length of the interarrival interval and denoting by

$A \equiv$ the event that of out $(i + 1)$ ongoing services, $(i + 1 - j)$ services will be completed in an interval of length t,

we get

$$p_{ij} = \int_0^\infty Pr\{A|t\} dA(t).$$

Now, $Pr\{$one service is completed in an interval of length $t\} = 1 - e^{-\mu t}$. Thus, A is a binomial RV with parameters $n = (i + 1)$ and $p = (1 - e^{-\mu t})$. We have

$$Pr\{A|t\} = \binom{i+1}{i+1-j}(1 - e^{-\mu t})^{i+1-j}(e^{-\mu t})^j$$

and as

$$\binom{i+1}{i+1-j} = \binom{i+1}{j},$$

$$p_{ij} = \int_0^\infty \binom{i+1}{j}(1 - e^{-\mu t})^{i+1-j}(e^{-\mu t})^j \, dA(t). \tag{8.7}$$

Case (iv): $i + 1 \geq c > j$. Here the arrival finds all the channels busy, and it can enter service only when $(i + 1 - c)$ services are completed and one of the channels becomes free and then the number of busy channels becomes c. Let the duration of this subinterval be H_c and its DF be $H(\cdot)$, $0 < H_c < (t_{n+1} - t_n)$. In order that the next arrival (arriving at t_{n+1}) finds j, services of $c - j$ must be completed during the remaining duration of the interarrival interval, that is, during the subinterval of length $T - H$.

Now H_c is the sum of $(i + 1 - c)$ IID exponential RVs with parameter $c\mu$ and is thus a gamma variable with parameters $(i + 1 - c)$ and $c\mu$. We have

$$dH(s) = e^{-c\mu s} \frac{(c\mu)(c\mu s)^{i-c}}{(i-c)!} ds. \tag{8.8}$$

Conditioning on the interarrival time T and then on the interval H, we get

$$\begin{aligned} p_{ij} &= \int_0^\infty Pr\{(i + 1 - j) \text{ service completions in time } t\} dA(t) \\ &= \int_0^\infty \int_0^t Pr\{(i + 1 - j) \text{ service completions in time } t | H = s\} dH(s) dA(t) \\ &= \int_0^\infty \int_0^t Pr\{B | H = s\} dH(s) dA(t) \end{aligned}$$

where

$$Pr\{B|H = s\} = Pr\{(c-j) \text{ service completions out of } c \text{ service in the (sub)interval of length } t-s\}$$

$$= \binom{c}{c-j}[1 - e^{-\mu(t-s)}]^{c-j}[e^{-\mu(t-s)}]^j.$$

Thus,

$$p_{ij} = \int_0^\infty \left\{ \int_0^t \binom{c}{j} [1 - e^{-\mu(t-s)}]^{c-j} [e^{-\mu(t-s)}]^j dH(s) \right\} dA(t) \tag{8.9}$$

$$= \binom{c}{j} \frac{(c\mu)^{i+1-c}}{(i-c)!} \int_0^\infty \left\{ \int_0^t \binom{c}{j} [1 - e^{-\mu(t-s)}]^{c-j} e^{-\mu(t-s)j} s^{i-c} e^{-c\mu s}\, ds \right\} dA(t). \tag{8.9a}$$

The limiting arrival point system size probabilities v_j are the unique solutions of

$$\mathbf{V} = \mathbf{VP}, \qquad \sum v_j = 1.$$

We note that for $i + 1 \geq j \geq c$, we get the same type of expression of p_{ij} (with g_r given by (8.6) for the c-server case in place of g_r given by (7.2) for the single-server case).

Thus, we can proceed in a similar manner as in the single server case to find v_j for $j \geq c$. We shall get

$$v_j = \mathrm{C} r_0^j, \qquad j \geq c, \tag{8.10}$$

where C is a constant and r_0 is the root of the equation

$$z - A^*(c\mu - c\mu z) = 0. \tag{8.11}$$

The constant C and the first c values, v_js (i.e., $v_0, v_1, \ldots, v_{c-1}$) are to be determined recursively from the first $(c-1)$ equations of $\mathbf{VP} = \mathbf{V}$, i.e., from

$$v_j = \sum_{i=0}^{\infty} v_i p_{ij}, \qquad j = 0, 1, 2, \ldots, c-1$$

and the normalizing relation $\sum_{j=0}^{\infty} v_j = 1$.

From the last relation, we get

$$\mathrm{C} = \frac{1 - \sum_{i=0}^{c-1} v_i}{\sum_{j=c}^{\infty} r_0^j} = \frac{1 - \sum_{i=0}^{c-1} v_i}{\dfrac{r_0^c}{(1-r_0)}}. \tag{8.12}$$

6.8.2.1. Waiting-Time Distribution of a G/M/c Queue. The distribution for the queueing time W_Q for a $G/M/c$ queue can be obtained in the same way as for a $G/M/1$ queue. We have

$$\begin{aligned} Pr\{W_Q = 0\} &= \sum_{i=0}^{c-1} v_i = 1 - \frac{Cr_0^c}{(1-r_0)} \quad \text{and} \\ Pr\{W_Q \neq 0\} &= \sum_{i=c}^{\infty} v_i = \frac{Cr_0^c}{(1-r_0)}. \end{aligned} \tag{8.13}$$

Since waiting-time distribution is a geometric confounding of exponential distributions, the distribution is modified exponential (similar to that of the simple queue). We shall have

$$\begin{aligned} W_Q(t) = Pr\{W_Q \le t\} &= 1 - Pr(W_Q \neq 0\} e^{-\mu c(1-r_0)t} \\ &= 1 - \frac{Cr_0^c}{(1-r_0)} e^{-\mu c(1-r_0)t}, \qquad t > 0. \end{aligned} \tag{8.14}$$

6.8.3. *The Model M/G/c*

This model cannot be analyzed in the same way as the model $G/M/c$. The difficulty arises because of the fact that an imbedded Markov chain cannot be extracted from it in the same way as it was done in case of the $G/M/c$ system. We cannot therefore get any tidy result about the queue-length distribution for this system.

However, Little's formula and its generalized version in terms of moments of higher order hold. As already mentioned in Remark (5) in Section 2.6, we get, for $r = 1, 2, 3, \ldots,$

$$E\{L^r\} = \frac{\lambda^r E\{W^r\}}{r!} \quad \text{and} \tag{8.15}$$

$$E\{L_{(r)}\} = \lambda^r E\{W^r\}, \tag{8.16}$$

where

$$L_{(r)} = L(L-1)\cdots(L-r+1).$$

The relation (8.16) connects the factorial moments of L with the moments of W. See Brumelle (1972) for derivation of the relations.

6.8.3.1. The Loss System M/G/c/c. We have discussed the $M/M/c/c$ loss system in Section 3.7 and obtained Erlang's loss formula as

$$p_c = Pr\{\text{an arrival finds all channels busy and is lost to the system}\}$$

$$= \frac{(\lambda/\mu)^c/c!}{\sum_{k=0}^{c} (\lambda/\mu)^k/k!}$$

Attempts have been made since Erlang's time to generalize the preceding formula (for example, by Palm, Kosten, Pollaczek, Vaulot, Sevast'yanov, and Fortet). It has been found that the formula holds for arbitrary distribution of the service time, i.e., for $M/G/c/c$. The first rigorous proof of the validity of the loss formula was given by Sevast'yanov (1956). Fortet (1956) obtained the result under the assumption that the distribution of service time is absolutely continuous. A proof based on the basic ideas of Sevast'yanov's proof is given in Gnedenko and Kovalenko (1968). A brief outline of the proof is given next.

If $N(t)$ is the queue length at epoch t (here the queue length equals the number of busy channels), then the process $\{N(t), t \geq 0\}$ is non-Markovian. Sevast'yanov uses the supplementary variable technique which consists of inclusion of additional variables.

Let us assume that at some epoch t, $N(t)$, the number of busy servers equals k, $1 \leq k \leq c$ and $N(t-0) \neq k$. Let the servers busy at epoch t be given serial numbers $1, 2, \ldots, k$ in random order and let $f_i(t)$ denote the time elapsed from epoch t until the server with serial number i completes the service. ($f_i(t)$ is the remaining service time from epoch t of server number i.) The vector stochastic process

$$\mathbf{V}(t) = \{N(t); f_1(t), \ldots, f_k(t)\} \tag{8.17}$$

is a Markov process. Denote

$$F_k(t; x_1, \ldots, x_k) = Pr\{N(t) = k; f_1(t) \leq x_1, \ldots, f_k(t) \leq x_k\}. \tag{8.18}$$

Then

$$F_k(t; \infty, \ldots, \infty) = Pr\{N(t) = k\} = p_k(t) \quad \text{and}$$

$$p_k = \lim_{t \to \infty} F_k(t; \infty, \ldots, \infty). \tag{8.19}$$

Denote

$$F_k(x_1, \ldots, x_k) = \lim_{t \to \infty} F_k(t; x_1, \ldots, x_k). \tag{8.20}$$

The Chapman–Kolmogorov equations are then obtained. It is shown that

$$F_k(x_1, \ldots, x_k) = \frac{\lambda^k}{k!} F_0 \prod_{i=1}^{k} \int_0^{x_i} [1 - B(u)]du \tag{8.21}$$

satisfies the equations ($B(.)$ being the service time DF). We get

$$p_k = \lim_{x_i \to \infty (1 \le i \le k)} F(x_1, \ldots, x_k) = \frac{\left(\frac{\lambda}{\mu}\right)^k}{k!} F_0.$$

Using the normalizing condition, one gets Erlang's formula. For details of Sevast'yanov's proof, refer to Gnedenko and Kovalenko (1968); also see Takács's (1969) paper for a general review.

Note. We have here a model having an invariant (or insensitivity) property discussed in the remark to Section 6.8.1.

Remark. The Erlang loss and delay formulas have been a subject of continued interest, especially problems of asymptotic analysis, approximations, inequalities, bounds, convexity of performance measures (of interest in the study of optimization models), and so on. These problems have been studied, among others, by Jagerman (1974), Sobel (1980), Akimaru and Takahashi (1981), Harel (1987), and Harel and Zipkin (1987).

Problems and Complements

6.1. Consider an $M/G/1$ queueing system in steady state. Show that $\{k_j\}$ is geometric *iff* $\{p_j\}$, $j = 0, 1, 2, \ldots$, is geometric. The result also holds for the zero-truncated geometric, i.e., for $j = 1, 2, \ldots$. (Rego and Szpankowski 1989).

6.2. Consider an $M/G/1$ system in steady state having service-time distribution $B(.)$ with finite first- and second-order moments. Show that any one of the following statements implies the other two:

(i) $B(.)$ is exponential.

(ii) $\{k_j\}$, $j = 0, 1, 2, \ldots$, is geometric with mass function

$$k_j = ab^j, \qquad a = 1 - b = \frac{\mu}{(\lambda + \mu)}.$$

(iii) $\{p_j\}$, $j = 0, 1, 2, \ldots$, is geometric with mass function

$$p_j = \left(\frac{2a-1}{a}\right)\left(\frac{1-a}{a}\right)^j.$$

(Rego and Szpankowski, 1989).

6.3. Suppose that the busy period T is initiated by $m(\geq 1)$ customers (i.e., there are m customers at the commencement of the busy period). Then the PGF of the number $N_m(T)$ served during the busy period is given by $[P(z)]^m$, where $P(z)$ is the PGF of the number served during the busy period initiated by a single customer.

Show that, for $n = m, m+1, \ldots$

(i) $$Pr\{N_m(T) = n\} = \frac{m}{n}\binom{2n-m-1}{n-1}\frac{\rho^{n-m}}{(1+\rho)^{2n-m}}$$

in the case of an $M/M/1$ queue (see Section 3.9.4.2);

(ii) $$Pr\{N_m(T) = n\} = \frac{m}{(n-m)!}\, n^{n-m-1}\rho^{n-m}e^{-n\rho}$$

in the case of an $M/D/1$ queue. This is called the Borel–Tanner distribution. Find the mean and variance of $N_m(T)$.

6.4. $M/G/1$ queue-length distribution (Willmott, 1988)

(a) Refer to Section 6.3 and the Pollaczek–Khinchin formula (3.7). Verify that $K(s)$ is a PGF; so also is

$$G(s) = \sum g_n s^n = \frac{K(s)-1}{\rho(s-1)} \tag{A}$$

where g_n is given by $(1 - \sum_{j=0}^{n} k_j)/\rho$, $n = 0, 1, 2, \ldots$. Show that the Eq. (3.7) can be decomposed into

$$V(s) = Q(s)K(s) \tag{B}$$

where

$$Q(s) = \sum q_n s^n = \frac{1-\rho}{1-\rho G(s)} \tag{C}$$

is the PGF of a compound geometric distribution. Hence show that

$$v_n = \sum_{j=0}^{\infty} q_j k_{n-j}, \qquad n = 0, 1, 2, \ldots, \tag{D}$$

and that q_n can be expressed as

$$q_n = \sum_{m=0}^{\infty} (1-\rho)\rho^m g_n^{(m)*} \tag{E}$$

where $g_n^{(m)*}$ is the coefficient of s^n in $[G(s)]^m$. Show that $Q(s)$ can also be expressed as

$$Q(s) = \frac{1-\rho_1}{1-\rho_1 G_1(s)} \tag{F}$$

where

$$\rho_1 = \frac{\rho(1-g_0)}{(1-\rho g_0)} \quad \text{and}$$

$$G_1(s) = \frac{G(s)-g_0}{1-g_0}, \tag{G}$$

whence

$$g_n = \sum_{m=0}^{n} (1-\rho_1)\rho_1^m g_1^{(m)*}, \qquad n = 0, 1, 2, \ldots. \tag{H}$$

(expressed as a finite sum as an alternate to (E)) where $g_1^{(m)*}$ = coefficient of s_n in $[G_1(s)]^m$.

(b) Obtain as a particular case the corresponding results for an $M/M/1$ queue.

(c) How would you generalize the results of (a) to bulk-arrival queues?

6.5. Consider an $M/G/1$ system with X_n as defined in Eq. (3.1) and $K(\ )$ as given in Eq. (2.2). Let $\mu(n)$ be defined by

$$\mu(n) = \begin{bmatrix} \inf\{m \geq 0 \colon X_{n+m} = 0\}, & \text{if this set is nonempty} \\ \infty & \text{otherwise,} \end{bmatrix}$$

$$n = 0, 1, 2, \ldots.$$

Show that for $0 \leq z \leq 1$, $n = 0, 1, 2, \ldots.$

$$E[Z^{X_n}] = E\left\{\left[\frac{z}{K(z)}\right]^{\mu(n)}\right\}.$$

(Baccelli and Makowski, 1989).

6.6. Show that (with notations as in Section 6.4.4)

$$E(T_b^2) = \frac{\lambda E(v^2)}{(1-\rho)^3} E(T_0) + \frac{\rho^2}{(1-\rho)^2} E(T_0^2)$$

$$E(T_c^2) = \frac{\lambda E(v^2)}{(1-\rho)^3} E(T_0) + \frac{E(T_0^2)}{(1-\rho)^2}$$

(Miller, 1975).

6.7. (a) Consider an $M/G/1/n$ finite queue with total space capacity n. Let T_0 be the delay with DF $H(t)$ and LST $H^*(s)$, let v be the service time with DF $B(t)$ and LST $B^*(s)$, and let $A_n(s)$ be the LST of ordinary busy period T and $A_n^d(s)$ be the LST of delay busy period T_d. Denote

$$u_k(s) = \int_0^\infty \frac{(\lambda t)^k}{k!} e^{-(\lambda+s)t}\, dB(t)$$

$$v_k(s) = \int_0^\infty \frac{(\lambda t)^k}{k!} e^{-(\lambda+s)t}\, dH(t).$$

Show that

$$A_n(s) = \frac{u_0(s)}{\left[1 - \sum_{k=1}^{n-1} u_k(s) \prod_{j=n-k+1}^{n-1} A_j(s) - \sum_{k=n}^{\infty} u_k(s) \prod_{j=1}^{n-1} A_j(s)\right]} \quad \text{and}$$

$$A_n^d(s) = v_0(s) + \sum_{k=1}^{n-1} v_k(s) \prod_{j=n-k+1}^{n} A_j(s) + \sum_{k=n}^{\infty} v_k(s) \prod_{j=1}^{n} A_j(s).$$

For an ordinary busy period T, writing

$$p_k = u_k(0), \qquad P_j = 1 - \sum_{k=0}^{j} p_k, \quad \text{and}$$

$$b_n = \text{mean busy period of } M/G/1/n,$$

one gets

$$b_1 = \frac{E(v)}{p_0}$$

$$b_n = \frac{\left[b_{n-1} - \sum_{j=1}^{n-1} b_j p_{n-j}\right]}{p_0}, \qquad n \geq 2.$$

Assume that $\rho < 1$, then $\lim_{n \to \infty} b_n$ exists. The generating function $B(z)$ of $\{b_n\}$ is given by

$$B(z) = \sum b_n z^n = \frac{z^{E(v)}}{B^*(\lambda - \lambda z) - z} \quad \text{and}$$

$$E(T) = \lim_{n \to \infty} b_n = \lim_{z \to 1} (1 - z) B(z)$$

(Miller, 1975; see also Harris, 1971).

(b) For an $M/M/1/n$ queue, deduce that

$$p_i = \frac{1}{1 + \rho} \left(\frac{\rho}{1 + \rho} \right)^i$$

$$P_i = \left(\frac{\rho}{1 + \rho} \right)^{i+1}$$

$$b_n = \frac{1}{\mu} \frac{1 - \rho^{n+1}}{1 - \rho}.$$

6.8. (a) Consider a $G/G/1$ system. Denote by R the arrival-epoch system size. Denote by N the general-time-system size. Denote for $n = 0$, 1, 2, 3, ...

$$a_n = v_n = Pr(R = n)$$

$$\hat{v}_n = Pr(R > n)$$

$$p_n = Pr(N = n),$$

$$\hat{p}_n = Pr(N > n), \quad \text{and}$$

V = the remaining service time of the service in progress provided the server is occupied

$$b_n = E(V \mid R = n), \qquad n = 1, 2, \ldots.$$

Let λ and μ be the arrival and service rates, respectively, and let

$$A_n(x) = \text{DF of } n\text{th interarrival interval}$$

$$B(x) = \text{DF of service time}$$

$$H(x) = \text{DF of actual waiting time.}$$

Then show that the following relations exist between the queue size and actual waiting time:

$$v_n = \int_0^\infty A_{n+1}(x)dH(x)$$

$$\hat{p}_n = \rho \int_0^\infty A_n(x)dH*B_1(x)$$

where

$$B_1(x) = \mu \int_0^\infty [1 - B(u)]du$$

and * denotes convolution (Mori, 1980).

(b) Show that the following relations hold good:

$$p_n = \lambda\left(b_n v_n + \frac{v_n}{\mu}\right), \qquad n = 1, 2, \ldots, \quad \text{and}$$

$$b_n = \frac{(\hat{p}_n - \rho\hat{v}_n)}{\lambda v_n}, \qquad n = 1, 2, \ldots .$$

Deduce the Pollaczek–Khinchin mean value formula for $M/G/1$. Also deduce the relation $p_n = \rho v_{n-1}$, $n = 1, 2, \ldots$, for the $G/M/1$ system (Fakinos, 1982).

6.9. Busy Period of an $M/G/\infty$ Queue

Define a busy period of an infinite-server queue with Poisson input as the interval during which at least one customer is present and is receiving service. Denote

$$B(.) = \text{DF of service-time distribution}$$

$$C(t) = 1 - \exp\left\{-\lambda \int_0^t [1 - B(x)]dx\right\}, \qquad t > 0$$

$$c(t) = C'(t) = \lambda[1 - B(t)][1 - C(t)]$$

$$H(t) = \text{DF of busy-period distribution.}$$

Show that

$$H(t) = 1 - \frac{1}{\lambda} \sum_{n=1}^{\infty} c^{(n)*}(t)$$

where $c^{(n)*}(t)$ is the n-fold convolution of $c(t)$ with itself (Stadje, 1985).

References

Akimaru, H., and Takahashi, H. (1981). Asymptotic expansion for Erlang loss function and its derivative, *IEEE Trans. Comm.* **29**, 1257–1260.

Asmussen, S. (1987). *Applied Probability and Queues.* Wiley, New York.

Baccelli, F., and Makowski, A. M. (1986). Martingale arguments for stability: The $M/GI/1$ case. *Systems Control Lett.* **6**, 181–186.

Baccelli, F., and Makowski, A. M. (1989). Dynamic, transient and stationary behavior of the $M/GI/1$ queue via martingales. *Annals Prob.* **17**, 1691–1699.

Bertsimas, D., and Papaconstantinou, X. (1988). On the steady-state solution of the $M/C_2(a, b)/s$ queueing system. *Transportation Sci.* **22**, 125–138.

Borthakur, A., and Medhi, J. (1974). A queueing system with arrival and service in batches of variable size. *Cah. du. Centre d'Et. de Rech. Oper.* **16**, 117–126.

Borthakur, A., Medhi, J., and Gohain, R. (1987). Poisson input queueing system with startup time and under control operating policy. *Comp. & Opns. Res.* **14**, 33–40.

Brandt, A. (1987). On stationary queue length distributions for $G/M/s/r$ queues. *Queueing Systems* **2**, 321–322.

Brumelle, S. L. (1972). A generalization of $L = \lambda W$ to moments of queue lengths and waiting times. *Opns. Res.* **20**, 1127–1136.

Burke, P. G. (1975). Delays in single server queues with batch input. *Opns. Res.* **23**, 830–833.

Cohen, J. W. (1982). *The Single Server Queue*, 2nd ed., North-Holland, Amsterdam.

Conway, R. W., Maxwell, W. L., and Miller, L. W. (1967). *Theory of Scheduling.* Addison-Wesley, Reading, MA.

Cosmetatos, G. P. (1976). Some approximate equilibrium results for the multiserver queue $M/G/r$. *Opnl. Res. Qrly*, **27**, 615–620.

Cox, D. R. (1955). The analysis of non-Markovian stochastic processes by the inclusion of supplementary variables. *Proc. Camb. Phil. Soc.* **51**, 433–441.

Disney, R. L., Farrell, R. L., and de Morais, P. R. (1973). A characterization of $M/G/1$ queues with renewal departure processes. *Mgmt. Sci.* **19**, 1222–1228.

Disney, R. L., and König, D. (1985). Queueing networks: a survey of their random processes. *SIAM Review* **27**, 335–403.

Dukhovny, I. M., and Koenigsberg, E. (1981). Invariance properties of queueing networks and their applications to computer communication systems. *INFOR* **19**, 185–204.

Enns, E. G. (1969). The trivariate distribution of the maximum queue length, the number of customers served and the duration of the busy period for the $M/G/1$ queueing system. *J. Appl. Prob.* **6**, 154–161.

Fabens, A. T. (1961). The solution of queueing and inventory models by semi-Markov processes. *J. Roy. Stat. Soc.* **B23**, 113–127.

Fabens, A. T., and Perera, A. G. A. D. (1963). A correction to Fabens (1961). *J.R.S.S.***B**. **25**, 455–456.

Fakinos, D. (1982). The expected remaining service time in a single server queue. *Opns. Res.* **30**, 1014–1017.

Feller, W. (1968; 1966). *An Introduction to Probability Theory and its Applications*, Vol. 1, 3rd ed., Vol. II, Wiley, New York.

Fischer, M. J. (1974). The waiting time in the $E_k/M/1$ queueing system. *Opns. Res.* **22**, 898–902.

Fortet, R. (1956). Random distributions with an application to telephone engineering. *Proc. Berk. Sym. Math. Stat. & Prob.* **2**, 81–88.

Gnendenko, B. V., and Kovalenko, I. N. (1968). *Introduction to Queueing Theory*, (2nd ed., 1980). Israel Prog. for Sci. Tran. Jerusalem.

Gross, D., and Harris, C. M. (1985). *Fundamentals of Queueing Theory*, 2nd ed. Wiley, New York.

Haight, F. A. (1961). A distribution analogous to Borel–Tanner. *Biometrika* **48**, 167–173.

Hajek, B. (1983). The proof of a folk theorem on queueing delay with applications to routing in networks. *J. Ass. Comp. Mach.* **30**, 834–851.

Hall, P. (1985). Heavy traffic approximations for busy period in an $M/G/\infty$ queue. *Stoch. Process & Appl.* **19**, 259–269.

Harel, A. (1987). Sharp bounds and simple approximations for the Erlang delay and loss formulas. *Mgmt. Sci.* (to appear).

Harel, A., and Zipkin, P. (1987). Strong convexity results for queueing systems. *Opns. Res.* **35**, 405–418.

Harel, A., and Zipkin, P. (1989). The convexity of a general performance measure for the $M/G/c$ queue. *J. Appl. Prob.* (to appear).

Harris, T. J. (1971). The remaining busy period of a finite queue. *Opns. Res.* **19**, 219–223.

Henderson, W. (1972). Alternative approaches to the analysis of $M/G/1$ and $G/M/1$ queues. *J. Oper. Res. Soc. Japan* **15**, 92–101.

Heyman, D. P. (1968). Optimal control policies for $M/G/1$ queueing systems. *Opns. Res.* **16**, 362–382.

Humblet, P. A. (1982). Determinism minimizes waiting time in queues. Technical Report, Dept. of Electrical Engineering and Computer Science, MIT, Cambridge, MA.

Jagerman, D. L. (1974). Some properties of the Erlang loss function. *Bell. Syst. Tech. J.* **53**, 525–551.

Kaufman, J. S. (1979). The busy probability in $M/G/N/N$ loss systems. *Opns. Res.* **27**, 204–206.

Keilson, J. (1965). *Green's Function Methods in Probability Theory*. Charles Griffin, London.

Keilson, J., and Kooharian, A. (1960). Time dependent queueing processes. *Ann. Math. Stat*, **31**, 104–112.

Kendall, D. G. (1951). Some problems in the theory of queues. *J. Roy. Stat. Soc.* **B13**, 151–185.

Kendall, D. G. (1953). Stochastic processes occurring in the theory of queues and their analysis by the method of Markov chains. *Ann. Math. Stat.* **24**, 338–354.

Kingman, J. F. C. (1963). Poisson counts for random sequence of events. *Ann. Math. Stat.* **34**, 1217–1232.

Kleinrock, L. (1975). *Queueing Systems*, Vol. I. Wiley, New York.

Lemoine, A. J. (1976). On random walks and stable $GI/G/1$ queues. *Maths. Opns. Res.* **1**, 159–164.

Louchard, G., and Latouche, G. (Eds. 1983). *Probability Theory and Computer Science*, Part II: *Queueing Models*. Academic Press, New York.

Lu, F. V., and Serfozo, R. F. (1984). Queueing decision processes with monotone hysteric optimal policies. *Opns. Res.* **32**, 1116–1132.

Medhi, J. (1984). *Recent Developments in Bulk Queueing Models*. Wiley Eastern, New Delhi, India.

Medhi, J., and Templeton, J. G. C. (1990). A Poisson input queue under N-policy and with general start-up time. *Comps. & Opns. Res.* (to appear).

Miller, L. (1975). A note on the busy period of an $M/G/1$ finite queue. *Opns. Res.* **23**, 1179–1182.

Mohanty, S. G. (1972). On queues involving batches. *J. Appl. Prob.* **9**, 430–435.

Mohanty, S. G. (1979). *Lattice Path Counting and Applications*. Academic Press, New York.

Mori, M. (1980). Relation between queue size and waiting time distributions. *J. Appl. Prob.* **17**, 822–830.

Neuts, M. F. (1966). An alternative proof of a theorem of Takács on $GI/M/I$ queue. *Opns. Res.* **14**, 313–317.

Palm, C. (1943). Intensität sschwankungen im Fernsprechverkehr. (Intensity fluctuations in telephone traffic.) *Erricsson Technics* **44**, 1–189.

Prabhu, N. U. (1960). Some results for the queue with Poisson arrivals. *J. Roy Stat. Soc.* **B22**, 104–107.

Prabhu, N. U. (1965). *Queues and Inventories*. Wiley, New York.

Prabhu, N. U., and Bhat, U. N. (1963a). Some first passage problems and their applications to queues. *Sankhyā* **A25**, 281–292.

Prabhu, N. U., and Bhat, U. N. (1963b). Further results for the queue with Poisson arrivals. *Opns. Res.* **11**, 380–386.

Rego, V. (1988). Characterizations of equilibrium queue length distributions in $M/G/1$ queues. *Comp. & Opns. Res.* **15**, 7–17.

Rego, V., and Szpankowski, W. (1989). The presence of exponentiality in entropy maximized queues. *Comp. & Opns. Res.* **16**, 441–449.

Rosenkrantz, W. A. (1983). Calculation of the L.T. of the length of the busy period for $M/G/1$ queue via martingales. *Ann. Prob.* **11**, 817–818.

Ross, S. M. (1980). *Introduction to Probability Models*, 2nd Ed., Academic Press, New York.

Scott, M., and Ulmer, M. B., Jr (1972). Some results for a simple queue with limited waiting room. *Zeit. F. Opns. Res.* **16**, 199–204.

Shanthikumar, J. G. (1988). DFR property of first-passage times and its preservation under geometric confounding. *Annals Prob.* **16**, 397–407.

Sobel, M. (1980). Some inequalities for multiserver queues. *Mgmt. Sci.* **26**, 951–956.

Sphicas, G. P., and Shimshak, D. G. (1978). Waiting time variability in some single server queueing systems. *J. Opnl. Res. Soc.* **29**, 65–70.

Stadje, W. (1985). The busy period of the queueing system $M/G/\infty$. *J. Appl. Prob.* **22**, 697–704.

Takács, L. (1962). *An Introduction to the Theory of Queues.* Oxford University Press, Oxford, U.K.

Takács, L. (1967). *Combinatorial Methods in the Theory of Queues.* Wiley, New York.

Takács, L. (1969). On Erlang's formula. *Ann. Math. Stat.* **40**, 71–78.

Takács, L. (1976). On the busy period of single-server queue with Poisson input and general service times. *Opns. Res.* **24**, 564–571.

Teghem, J., Loris-Teghem, J., and Lambotte, J. P. (1969). *Modèles d'attente $M/G/1$ et $GI/M/1$ à Arrivées et Services en Groupes.* Lecture Notes on O.R. 8, Springer-Verlag, Berlin.

Van de Liefvoort, A., and Medhi, J. (1990). The waiting time distribution of the $M^X/G/1$ queue (under submission).

Willmot, G. E. (1988). A note on the equilibrium $M/G/1$ queuelength. *J. Appl. Prob.* **25**, 228–231.

Wolff, R. W., and Wrightson, C. W. (1976). An extension of Erlang's formula. *J. Appl. Prob.* **21**, 628–632.

Yadin, M., and Naor, P. (1963). Queueing systems with a removable service station. *Opnl. Res. Qrly.* **14**, 393–405.

7 Queues with General-Arrival-Time and General-Service-Time Distributions

7.1. The $G/G/1$ Queue with General-Arrival-Time and General-Service-Time Distributions

In this system, one server provides service to customers one by one according to FIFO service discipline. Let

$t_n \equiv$ instant of arrival of the nth customer

$u_n \equiv$ interarrival time between the nth and $(n + 1)$th customer

$= t_{n+1} - t_n$

$v_n \equiv$ service time of the nth customer

$X_n \equiv v_n - u_n$

$W_n =$ waiting time in queue of the nth customer

$D_n \equiv$ instant of departure of the nth customer (and instant of commencement of service of the $(n + 1)$th customer, if already in queue)

$\tau_n \equiv$ time between the nth and $(n + 1)$th departures $= D_{n+1} - D_n$

$I_{n-1} \equiv$ idle time (if any) preceding the nth arrival (idle period, if any, preceding the nth arrival)

$J_n \equiv$ total idle period up to the instant of the nth arrival

$\equiv I_1 + I_2 + \cdots + I_n$

$I \equiv$ Idle period with DF $H(x) = Pr\{I \leq x\}$

$B =$ Busy period with DF $\beta(x) = Pr\{B \leq x\}$

$N \equiv$ Number served during a busy period

Assume further

(i) that $u_1, u_2, \ldots$ are IID random variables with a common DF $A(u) = P\{u_i \leq u\}$ with mean $E(u) = 1/\lambda$ and $\text{var}(u) = \sigma_u^2$
(ii) that $v_1, v_2, \ldots$ are IID RV with a common DF $B(v) = P\{v_i \leq v\}$ with mean $E(v) = 1/\mu$ and $\text{var}(v) = \sigma_v^2$ and
(iii) that u_n and v_n are mutually independent. The RVs $X_n = v_n - u_n$ are IID. Let $K(x) = P\{X_n \leq x\}$ be its DF. Then

$$\alpha = E(X_n) = \frac{1}{\mu} - \frac{1}{\lambda} \quad \text{and } \text{var}(X_n) = \sigma_u^2 + \sigma_v^2.$$

We shall assume that the system is in steady state. This happens *iff* $\rho = \lambda/\mu < 1$, which implies that

$$\alpha = \frac{1}{\mu} - \frac{1}{\lambda} < 0.$$

Let $S_n = X_1 + \cdots + X_n$. The random walk $\{S_n, n \geq 1\}$ is the basic process underlying the queueing model. We shall here be mainly interested in the stochastic process $\{W_n, n \geq 1\}$, a process in discrete time with continuous state space. Two cases arise.

First, a customer arrives to find the server busy. Suppose that the $(n+1)$th customer is such a customer. Its waiting time $W_{n+1}(\geq 0)$ is given by

$$\begin{aligned} W_{n+1} &= D_n - t_{n+1} \\ &= v_n + (D_{n-1} - t_{n+1}) \\ &= v_n + W_n - u_n \\ &= W_n + (v_n - u_n) = W_n + X_n(\geq 0). \end{aligned}$$

Second, a customer arrives to find the server idle. Suppose that the $(n+3)$th customer is such a customer. Its waiting time $W_{n+3} = 0$ since $D_{n+2} - t_{n+3} < 0$ or $W_{n+2} + v_{n+2} - u_{n+2} < 0$.

Thus, we can write

$$\begin{aligned} W_{n+1} &= W_n + v_n - u_n \quad \text{if } W_n + v_n - u_n \geq 0 \\ &= 0 \qquad\qquad\qquad\;\; \text{if } W_n + v_n - u_n < 0. \end{aligned} \tag{1.1}$$

Another expression of W_{n+1} is as follows:

$$W_{n+1} = \max(0, W_n + v_n - u_n). \tag{1.2}$$

The RV's v_n and u_n of the sequences $\{v_n\}$ and $\{u_n\}$ are independent among themselves and each other. The value of W_{n+1} depends on the sequence of RV's W_i, $i = 1, 2, \ldots, n$, only through its most recent value W_n plus a RV $X_n = v_n - u_n$ which is independent of all W_i for $i \leq n$. Thus, $\{W_n, n \geq 1\}$ is a Markov process with stationary transition probabilities.

Now $I_n = -\min(0, W_n + v_n - u_n)$; so that $I_n = 0$ or $I_n > 0$, and

$$I_n > 0 \Rightarrow I_n = I,$$

the idle period.

7.1.1. Lindley's Integral Equation

Let us denote the stationary distribution of W_n by

$$\lim_{n \to \infty} \{W_n \leq x\} = W(x). \tag{1.3}$$

We have

$$K(x) = P\{X_n \leq x\} = P\{v_n - u_n \leq x\}.$$

Conditioning on u_n we get

$$\begin{aligned} K(x) &= \int_{u=0}^{\infty} P\{v_n \leq x + u \mid u_n = u\} dA(u) \\ &= \int_{u=0}^{\infty} B(x+u) dA(u). \end{aligned} \tag{1.4}$$

We have, for $x \geq 0$,

$$\begin{aligned} W_{n+1}(x) &= P\{W_{n+1} \leq x\} \\ &= P\{W_n + X_n \leq x\}. \end{aligned}$$

Conditioning on W_n, we get

$$W_{n+1}(x) = \int_0^\infty P\{X_n \le x - t \mid W_n = t\} dW_n(t). \tag{1.5}$$

Now since X_n is independent of W_n, we have

$$P\{X_n \le x - t \mid W_n = t\} = P\{X_n \le x - t\} = K(x - t).$$

Thus, for $x \ge 0$,

$$W_{n+1}(x) = \int_{0-}^\infty K(x - t) dW_n(t).$$

Taking limits as $n \to \infty$ and noting that $\lim_{n\to\infty} \{W_n \le x\} = W(x)$, we get

$$W(x) = \int_{0-}^\infty K(x - t) dW(t), \qquad x \ge 0.$$

Further, we have

$$W(x) = 0, \qquad x < 0.$$

Thus, we get Lindley's integral equation

$$\begin{aligned} W(x) &= \int_{0-}^\infty K(x - t) dW(t), \qquad 0 \le x < \infty \\ &= 0, \qquad x < 0. \end{aligned} \tag{1.6}$$

The equation can be written in two other alternative forms as follows:

We have for $x \ge 0$,

$$\begin{aligned} W(x) &= K(x - t)W(t)\Big|_{t=0}^{\infty} - \int_{0-}^\infty dK(x - t)W(t) \\ &= \lim_{t\to\infty} K(x - t)W(t) - K(x - t)W(0^-) \\ &\quad - \int_{0-}^\infty W(t) dK(x - t) \\ &= -\int_{0-}^\infty W(t) dK(x - t). \end{aligned}$$

Thus,

$$W(x) = -\int_{0-}^{\infty} W(t)dK(x-t), \qquad x \geq 0$$
$$= 0, \qquad x < 0. \tag{1.7}$$

Changing the variable in the integral by putting $x - t = u$, we get

$$W(x) = \int_{-\infty}^{x} W(x-u)dK(u), \qquad x \geq 0$$
$$= 0, \qquad x < 0. \tag{1.8}$$

Equations (1.6)–(1.8), which are all called Lindley's integral equation, are Wiener–Hopf integral equations. From the integral equation it is clear that the waiting-time distribution function $W(x)$ depends only on the distribution function $K(x) = Pr\{v_n - u_n \leq x\}$, i.e., on the DF of the difference of the service-time and interarrival-time distributions, rather than on the DFs of the individual distributions. This basic equation describes the waiting-time distribution of the $G/G/1$ queue. The Lindley integral equation, which looks like a convolution integral, is not exactly of convolution type. The Lindley integral equation holds only for nonnegative values of the variables, while the distribution function vanishes for negative values of the variable. The integral equation can be solved using techniques of complex variable theory.

7.1.2. Laplace Transform of W

We shall obtain next the Laplace Transform of the waiting-time distribution as given in (1.8).

As the integral on the RHS does not have a numerical value (other than zero) for negative x, ($x < 0$), we shall define an integral as follows for negative values of x.

$$W^{-}(x) = \int_{-\infty}^{x} W(x-u)dK(u) \quad \text{when } x < 0$$
$$= 0 \qquad \text{when } x \geq 0. \tag{1.9}$$

Combining (1.8) and (1.9) we get

$$W^{-}(x) + W(x) = \int_{-\infty}^{x} W(x-u)dK(u), \qquad -\infty < x < \infty, \tag{1.10}$$

which holds for all real x.

We define the two-sided Laplace Transforms of $W(t)$ and $W^-(t)$ as follows:

$$\bar{W}(s) = \int_{-\infty}^{\infty} e^{-st} W(t)dt = \int_{0}^{\infty} e^{-st} W(t)dt$$

$$\bar{W}^-(s) = \int_{-\infty}^{\infty} e^{-st} W^-(t)dt = \int_{-\infty}^{0} e^{-st} W^-(t)dt.$$

Let $A^*(s)$ and $B^*(s)$ be the LST of u_n and v_n, respectively. Then since K is the distribution function of $X_n = v_n - u_n$, the two-sided LST of $K(u)$ is given by

$$K^*(s) = \int_{-\infty}^{\infty} e^{-st}\, dK(t) = B^*(s)A^*(-s). \tag{1.11}$$

Taking the two-sided LT of the RHS of (1.10) we get

$$\begin{aligned}
&\int_{-\infty}^{\infty}\int_{-\infty}^{x} e^{-sx} W(x-u)dK(u) \\
&= \int_{-\infty}^{\infty}\int_{-\infty}^{x} e^{-(x-u)s} W(x-u)e^{-us}\, dK(u) \\
&= \left[\int_{-\infty}^{\infty} e^{-(x-u)s} W(x-u)du\right]\left[\int_{-\infty}^{\infty} e^{-us}\, dK(u)\right] \\
&\qquad\qquad\qquad\qquad (\text{since } W(x-u) = 0 \quad \text{for } u \geq x) \\
&= \left[\int_{-\infty}^{\infty} e^{-ts} W(t)dt\right]\left[\int_{-\infty}^{\infty} e^{-us}\, dK(u)\right] \\
&= \bar{W}(s)K^*(s).
\end{aligned} \tag{1.12}$$

Using (1.11) and (1.12)

$$\begin{aligned}
\bar{W}(s) + \bar{W}^-(s) &= \bar{W}(s)K^*(s) \\
&= \bar{W}(s)B^*(s)A^*(-s)
\end{aligned}$$

whence, we get

$$\bar{W}(s) = \frac{\bar{W}^-(s)}{B^*(s)A^*(-s) - 1}. \tag{1.13}$$

Thus, the LT $\bar{W}(s)$ of the waiting time W is obtained in terms of the LST of the RVs u_n and v_n and the unknown $\bar{W}^-(s)$. Solution of Lindley's equation becomes difficult because determination of the unknown $\bar{W}^-(s)$ requires techniques of complex variable theory.

Example 7.1. The $M/M/1$ queue
Here

$$A^*(s) = \frac{\lambda}{\lambda + s}, \qquad B^*(s) = \frac{\mu}{\mu + s}$$

so that the Eq. (1.13) becomes

$$\bar{W}(s) = \frac{\bar{W}^-(s)}{\dfrac{\lambda}{\lambda - s}\dfrac{\mu}{\mu + s} - 1}$$

$$= \frac{(\lambda - s)(\mu + s)\bar{W}^-(s)}{s(\mu - \lambda + s)}. \tag{1.14}$$

Now to find $\bar{W}^-(s)$. We note that

$$W^*(\lambda) = \int_0^\infty e^{-\lambda t}\, dW(t) = e^{-\lambda t} W(t)\Big|_{t=0}^{\infty} + \lambda \int_0^\infty e^{-\lambda t} W(t) dt.$$

We are concerned with computation of $W(t)$ for $t > 0$; we may assume for the time being that $W(0) = 0$. We can ultimately put $W(0) = p_0 = 1 - \rho$, which holds for all $M/G/1$ queues. Thus,

$$W^*(\lambda) = \lambda \bar{W}(\lambda).$$

Now $\int_0^\infty e^{-\lambda t}\, dW(t)$ is the probability that no customer arrives during the waiting time of an arbitrary customer. This is the probability that an arrival finds no other customers waiting for service or that the arrival finds the system idle or with one customer in service.

Thus, for an $M/M/1$ system,

$$W^*(\lambda) = \int_0^\infty e^{-\lambda t}\, dW(t) = p_0 + p_1 = (1 - \rho) + \rho(1 - \rho)$$
$$= (1 - \rho)(1 + \rho)$$

so that

$$\bar{W}(\lambda) = \frac{(1 - \rho)(1 + \rho)}{\lambda} \neq 0. \tag{1.15}$$

Since $\bar{W}(\lambda) \neq 0$, it is clear from (1.14) that $\bar{W}^{-}(s)$ is of the form

$$\bar{W}^{-}(s) = \frac{c}{\lambda - s} \text{ where } c \text{ is a constant or}$$

$$(\lambda - s)\bar{W}^{-}(s) = c.$$

From (1.14) we find

$$\bar{W}(\lambda) = \lim_{s \to \lambda} \left[\frac{c(\mu + s)}{s(\mu - \lambda + s)} \right]$$

$$= c\,\frac{\mu + \lambda}{\lambda\mu} = \frac{c}{\lambda}(1 + \rho).$$

Using (1.15), we get

$$\frac{(1 - \rho)(1 + \rho)}{\lambda} = \frac{c}{\lambda}(1 + \rho) \quad \text{or}$$

$$c = 1 - \rho.$$

Thus,

$$\bar{W}^{-}(s) = \frac{1 - \rho}{\lambda - s}. \tag{1.16}$$

Substituting this value of $\bar{W}^{-}(s)$ in (1.14), we get

$$\bar{W}(s) = \frac{(\lambda - s)(\mu + s)}{s(\mu - \lambda + s)} \frac{1 - \rho}{\lambda - s}$$

$$= \frac{(\mu + s)(1 - \rho)}{s(\mu - \lambda + s)} \tag{1.17}$$

$$= \frac{1 - \rho}{s} + \frac{\lambda(1 - \rho)}{s(\mu - \lambda + s)}.$$

Inverting the LT, we get

$$W(x) = (1 - \rho) + \frac{\lambda(1 - \rho)}{\mu - \lambda} [1 - e^{-(\mu - \lambda)x}]$$

$$= 1 - \rho e^{-\mu(1 - \rho)x}, \qquad (x > 0). \tag{1.18}$$

We now note that

$$W(0) = p_0 = 1 - \rho.$$

Note that

$$\bar{W}(s) = \frac{(1-\rho)(\mu+s)}{s(\mu-\lambda+s)}$$

is the Pollaczek–Khinchine formula for $M/G/1$ for the special case of exponential service time, $G \equiv M$. Result (1.18) was earlier obtained directly.

7.1.3. Generalization of the Pollaczek–Khinchine Transform Formula

We shall now obtain the LST of W in terms of the LST of the idle time I and the LST of X_n. We have

$$W_{n+1} - I_n = W_n + X_n$$

with $W_{n+1} I_n = 0$, and $I_n = I$ whenever $I_n > 0$. Now since W_n and X_n are independent, we have

$$E[e^{-s(W_{n+1}-I_n)}] = E[e^{-sW_n}]E[e^{-sX_n}]. \tag{1.19}$$

Again, since $W_{n+1} = 0$, when $I_n > 0$ and $W_n \neq 0$, when $I_n = 0$

$$\begin{aligned} E[e^{-s(W_{n+1}-I_n)}] &= E[e^{-s(-I_n)} | I_n > 0] Pr(I_n > 0) \\ &\quad + E[e^{-sW_{n+1}} | I_n = 0] Pr(I_n = 0). \end{aligned} \tag{1.20}$$

Now

$$\begin{aligned} E[e^{-sW_{n+1}}] &= E[e^{-sW_{n+1}} | I_n = 0] Pr(I_n = 0) \\ &\quad + E[e^{-sW_{n+1}} | I_n > 0] Pr(I_n > 0). \end{aligned} \tag{1.21}$$

Since $W_{n+1} = 0$ when $I_n > 0$,

$$E[e^{-sW_{n+1}} | I_n > 0] = 1,$$

and since $Pr(I_n > 0) = Pr\,\{\text{arrivals find system idle}\} = a_0$

$$E[e^{-sW_{n+1}}] = E[e^{-sW_{n+1}} | I_n = 0] Pr(I_n = 0) + a_0$$

and from (1.20), we get

$$\begin{aligned} E[e^{-s(W_{n+1}-I_n)}] &= E[e^{-s(I_n)} | I_n > 0] Pr(I_n > 0) \\ &\quad + E[e^{-sW_{n+1}}] - a_0. \end{aligned} \tag{1.22}$$

Thus, from (1.19) and (1.22), we get

$$E[e^{-sW_n}]E[e^{-sX_n}] = E[e^{-s(-I_n)} | I_n > 0] a_0 + E[e^{-sW_{n+1}}] - a_0. \tag{1.23}$$

Note that when $I_n > 0$, $I_n = I$ and in steady state

$$\lim E[e^{-sW_n}] = \lim E[e^{-sW_{n+1}}] = E[e^{-sW}].$$

Thus, taking limit of (1.23), we get

$$E[e^{-sW}]E[e^{-sX_n}] = E[e^{-s(-I)}]a_0 + E[e^{-sW}] - a_0. \tag{1.24}$$

Now

$$E[e^{-sW}] = W^*(s)$$

is the LST of W. $E[e^{-sX_n}] = K^*(s)$ is the two-sided Laplace–Stieltjes Transform of

$$K(u) = P(X_n = v_n - u_n \leq u);$$

$$E[e^{-s(I)}] = I^*(s)$$

is the Laplace–Stieltjes Transform of I. Thus, from (1.24) we get

$$W^*(s) = \frac{a_0[1 - I^*(-s)]}{1 - K^*(s)}. \tag{1.25}$$

The preceding generalization of Pollaczek–Khinchine (Marshall, 1968c) formula holds for the $G/G/1$ system.

Example 7.2. **The *M/G/1* system.** For a Poisson arrival system, the idle-time distribution is the same as the interarrival-time distribution, so that $A^*(s) = I^*(s) = \lambda/(\lambda + s)$. Again, $a_0 = p_0 = 1 - \rho$. Further,

$$\begin{aligned} K^*(s) &= E[e^{-s(v_n - u_n)}] \\ &= E[e^{-sv_n}]E[e^{su_n}] \\ &= B^*(s)\frac{\lambda}{\lambda - s} \end{aligned}$$

where $B^*(s)$ is the LST of the service-time distribution. Substituting in (1.25) we get

$$\begin{aligned} W^*(s) &= \frac{(1-\rho)\left[\dfrac{-s}{(\lambda - s)}\right]}{1 - \left[\dfrac{\lambda}{(\lambda - s)}\right]B^*(s)} \\ &= \frac{s(1-\rho)}{s - \lambda + \lambda B^*(s)} \end{aligned} \tag{1.26}$$

which is exactly the Pollaczek–Khinchine Transform formula for the LST of waiting time.

Noting that $s\bar{W}(s) = W^*(s)$, we get the P–K formula for the LT of the waiting time.

Example 7.3. **The *M/M/1* system.** For an $M/M/1$ system also

$$p_0 = 1 - p \quad \text{and}$$

$$A^*(s) = I^*(s) = \frac{\lambda}{(\lambda + s)}.$$

Since $B^*(s) = \mu/(\mu + s)$, we get from (1.26)

$$W^*(s) = \frac{s(1 - \rho)}{s - \lambda + \dfrac{\lambda\mu}{(\mu + s)}}$$

$$= \frac{(1 - \rho)(\mu + s)}{\mu + s - \lambda}. \tag{1.27}$$

Noting that $s\bar{W}(s) = W^*(s)$, we get the same result as obtained in (1.17).

7.2. Mean and Variance of Waiting Time W

7.2.1. Mean of W (Single Server Queue)

Marshall (1968a) obtained the mean and the variance of W. The mean of the waiting time can be obtained in terms of the mean and variance of the interarrival, service, and idle-time distributions. This is given shortly in Theorem 7.1.

First we prove the following:
For a $G/G/1$ queue with $\rho < 1$, the mean idle time $E(I)$ is given by

$$E(I) = \frac{\left(\dfrac{1}{\lambda} - \dfrac{1}{\mu}\right)}{a_0} = \frac{(1 - \rho)}{\lambda a_0} \tag{2.1}$$

where a_0 is the probability that an arrival finds the system empty.

We have

$$W_{n+1} - I_n = W_n + X_n. \tag{2.2}$$

Since the queue is stationary, $E(W_{n+1}) = E(W_n)$. Whenever $I_n > 0$, I_n equals the idle period I. We have

$$E(I_n) = Pr\{\text{system is found empty}\} \times E\{\text{idle time}\}$$

$$= a_0 E(I).$$

Thus,

$$-a_0 E(I) = E(X_n) = E(v_n) - E(u_n)$$

whence

$$E(I) = \frac{\left(\frac{1}{\lambda} - \frac{1}{\mu}\right)}{a_0} = \frac{(1-\rho)}{\lambda a_0}.$$

Notes.

(1) Either W_{n+1} or I_n is always zero, so that

$$W_{n+1} I_n = 0, \qquad W_{n+1}^2 I_n = 0, \quad \text{and}$$
$$W_{n+1} I_n^2 = 0.$$

(2) For Poisson input system, $E(I) = I/\lambda$.

Theorem 7.1. *For all $G/G/1$ queues with $\rho < 1$,*

$$E(W) = \frac{\lambda^2(\sigma_u^2 + \sigma_v^2) + (1-\rho)^2}{2\lambda(1-\rho)} - \frac{\upsilon_h^{(2)}}{2\upsilon_h} \tag{2.3}$$

where υ_h and $\upsilon_h^{(2)}$ are the first and second moments of the idle period I.

Proof. Squaring (2.2) and noting that

$$W_{n+1} \cdot I_n = 0,$$

we get

$$W_{n+1}^2 + I_n^2 = W_n^2 + 2W_n X_n + X_n^2. \tag{2.4}$$

We have

$$E(I_n^2) = Pr\{\text{system is empty}\} E(I^2)$$
$$= a_0 E(I^2) \quad \text{and}$$
$$E(W_n X_n) = E(W_n) E(X_n)$$

since W_n, X_n are independent. Thus, taking the expectation of (2.4) and noting that in steady state $E(W_n) = E(W)$ and $E(W_{n+1}^2) = E(W_n^2)$, we get

$$a_0 E(I^2) = E(X_n^2) + 2E(W)E(X_n)$$

so that

$$E(W) = \frac{a_0 E(I^2) - E(X_n^2)}{2E(X_n)}. \tag{2.5}$$

Now

$$E(X_n) = \frac{1}{\mu} - \frac{1}{\lambda} = \frac{1}{\lambda}(\rho - 1),$$

$$E(X_n^2) = E(v_n^2 + u_n^2 - 2u_n v_n) = E(v_n^2) + E(u_n^2) - 2E(u_n)E(v_n)$$

(since u_n, v_n are independent)

$$= \sigma_v^2 + \frac{1}{\mu^2} + \sigma_u^2 + \frac{1}{\lambda^2} - 2\left(\frac{1}{\lambda}\right)\left(\frac{1}{\mu}\right) = \sigma_u^2 + \sigma_v^2 + \frac{(1-\rho)^2}{\lambda^2},$$

$$v_h = E(I) = \frac{\left(\frac{1}{\lambda} - \frac{1}{\mu}\right)}{a_0} = \frac{1}{\lambda a_0}(1 - \rho) = -\frac{E(X_n)}{a_0} \quad \text{and}$$

$$v_h^{(2)} = E(I^2).$$

From (2.5) we get

$$E(W) = \frac{E(X_n^2)}{-2E(X_n)} - \frac{E(I^2)}{2E(I)} \tag{2.5a}$$

and using the preceding relations, we get

$$E(W) = \frac{a_0 v_h^{(2)} - \left[\sigma_u^2 + \sigma_v^2 + \frac{(1-\rho)^2}{\lambda^2}\right]}{(2/\lambda)(\rho - 1)}$$

$$= \frac{\lambda^2(\sigma_u^2 + \sigma_v^2) + (1-\rho)^2}{2\lambda(1-\rho)} - \frac{v_h^{(2)}}{2v_h}, \tag{2.5b}$$

which is the relation (2.3).

Notes.

(1) Denoting

$$C_g \equiv \text{coeff. of variation of } X_n$$

we can put (2.3) in an alternative form in terms of C_g.

(2) The quantity $v_h^{(2)}/2v_h$ is the expectation of the RV Z with distribution function

$$\frac{1}{v_h}\int_0^t [1 - H(u)]du$$

where $H(u)$ is the DF of the idle period I. Z is the equilibrium excess idle distribution.

(3) For a queue with Poisson arrivals, we have

$$v_h = E(I) = \frac{1}{\lambda} \quad \text{and} \quad v_h^{(2)} = E(I^2) = \frac{2}{\lambda^2}, \qquad \sigma_u^2 = \frac{1}{\lambda^2},$$

so that for an $M/G/1$ queue, Eq. (2.3) becomes

$$E(W) = \frac{1 + \lambda^2\sigma_v^2 + 1 - 2\rho + \rho^2}{2\lambda(1-\rho)} - \frac{\frac{2}{\lambda^2}}{\frac{2}{\lambda}}$$

$$= \frac{\rho^2 + \lambda^2\sigma_v^2}{2\lambda(1-\rho)} \tag{2.6}$$

which is the Pollaczek–Khinchine formula for the mean waiting time of the $M/G/1$ queue.

(4) For the $D/D/1$ queue,

$$\sigma_u^2 = \sigma_v^2 = 0 \quad \text{and}$$

$$v_h^{(2)} = (v_h)^2 = \left(\frac{1}{\lambda} - \frac{1}{\mu}\right)^2,$$

so that (2.3) reduces to

$$E(W) = 0.$$

7.2.1.1. Mean of W (Multiserver Queues). *M/G/c queue.* Though no closed form expression for $E\{W_q\}$ is known, several authors (for example, Hokstad (1978) and Stoyan (1976); also Lee and Longton (*Opnl. Res. Qrly*, **10** (1957) 56), Nozaki and Ross (*J. Appl. Prob.* **15** (1978) 826), Maaløe (*Mgmt. Sci.* **19** (1973) 703) have independently suggested the approximation:

$$E\{W_q\} = \frac{1 + c_v^2}{2\lambda(1-\rho)}\rho C$$

where $C = Pr\{N \geq c\} = p_c/(1-\rho)$ is the blocking probability (see Eq. (6.4) Chapter 3) in an $M/M/c$ system with the same traffic intensity $\rho = \lambda/c\mu$, (c_v

is the coefficient of variation of service time distribution). The formula is exact for $M/M/c$ and $M/G/1$ queues. The approximation is quite satisfactory for large ρ, as well as for

$$E\{W_s\} = E\{W_q\} + 1/\mu.$$

Harel and Zipkin (1987) show that $f(\rho) = 1/E\{W_s\}$ is strictly concave for: (*i*) $M/G/1$ *iff* $c_v^2 < 1$; (ii) $M/G/c$, $c \geq 2$ for sufficiently light traffic. Refer to Harel and Zipkin for further details as well as for references of the approximation of $E\{W_q\}$.

Some other approximations have also been advanced, for example, Boxma *et al.* (1979), Tijms *et al.* (1981), and Kimura (1986).

G/G/c queue. Kimura (1986) suggests a simple two-moment approximate formula for the mean waiting time (in the queue) $E\{W(G/G/c)\}$ of a $G/G/c$ system as a weighted harmonic mean of the mean waiting times of three systems. It is as follows:

$$[E\{W(G/G/c)\}]^{-1} \simeq (c_u^2 + c_v^2)^{-1}\left[\frac{2(c_u^2 + c_v^2 - 1)}{E\{W(M/M/c)\}} + \frac{1 - c_v^2}{E\{W(M/D/c)\}} + \frac{1 - c_u^2}{E\{W(D/M/c)\}}\right].$$

[$c_u(c_v)$ is the coefficient of variation of $u(v)$].

The numerical values of the building block systems $E\{W(M/M/c)\}$, $E\{W(M/D/c)\}$ and $E\{W(D/M/c)\}$ can be obtained, for specific values of c and ρ, from Hillier and Yu (1981) and Page (1982). See Kimura (1986) for an account of other approximation formulas and for some numerical comparisons.

7.2.2 Variance of W

Theorem 7.2. *For a $G/G/1$ queue with $\rho < 1$ with FIFO service discipline, the variance σ_W^2 of W is given by*

$$\sigma_W^2 = \frac{3E(X_n^3)}{-3E(X_n)} + \left[\frac{E(X_n^2)}{-2E(X_n)}\right]^2 + \frac{E(I^3)}{3E(I)} - \left[\frac{E(I^2)}{2E(I)}\right]^2 \tag{2.7}$$

$$= \frac{[\lambda(v_u^{(3)} - v_v^{(3)}) - 3(\rho v_u^{(2)} - v_v^{(2)})]}{3(1 - P)} + \left[\frac{\{\lambda^2(\sigma_u^2 + \sigma_v^2) + (1 - \rho)\}^2}{2\lambda(1 - \rho)}\right]^2 + \frac{v_h^{(3)}}{3v_h} - \left[\frac{v_h^{(2)}}{2v_h}\right]^2 \tag{2.8}$$

where $v_u^{(r)}$, $v_v^{(r)}$, and $v_h^{(r)}$ are the rth moments ($r = 1, 2, 3$) of the interarrival, service-time, and idle-time distributions respectively, and $v^{(1)} = v$.

Proof. Cubing both sides of (2.2)

$$W_{n+1} - I_n = W_n + X_n,$$

noting that $I_n^2 W_{n+1} = I_n W_{n+1}^2 = 0$ and taking expectations, we get

$$E[W_{n+1}^3 - I_n^3] = E[W_n^3 + X_n^3 + 3W_n^2 X_n + 3W_n X_n^2]. \tag{2.9}$$

In steady state

$$E[W_{n+1}^r] = E[W_n^r] = E[W^r], \qquad r = 1, 2, 3.$$

Noting that W_n and X_n are independent, we get from (2.9)

$$E[-I_n^3] = E[X_n^3] + 3E[X_n]E[W^2] + 3E[W]E[X_n^2]. \tag{2.10}$$

Using $E(I_n^3) = a_0 E(I^3)$ and $E(X_n) = -a_0 E(I)$, we get

$$E(W^2) = \frac{E(X_n^3)}{-3E(X_n)} + \frac{E(X_n^2)}{-E(X_n)} E(W) + \frac{E(I^3)}{3E(I)}. \tag{2.11}$$

Putting the value of

$$E(W) = \frac{E(X_n^2)}{-2E(X_n)} - \frac{E(I^2)}{2E(I)}$$

from (2.5a), we get

$$\begin{aligned}
\sigma_W^2 &= E(W^2) - [E(W)]^2 \\
&= \frac{E(X_n^3)}{-3E(X_n)} + \frac{E(I^3)}{3E(I)} + \left[\frac{E(X_n^2)}{-E(X_n)} + \frac{E(X_n^2)}{2E(X_n)} + \frac{E(I^2)}{2E(I)}\right] \\
&\qquad \times \left[\frac{E(X_n^2)}{-2E(X_n)} - \frac{E(I^2)}{2E(I)}\right] \\
&= \frac{E(X_n^3)}{-3E(X_n)} + \left[\frac{E(X_n^2)}{-2E(X_n)}\right]^2 + \frac{E(I^3)}{3E(I)} - \left[\frac{E(I^2)}{2E(I)}\right]^2
\end{aligned}$$

which gives (2.7). Now

$$E(X_n) = -\frac{1}{\lambda}(1 - \rho)$$

$$E(X_n^2) = \frac{[\lambda^2(\sigma_u^2 + \sigma_v^2) + (1 - \rho)^2]}{\lambda^2}$$

$$E(X_n^3) = v_u^{(3)} - v_v^{(3)} - 3v_u^{(2)}\left(\frac{1}{\mu}\right) + 3\left(\frac{1}{\lambda}\right)v_v^{(2)}$$
$$= \frac{[\lambda(v_u^{(3)} - v_v^{(3)}) - 3(\rho v_u^{(2)} - v_v^{(2)})]}{\lambda}.$$

Using these expressions in (2.7), we at once get (2.8).

Note. It is tacitly assumed in the preceding derivations that all moments up to the required order (up to order two for validity of (2.4) and up to order three for validity of (2.8)) of the three distributions exist. It is shown by Marshall (1968) that the necessary and sufficient condition for the existence of all the moments up to order three is that the first three moments of the interarrival- and service-time distributions exist.

7.3. Queues with Batch Arrivals $G^{(x)}/G/1$

Marshall (1968b) extended his results to queues with batch arrival and batch service.

Suppose that the arrivals in batches occur in a renewal process with rate λ and each arrival consists of a batch of customers of random size X with probability mass function

$$P(X = m) = a_m, \qquad m = 0, 1, 2, \ldots$$

with rth moment

$$v_X^{(r)} \qquad (v_X^{(1)} = v_X), \qquad r = 1, 2, \ldots.$$

Customers arriving in a batch are assumed to be numbered in some way to denote the order of service, and service discipline is FIFO.

Let $V_{k,n}$ be the service time of the kth customer in the nth batch and let $V_{k,n}$ be IID RVs with common DF $A(\cdot)$ for all n and $k \geq 1$ and $V_{0,n} \equiv 0$. Let $W_{1,n}$ be the waiting time in queue of the first customer served of the nth batch. Let

$$V_n^* = V_{0,n} + V_{1,n} + \cdots + V_{X,n} \tag{3.1}$$

be the total service time of all arrivals of the nth batch and let its DF be $B^*(v) = P(V_n^* \leq v)$.

Now $\{W_{1,n+1}\}$ satisfies the relation

$$W_{1,n+1} = \max[0, W_{1,n} + X_n^*] \tag{3.2}$$

where $X_n^* = V_n^* - T_n^*$, T_n^* being the interarrival time between the arrival of the nth and $(n+1)$th batches. Assume that

$$\lambda v_X < \mu \left(\rho = \frac{\lambda v_X}{\mu} < 1 \right),$$

the queue is stationary, and

$$E[W_{1,n}] \to E[W_{1,.}],$$

the mean waiting time in queue of the first customer of an arbitrary batch.

We shall use the relation for the $G/G/1$ queue with single-arrival and individual service,

$$E[W] = \frac{E[X_n]^2}{-2E(X_n)} - \frac{E(I^2)}{2E(I)}$$

to find the mean waiting time in queue for the first customer in each batch.

For the batch-arrival case, we have V_n^* as the sum of a random number of IID RV each with mean v_v and variance σ_v^2. Thus,

$$\begin{aligned}\sigma_{V^*}^2 &= E(X)\ \text{var}(v) + \text{var}(X)[E(v)]^2 \\ &= v_X \sigma_v^2 + \sigma_X^2 \left(\frac{1}{\mu^2} \right).\end{aligned} \tag{3.3}$$

Thus, applying the result of Theorem 7.1, we get

$$E[W_{1,.}] = \frac{\sigma_u^2 + \sigma_{V^*}^2}{\dfrac{2}{\lambda}\left(1 - \dfrac{\lambda v_X}{\mu}\right)} + \frac{1 - \dfrac{\lambda v_X}{\mu}}{2\lambda} - \frac{v_h^{(2)}}{2v_h}. \tag{3.4}$$

We now proceed to find the mean waiting time in queue of an *arbitrary* customer in a batch. The mean total additional waiting time of all customers in an average batch is obtained and it is divided by the average number of customers per batch. Let Z_n be the total additional waiting time of all customers in some batch with n arrivals. Then $Z_n = 0$ if $X = 0$, 1 and for $X \geq 2$

$$Z_n = V_{1,n} + (V_{1,n} + V_{2,n}) + \cdots + (V_{1,n} + V_{2,n} + \cdots + V_{(X-1),n})$$

so that

$$E(Z_n|X = k) = E[(V_{1,n}) + (V_{1,n} + V_{2,n}) + \cdots + (V_{1,n} + \cdots + V_{k-1,n})]$$
$$= \frac{k(k-1)}{2} E(V) = \frac{k(k-1)}{2} \frac{1}{\mu} \quad \text{for } k \geq 0$$

and, thus,

$$E(Z_n) = \sum_{k=1}^{\infty} E(z_n|X = k)Pr(X = k)$$
$$= \sum_{k=1}^{\infty} \frac{k(k-1)}{2} \frac{1}{\mu} a_k$$
$$= \frac{1}{2\mu}\left[\sum_{k=1}^{\infty} k^2 a_k - \sum_{k=1} k a_k\right]$$
$$= \frac{1}{2\mu}[\nu_X^{(2)} - \nu_X] \quad \text{and}$$

$$\frac{E(Z_n)}{\text{average number of customers per batch}} = \frac{E(Z_n)}{\nu_X} = \frac{1}{2\mu}\left[\frac{\nu_X^{(2)}}{\nu_X} - 1\right]. \tag{3.5}$$

Thus, the expected wait of any unspecified customer $E(W)$ is then obtained by using (3.4) and (3.5). We have

$$E(W) = E(W_{1,.}) + \frac{1}{2\mu}\left[\frac{\nu_X^{(2)}}{\nu_X} - 1\right]. \tag{3.6}$$

7.4. The Output Process of a $G/G/1$ System

The interdeparture interval τ_n between the nth and $(n+1)$ departures is given by

$$\tau_n = D_{n+1} - D_n$$
$$= t_{n+1} + W_{n+1} + v_{n+1} - (t_n + W_n + v_n) \tag{4.1}$$
$$= (t_{n+1} - t_n) + W_{n+1} - W_n + v_{n+1} - v_n.$$

In steady state, $E(W_{n+1}) = E(W_n)$ so that the expected interdeparture interval is given by

$$\begin{aligned} E(\tau_n) &= E(t_{n+1} - t_n) \\ &= E(u_n) = \frac{1}{\lambda}. \end{aligned} \tag{4.2}$$

Theorem 7.3. *The variance of the interdeparture interval $\tau_n \equiv \tau$ in a $G/G/1$ queue is given by*

$$\operatorname{var}(\tau) = \sigma_v^2 - \frac{(1-\rho)^2}{\lambda^2} + \left[\frac{(1-\rho)}{\lambda}\right]\left[\frac{v_h^{(2)}}{v_h}\right].$$

Proof. We have

$$\tau_n = v_{n+1} + I_n \tag{4.3}$$

where v_{n+1} and I_n are independent. Hence

$$\operatorname{var}(\tau_n) = \operatorname{var}(v_{n+1}) + \operatorname{var}(I_n). \tag{4.4}$$

Again,

$$W_{n+1} - I_n = W_n + X_n = W_n + u_n - v_n \quad \text{and}$$

$$\begin{aligned} \operatorname{var}(W_{n+1} - I_n) &= \operatorname{var}(W_n) + \operatorname{var}(u_n) + \operatorname{var}(v_n) \\ &= \sigma_W^2 + \sigma_u^2 + \sigma_v^2. \end{aligned} \tag{4.5}$$

But

$$\begin{aligned} \operatorname{var}(W_{n+1} - I_n) &= \operatorname{var}(W_{n+1}) + \operatorname{var}(I_n) \\ &\quad -2\operatorname{cov}(W_{n+1}I_n). \end{aligned} \tag{4.6}$$

Now $W_{n+1}I_n = 0$, and hence

$$\begin{aligned} \operatorname{cov}(W_{n+1}I_n) &= E(W_{n+1}I_n) - E(W_{n+1})E(I_n) \\ &= -E(W)\left[\frac{1}{\lambda} - \frac{1}{\mu}\right]. \end{aligned} \tag{4.7}$$

From Eqs. (4.5)–(4.7), we get

$$\sigma_W^2 + \sigma_u^2 + \sigma_v^2 = \sigma_W^2 + \operatorname{var}(I_n) + 2E(W)\left(\frac{1}{\lambda} - \frac{1}{\mu}\right)$$

so that

$$\operatorname{var}(I_n) = \sigma_u^2 + \sigma_v^2 - 2\left(\frac{1}{\lambda} - \frac{1}{\mu}\right)E(W) \tag{4.8}$$

and putting the value of (4.8) in (4.4), we get

$$\operatorname{var}(\tau_n) = \sigma_u^2 + 2\sigma_v^2 - \frac{2}{\lambda}(1-\rho)E(W). \tag{4.9}$$

Using the value of $E(W)$ from (2.3), we get

$$\begin{aligned}\operatorname{var}(\tau_n) &= \sigma_u^2 + 2\sigma_v^2 - \left[\sigma_u^2 + \sigma_v^2 + \frac{1}{\lambda^2}(1-\rho)^2\right] + \frac{1}{\lambda}(1-\rho)\left(\frac{v_h^{(2)}}{v_h}\right)\\ &= \sigma_v^2 - \frac{(1-\rho)^2}{\lambda^2} + \left[\frac{(1-\rho)}{\lambda}\right]\left[\frac{v_h^{(2)}}{v_h}\right].\end{aligned} \tag{4.10}$$

7.4.1. Particular Case

For an $M/G/1$ queue, $v_h^{(2)}/v_h = 2/\lambda$ so we get

$$\begin{aligned}\operatorname{var}(\tau_n) &= \sigma_v^2 - \frac{(1-\rho)^2}{\lambda^2} + \frac{(1-\rho)}{\lambda}\frac{2}{\lambda}\\ &= \sigma_v^2 + \frac{1-\rho^2}{\lambda^2}.\end{aligned} \tag{4.11}$$

The variance of the output process of an $M/G/1$ system in steady state is exactly known when the mean and variance of the service-time distribution are exactly known. When the service time is exponential, i.e., when the system is $M/M/1$, $\sigma_v^2 = 1/\mu^2$ so that

$$\operatorname{var}(\tau_n) = \frac{1}{\mu^2} + \frac{1-\rho^2}{\lambda^2} = \frac{1}{\lambda^2}.$$

As is well known, the output process of an $M/M/1$ queue in steady state is Poisson with the same rate as the arrival process.

7.4.2. Output Process of a $G/G/c$ System

It is well known that the output process of $M/M/c$ and $M/G/\infty$ stationary queueing systems are Poisson. Whitt (1984b) examines the output process of a $G/G/c$ system. He shows that the output process in a large class of stationary $G/G/c$ systems is *approximately Poisson*, when there are many busy slow servers. Refer to Whitt for details and for limit theorems for the case when c and ρ increase.

7.5. Some Bounds for the $G/G/1$ System

7.5.1. Bound for $E(I)$

We have from (2.1) as $a_0 \leq 1$

$$E(I) \geq \frac{1-\rho}{\lambda} = \frac{1}{\lambda} - \frac{1}{\mu} \tag{5.1}$$

which gives a lower bound for $E(I)$. The equality holds for the $D/D/1$ queue.

7.5.2. Bounds for $E(W)$

7.5.2.1. *Upper Bound.* We have

$$v_h^{(2)} = E(I^2) = \text{var}(I) + [E(I)]^2 \geq E(I)^2$$

(from (5.1)) and so

$$\frac{v_h^{(2)}}{v_h} \geq E(I) \geq \frac{1}{\lambda}(1-\rho).$$

From (2.3), we get

$$\begin{aligned} E(W) &\leq \frac{\lambda^2(\sigma_u^2+\sigma_v^2)+(1-\rho)^2}{2\lambda(1-\rho)} - \frac{1}{2\lambda}(1-\rho) \\ &= \frac{\lambda(\sigma_u^2+\sigma_v^2)}{2(1-\rho)} \end{aligned} \tag{5.2}$$

which is an upper bound for $E(W)$ for a $G/G/1$ system.

The equality holds for the $D/D/1$ queue. The importance of the bounds (5.1) and (5.2) is that they involve the first two moments of the arrival- and service-time distributions.

7.5.2.2. *Lower Bound.* We shall discuss Marshall's lower bound as given next.

Theorem 7.4. *For a $G/G/1$ queue, $E(W) \geq r$ where r is the unique nonnegative root of the equation*

$$x = \int_{-x}^{\infty} \{1 - K(u)\}du; \tag{5.3}$$

the root is unique iff $\rho < 1$.

Proof. Let

$$f(x) = x - \int_{-x}^{\infty} \{1 - K(u)\} du;$$

we are to show that r is the unique nonnegative root of $f(x) = 0$. We have

$$\begin{aligned} f'(x) &= 1 - \{1 - K(-x)\} \\ &= K(-x) \geq 0 \quad \text{for } x \geq 0 \end{aligned}$$

so that $f(x)$ is monotonically increasing for $x \geq 0$. For $x = 0$,

$$f(0) = 0 - \int_{0}^{\infty} \{1 - K(u)\} du < 0.$$

For large x, say $x \to A(>0)$, we have

$$\begin{aligned} f(A) &= A - \int_{-A}^{\infty} \{1 - K(u)\} du \\ &= A - \int_{-A}^{\infty} \left\{ \int_{u}^{\infty} dK(t) \right\} du \\ &= A - \int_{-A}^{\infty} \left\{ \int_{-A}^{t} du \right\} dK(t) \\ &= A - \int_{-A}^{\infty} (t + A) dK(t) \\ &\geq A - \int_{-\infty}^{\infty} (t + A) dK(t) \\ &= A - \int_{-\infty}^{\infty} t dK(t) - A \int_{-\infty}^{\infty} dK(t) \\ &= A - \left(\frac{1}{\mu} - \frac{1}{\lambda} \right) - A \\ &= \frac{1 - \rho}{\lambda} > 0 \quad \text{when } \rho < 1. \end{aligned}$$

Thus, we have $f(0) < 0$ and $f(A) > 0$ where A is large and $\rho < 1$. Thus, $f(x) = 0$ has a unique nonnegative root when $\rho < 1$ i.e. Eq. (5.3) has a unique nonnegative root when $\rho < 1$. Denoting this root by r, we shall show that $E(W) \geq r$. Let

$$g(x) = \int_{-x}^{\infty} \{1 - K(u)\} du \tag{5.4}$$

so that

$$f(x) = x - g(x).$$

As $f(0) < 0$, $f(r) = 0$, and $f(A) > 0$ for large A, we get

$$\begin{aligned} g(x) &> x \quad \text{when } x < r \\ &\leq x \quad \text{when } x \geq r. \end{aligned} \tag{5.5}$$

Thus, the function $g(x)$ is continuous and convex for $x \geq 0$. Again, since

$$W_{n+1} = \max(0, W_n + X_n)$$

we have that

$$Z = (W_{n+1} | W_n = x) = \max(0, X_n + x)$$

is positive, so that

$$E(Z) = E(W_{n+1} | W_n = x) = \int_0^\infty \{1 - G(t)\} dt$$

where $G(t) = P(Z \leq t)$ is the DF of Z. Now

$$\begin{aligned} G(t) &= P\{Z \leq t\} \\ &= P\{0 \leq t\} P\{X_n + x \leq t\} \\ &= P\{X_n \leq t - x\} \\ &= K(t - x). \end{aligned}$$

Thus, for $x \geq 0$,

$$\begin{aligned} E(Z) &= \int_0^\infty \{1 - K(t - x)\} dt \\ &= \int_{-x}^\infty \{1 - K(v)\} dv, \text{ putting } v = t - x \\ &= g(x). \end{aligned} \tag{5.6}$$

Again using the relation

$$\begin{aligned} E(X) &= E\{E(X | Y = y)\} \\ &= \int_0^\infty E(X | Y = y) dF(y), \end{aligned} \tag{5.7}$$

where F is the DF of Y; we have, using (5.6),

$$\begin{aligned} E(W_{n+1}) &= E[E(W_{n+1} \mid W_n = x)] \\ &= \int_0^\infty E(W_{n+1} \mid W_n = x) dW(x) \\ &= \int_0^\infty g(x) dW(x) \end{aligned} \tag{5.8}$$

where $g(x)$ is a continuous convex function.

Using Jensen's inequality for the expected value of a convex function of a nonnegative RV, we get

$$E(W_{n+1}) \geq g[E(W_n)]. \tag{5.9}$$

In steady state, $E(W_n) = E(W_{n+1}) = E(W)$, so that

$$\begin{aligned} E(W) &\geq g(E(W)) \\ &= \int_{-E(W)}^\infty \{1 - K(u)\} du. \end{aligned} \tag{5.10}$$

Assume, if possible, that $E(W) < r$; then from (5.5)

$$g[E(W)] = \int_{-E(W)}^\infty \{1 - K(u)\} du > E(W) \quad \text{for } E(W) < r$$

which contradicts (5.10). This contradiction is due to our assumption that $E(W) < r$. Thus, we must have

$$E(W) \geq r.$$

Finally, putting the upper and lower bounds together (from Eqs. (5.2) and (5.3)) we get

$$r \leq E(W) \leq \frac{\lambda(\sigma_u^2 + \sigma_v^2)}{2(1 - \rho)}. \tag{5.11}$$

Remarks.

(1) We can write Eq. (5.3) as

$$\begin{aligned} x &= \alpha + \int_{-x}^0 \{1 - K(u)\} du \\ \text{where} \quad \alpha &= \int_0^\infty \{1 - K(u)\} du. \end{aligned} \tag{5.12}$$

Equation (5.3) has a solution *iff* the curves $y = x$ and

$$y = \alpha + \int_{-x}^{0} \{1 - K(u)\} du$$

intersect.

If $\alpha = 0$, then the second curve also passes through the origin and $x = 0$ is a solution.

If $\alpha > 0$, then $y = \alpha > 0$ when $x=0$, so that the second curve will lie above the first curve at the origin. Thus, the two curves will intersect if and only, for x sufficiently large, the first curve will lie above the second curve, that is, *iff*

$$\begin{aligned} x &> \alpha + \int_{-x}^{0} \{1 - K(u)\}\, du \\ &= \alpha + x - \int_{-x}^{0} K(u) du, \end{aligned}$$

or *iff* $\int_{-x}^{0} K(u)du > \alpha = \int_{0}^{\infty} \{1 - K(u)\}du$, in other words, *iff*

$$\beta = \int_{-\infty}^{0} K(u)du > \int_{0}^{\infty} \{(1 - K(u)\}du = \alpha. \tag{5.13}$$

Now write

$$\begin{aligned} Z_1 &= \max(0, X_n) \\ Z_2 &= -\min(0, X_n) \\ F_1(t) &= Pr(Z_1 \leq t) = Pr(0 \leq t)Pr(X_n \leq t) = K(t), \\ F_2(t) &= Pr(Z_2 \leq t) = Pr\{-\min(0, X_n) \leq t\} \\ &= Pr\{\min(0, X_n) \geq -t\} \\ &= Pr\{0 \geq -t\}Pr\{X_n \geq -t\} \\ &= 1 - K(-t). \end{aligned}$$

Since Z_1 and Z_2 are positive RVs,

$$E(Z_1) = \int_{0}^{\infty} \{1 - F_1(t)\}dt = \int_{0}^{\infty} \{1 - K(t)\}dt = \alpha \quad \text{and}$$

$$E(Z_2) = \int_{0}^{\infty} \{1 - F_2(t)\}dt = \int_{0}^{\infty} K(-t)dt = -\int_{-\infty}^{0} K(t)dt = -\beta$$

so that (5.13) implies

$$E\{\min(0, X_n)\} > E\{\max(0, X_n)\}.$$

This leads to

$$\begin{aligned} E\{\max(0, X_n) - \min(0, X_n)\} &< 0 \quad \text{or} \\ E\{X_n\} &< 0 \quad \text{or} \\ E(v_n - u_n) &< 0 \quad \text{or} \\ \frac{1}{\mu} - \frac{1}{\lambda} &< 0 \quad \text{or} \\ \rho &< 1. \end{aligned}$$

Thus, the two curves will intersect *iff* $\rho < 1$. The x-coordinate of the point of intersection will be the lower bound of the expected waiting time. The situation is shown in Fig. 7.1.

(2) Since $\sigma_u^2 + \sigma_v^2 > 0$ for all systems except for $D/D/1$, both the bounds tend to infinity if $\rho \to 1$.

(3) Rosberg (1987) derives new lower and upper bounds. The bounds obtained are functions of moments of order $r(r > 2)$ of u_n and v_n. It is claimed that these moments are better for low traffic intensity.

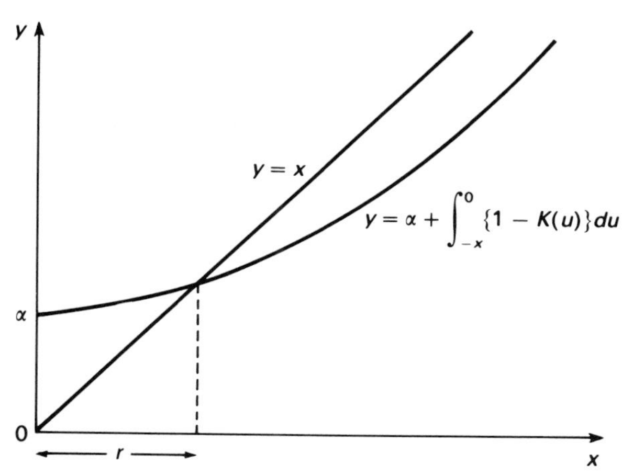

FIGURE 7.1. Graphical determination of the lower bound of the expected waiting time in a $G/G/1$ queue.

Example 7.4. The value of r for an $M/M/1$ queue is given by

$$r = \left(-\frac{1}{\lambda}\right)\log(1-\rho^2).$$

Proof. For an $M/M/1$ queue

$$A(x) = P(u_n \leq x) = 1 - e^{-\lambda x} \quad \text{and}$$
$$B(x) = P(v_n \leq x) = 1 - e^{-\mu x}$$

so that from (1.4), for $x \geq 0$

$$\begin{aligned} K(x) &= \int_0^\infty \{1 - e^{-\mu(x+u)}\}\lambda e^{-\lambda u}\,du \\ &= 1 - \lambda\int_0^\infty e^{-(\lambda+\mu)u}e^{-\mu x}\,du. \end{aligned}$$

Thus,

$$K(x) = 1 - \frac{\lambda e^{-\mu x}}{\lambda+\mu}, \qquad x \geq 0. \tag{5.14a}$$

For $x < 0$,

$$\begin{aligned} K(x) &= \int_0^\infty B(x+u)dA(u) \\ &= \int_0^\infty B(v)dA(v-x), \quad \text{since } x < 0 \\ &= \int_0^\infty (1 - e^{-\mu v})\lambda e^{-(v-x)\lambda}\,dv \\ &= \lambda\left[\int_0^\infty e^{-(v-x)\lambda}\,dv - \int_0^\infty e^{-(\mu+\lambda)v}e^{x\lambda}\,dv\right]. \end{aligned}$$

Thus,

$$K(x) = e^{\lambda x} - \frac{\lambda e^{\lambda x}}{\lambda+\mu} = \frac{\mu e^{\lambda x}}{\lambda+\mu}, \qquad x < 0. \tag{5.14b}$$

The lower bound r is the unique positive root of

$$\begin{aligned}
0 &= x - \int_{-x}^{\infty} \{1 - K(t)\} dt \\
&= x - \int_{-x}^{0} \{1 - K(t)\} dt - \int_{0}^{\infty} \{1 - K(t)\} dt \\
&= x - \int_{-x}^{0} \left\{1 - \frac{\mu e^{\lambda t}}{\lambda + \mu}\right\} dt - \int_{0}^{\infty} \frac{\lambda}{\lambda + \mu} e^{-\mu t} dt \\
&= x - x + \frac{\mu}{\lambda(\lambda + \mu)} (1 - e^{-\lambda x}) - \frac{\lambda}{\lambda + \mu} \frac{1}{\mu} \\
&= \frac{\mu^2 - \lambda^2 - \mu^2 e^{-\lambda x}}{\lambda \mu (\lambda + \mu)} \\
&= \frac{\mu - \lambda}{\lambda \mu} - \frac{\mu e^{-\lambda x}}{\lambda(\lambda + \mu)}
\end{aligned}$$

whence

$$e^{-\lambda x} = (1 - \rho^2)$$

so that the root r is given by

$$r = -\frac{1}{\lambda} \log_e(1 - \rho^2) = -\frac{1}{\lambda} \ln(1 - \rho^2). \tag{5.15}$$

Note. The lower bound tends to ∞ as $\rho \to 1$. The upper bound is

$$\frac{\lambda\left(\frac{1}{\lambda^2} + \frac{1}{\mu^2}\right)}{2\left(1 - \frac{\lambda}{\mu}\right)} = \frac{1}{\lambda} \frac{1 + \rho^2}{2(1 - \rho)}$$

which also tends to ∞ as $\rho \to 1$. Thus, as $\rho \to 1$, both the bounds tend to ∞. As $\rho \to 0$, the lower bound tends to 0 and the upper bound tends to $1/2\lambda$. The upper and lower bounds and the true value of $E(W)$ are shown in Fig. 7.2 for fixed $\lambda = 1$ and varying μ, i.e., for $\rho = 0$ to $\rho = 1$.

Another lower bound has been put forward by Marchal (1978), which we discuss shortly.

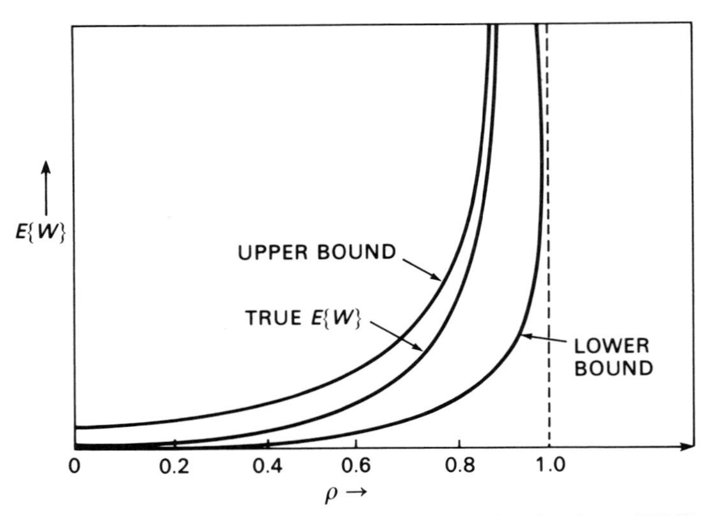

FIGURE 7.2. Upper and lower bounds of the expected waiting time in an $M/M/1$ queue.

Figures 7.1 and 7.2 with kind permissions from the author K. T. Marshall and the publishers of *Operations Research.*

Theorem 7.5. *For a $G/G/1$ queue*

$$E(W) \geq \frac{\lambda^2\sigma_v^2 + \rho(\rho - 2)}{2\lambda(1 - \rho)}.$$

Marchal derives the preceding lower bound from the lower bound obtained by Kingman (1962), which is as follows:

$$E\{W\} \geq \frac{\lambda}{2(1 - \rho)} E\{[\max(0, v - u)]^2\}. \tag{5.16}$$

For details of proof and other results, such as bounds of $G/G/c$ and of $M/G/c$ systems, refer to Marchal (1978).

Notes.

(1) The lower bound depends on the mean of u and the first two moments of v.

(2) Marchal's lower bound is positive *iff*

$$\sigma_v^2 \geq \frac{2 - \rho}{\lambda\mu}.$$

Thus, this lower bound is of use only when this inequality is satisfied.

Example 7.5. The bounds of $\text{var}(\tau_n)$ are given by

$$\sigma_v^2 \le \text{var}(\tau_n) \le \sigma_u^2 + 2\sigma_v^2 - 2r\left(\frac{1}{\lambda} - \frac{1}{\mu}\right).$$

We have from (4.9)

$$\text{var}(\tau_n) = \sigma_u^2 + 2\sigma_v^2 - \left(\frac{2}{\lambda}\right)(1 - \rho)E(W).$$

Using (5.11) we get at once

$$\sigma_v^2 \le \text{var}(\tau_n) \le \sigma_u^2 + 2\sigma_v^2 - 2r\,\frac{(1-\rho)}{\lambda}.$$

Remarks.

(1) It may be noticed that the moments of the idle-time distribution figured in many of the expressions obtained. The idle-time distribution is some complicated tail distribution of interarrival time. Marshall (1968a) observed that by placing some restrictions on the interarrival-time distribution, it is possible to obtain some desirable properties of the moments of the idle-time distribution. Marshall considered some special type of interarrival-time distribution (having bounded mean residual life) and obtained bounds for $E(I^2)/2E(I)$. In another paper (1968b) he considered further generalizations. (See Problems and Complements.)

(2) Another approach to a $G/G/1$ queue through Wiener–Hopf factorization is considered in detail by Prabhu (1980): ladder processes and the random walk $\{S_n\}$, where

$$S_n = X_1 + \cdots + X_n,$$

prove very important in this approach. Denote

$$\bar{N} = \min\{n > 0;\ S_n \le 0\} \quad \text{and}$$

$$\begin{aligned} g_n(x) &= Pr\{\bar{N} = n, \quad S_{\bar{N}} \le x\} \\ &= Pr\{S_1 > 0, S_2 > 0, \ldots, S_{n-1} > 0, S_n \le x < 0\}. \end{aligned}$$

Then $\bar{N}$ is the number served during a busy period and $I = -S_{\bar{N}}$ is the idle period. The distribution of $\{W_n\}$ is completely solved, and the joint distribution of $\bar{N}$ and I has been obtained in explicit form for $M/M/1$ and $G/M/1$

systems. (See Problems and Complements.) Ladder processes, which were primarily used for the study of waiting times and idle periods of a $G/G/1$ queue, have also been found to be important in the study of queue-length processes in some special cases. (See Prabhu (1980) for further results for the $G/G/1$ queue; also Whitt (1980, 1984).)

Problems and Complements

7.1. Using (5.14a) and (5.14b), show that the (two-sided) LST of the RV $X_n = v_n - u_n$ for an $M/M/1$ system is given by

$$K^*(s) = \frac{\lambda\mu}{(\lambda - s)(\mu + s)}.$$

Verify that $K^*(s) = B^*(s)A^*(-s)$.

7.2. Show that the DF $K(x) = P\{X_n \le x\}$, $X_n = v_n - u_n$ for the system $G/M/1$ is given by

$$K(x) = \int_{-x}^{\infty} dA(u)\{1 - e^{-\lambda(u+x)}\}, \qquad x \le 0, \quad \text{and}$$

$$= 1 - ce^{-\lambda x}, \qquad x \ge 0,$$

where $c = \psi(\theta) = \int_0^\infty e^{-\theta x}\, dA(x)$ is the LST of the interarrival-time distribution. Further show that $K(x)$ has the partial lack of memory

$$P\{X_n \le y | X_n > x\} = 1 - e^{-\lambda(y-x)}, \qquad y > x \ge 0.$$

(Prabhu, 1980).

7.3. For an $M/D/1$ system find $K^*(s)$, $K(u)$; find $E(X_n)$ from $K^*(s)$.

7.4. Obtain the upper and lower bounds of $E(W)$ for a $D/M/1$ queue.

7.5. Using Eq. (1.25) find the waiting-time distribution for the $D/D/1$ queue.

7.6. For the sample sequence $\{W_n, n \ge 0\}$ with first eight elements

$$A = \{W_0 = 0,\ W_1 > 0,\ W_2 = 0,\ W_3 > 0,\ W_4 > 0,\ W_5 = 0,$$
$$W_6 > 0,\ W_7 > 0\}$$

compute the W_ns in terms of S_ns, and the $I_n s$ in terms of S_ns, and the number served during the busy periods corresponding to A. (With notations as in Section 7.1).

7.7. Show that

$$W_n = \max(0, X_n + X_{n-1} + \cdots + X_{n-r+1}, 1 \le r \le n).$$

7.8. Find the expressions for the mean and the variance of the waiting time for the systems (i) $E_2/M/1$ and (ii) $M/G/1$.

7.9. Batch service with fixed batch size k. Let $W_{k,n}$ be the waiting time in queue of the last (kth) person who arrived and formed the nth batch of k customers. Then show that the expected waiting time for an unspecified customer $E(W)$ is given by

$$E(W) = E(W_k, .) + \frac{(k-1)}{2\lambda}.$$

Deduce that Little's formula is satisfied for $M/G^k/1$ with services in batches of size k (Marshall, 1968b).

7.10. Suppose that there is a random start-up time R after every idle period so that the first customer in every busy period suffers a random delay R before his service commences, and assume that the idle time WRT the server is $I + R$. Then

$$E(W) = \frac{E(U^2)}{-2E(U)} + \frac{E(R^2) - E(I^2)}{2[E(R) + E(I)]} + \frac{\operatorname{cov}(W_n, U_n)}{E(U)}$$

where $U_n = v_n - u_n$; and assuming independence of I and R we get that $E(W)$ is given by the first two terms on the RHS. (Marshall, 1968b).

Note. A specific example of the preceding class of queues is the single-service queue with N-policy, where the server starts his first service in a busy period only after N customers arrive and then serves one by one until all the customers are served. See Sections 6.45 and 8.3.3 for a queue under N-policy.

7.11. Marchal's weighting factors.
As an approximation for $E(W)$, Marchal suggests some weighting factors to be used with the upper bound. The weighting factor

$$\frac{\rho^2 + \lambda^2\sigma_v^2}{1 + \lambda^2\sigma_v^2}$$

(which tends to 1 as ρ tends to 1) used to scale down the upper bound leads to the approximation

$$E(W) \simeq \frac{\lambda(\sigma_u^2 + \sigma_v^2)}{2(1-\rho)} \frac{\rho^2 + \lambda^2\sigma_v^2}{1 + \lambda^2\sigma_v^2}.$$

When the arrival process is Poisson, show that this leads to the

Pollaczek–Khinchine formula, i.e., it is exact for $M/G/1$. Show that it works well with a $G/M/1$ system also.

Another weighting factor suggested is

$$\frac{\rho^2\sigma_u^2 + \sigma_v^2}{\sigma_u^2 + \sigma_v^2}$$

(which also tends to 1 as ρ tends to 1). Using this to scale down the upper bound, show that one can get as an approximation

$$E(W) \simeq \frac{\rho(\lambda^2\sigma_u^2 + \mu^2\sigma_v^2)}{2\mu(1-\rho)}.$$

Show that it is exact for $M/G/1$ and $D/D/1$ (Marchal, 1978).

7.12. Distribution with bounded mean residual life. A nondiscrete distribution F is said to have its mean residual life bounded above (below) by γ, denoted by γ-MRLA (γ-MRLB) *iff*

$$\int_t^\infty \frac{\{1-F(u)\}du}{\{1-F(t)\}} \underset{(\geq)}{\leq} \gamma \quad \text{for all } t \geq 0,\ \gamma < \infty.$$

For a $G/G/1$ queue where the interarrival-time distribution is γ-MRLA (γ-MRLB), show that

$$\frac{E(I^2)}{2E(I)} = \frac{v_h^{(2)}}{2v_h} \underset{(\geq)}{\leq} \gamma$$

(Marshall, 1968a).

7.13. Show that for an $M/M/1$ queue

$$E\{z^{\bar{N}}e^{itI}\} = \frac{\lambda\zeta(z)}{\lambda - it}.$$

where

$$\zeta(z) = \frac{[(\lambda+\mu) - \sqrt{\{(\lambda+\mu)^2 - 4\lambda\mu z\}}]}{2\lambda}.$$

Write down the PGF of $\bar{N}$ and the PDF of I. (Prabhu, 1980).

7.14. Show that for a $G/M/1$ queue

$$E\{z^{\bar{N}}e^{-itI}\} = \frac{\mu\xi - \mu z\phi_1(-t)}{\mu\xi - \mu + it}, \qquad 0 < z < 1, \qquad t \text{ real}$$

where ϕ_1 is the characteristic function of the RV having DF $A(u) =$

$Pr\{u_n \leq u\}$ and $\xi = \xi(z)$ is the unique continuous solution of the equation

$$\xi = z\psi(\mu - \mu\xi)$$

in the interval $0 < z < 1$, $\psi(\theta)$ being the LST of the RV having DF $A(u)$ (Prabhu 1980).

References

Boxma, O. J. Cohen, J. W., and Huffels, N. (1979). Approximations of the mean waiting time in an $M/G/s$ queueing system. *Opns. Res.* **27**, 1115–1127.

Brumelle, S. L. (1973). Bounds on the wait in a $GI/M/k$ Queue. *Mgmt. Sci.* **19**, 773–777.

Cohen, J. W. (1982). *The Single Server Queue*, 2nd Ed. North-Holland, Amsterdam.

Gross, D., and Harris, C. M. (1985). *Fundamentals of Queueing Theory*, 2nd Ed. Wiley, New York.

Harel, A. and Zipkin, P. (1987). Strong convexity results for queueing systems. *Opns. Res.* **35**, 405–418.

Heyman, D. P. (1968). Optimal control policies for $M/G/1$ queueing systems. *Opns. Res.* **16**, 362–382.

Hillier, F. S., and Yu, O. S. (1981). *Queueing Tables and Graphs.* North-Holland, New York.

Hokstad, P. (1978). Approximations for the $M/G/m$ queue. *Opns. Res.* **26**, 510–523.

Jagerman, D. (1987). Approximations for waiting time in $GI/G/1$ systems. *Queueing Systems* **2**, 351–362.

Jain, J. L., and Grassman, W. K. (1989). Numerical solution for the departure process from $GI/G/1$ queue. *Comp. & Opns. Res.* **15**, 293–296.

Kimura, T. (1986). A two-moment approximation for the mean waiting time in $GI/G/s$ queue. *Mgmt. Sci.* **32**, 751–763.

Kingman, J. F. C. (1962). Some inequalities for the queue $GI/G/1$. *Biometrika* **49**, 315–324.

Kingman, J. F. C. (1970). Inequalities in the theory of queues. *J.R.S.S.* **B32**, 102–110.

Kleinrock, L. (1976). *Queueing Systems, Vol. II: Computer Applications.* Wiley, New York.

Lindley, D. V. (1952). The theory of queues with a single server, *Proc. Camb. Phil, Soc.* **48**, 277–289.

Marchal, W. G. (1978). Some simpler bounds on the mean queueing time. *Opns. Res.* **26**, 1083–1088.

Marshall, K. T. (1968a). Some inequalities in queueing *Opns. Res.* **16**, 651–665; comments (by R. V. Evans), 666–668.

Marshall, K. T. (1968b). Bounds for some generalizations of the $GI/G/1$ queue. *Opns. Res.* **16**, 841–848.

Marshall, K. T. (1968c). Some relationships between the distributions of waiting time, and inter-output time in the $GI/G/1$ queue. *SIAM J. Appl. Math.* **16**, 324–327.

Newell, G. F. (1971). *Applications of Queueing Theory.* (2nd ed., 1982) Chapman and Hall, London.

Page, E. (1982). Tables of waiting times for $M/M/n$, $M/D/n$ and $D/M/n$ and their use to give approximate waiting times in more general queues. *J. Opnl. Res. Soc.* **33**, 453–473.

Prabhu, N. U. (1980) *Stochastic Storage Processes.* Springer-Verlag, New York.

Rosberg, Z. (1987). Bounds on the expected waiting time in a $GI/G/1$ queue: upgrading for lower traffic intensity. *J. Appl. Prob.* **24**, 749–757.

Ross, S. M. (1974). Bounds on the delay distribution in $GI/G/1$ queues. *J. Appl. Prob.* **24**, 749–757.

Schassberger, R. (1970). On the waiting time in queueing system $GI/G/1$. *Ann. Math. Statist.* **41**, 182–187.

Sobel, M. (1980). Simple inequalities for multiserver queues. *Mgmt. Sci.* **26**, 951–956.

Stoyan, D. (1976). Approximations for $M/G/s$ queues. *Math. Oper. Statist.* **7**, 587–594.

Stoyan, D. (1977). Bounds and approximations in queueing through monotonicity and continuity. *Opns. Res.* **25**, 851–863.

Tijms, H. C., Van Hoorn, M. H., and Federgruen, A. (1981). Approximations for the steady-state probabilities in the $M/G/c$ queue, *Adv. Appl. Prob.* **13**, 186–206.

Weber, R. R. (1983). A note on the waiting times in single server queues. *Opns. Res.* **31**, 950–951.

Whitt, W. (1980). The effect of variability in $GI/G/s$ queue. *J. Appl. Prob.* **17**, 1062–1071.

Whitt, W. (1984a). Minimizing delays in the $GI/G/1$ queue. *Opns Res.* **32**, 41–51.

Whitt, W. (1984b). Departures from a queue with many servers. *Math. Opns. Res.* **9**, 534–544.

8 Miscellaneous Topics

8.1. Heavy-Traffic Approximation for Waiting-Time Distribution

8.1.1. Kingman's Heavy-Traffic Approximation for a G/G/1 Queue

A queue with traffic intensity barely less than unity is called a heavy-traffic queue. The behavior of a queueing system $G/G/1$ in the heavy-traffic case was first investigated by Kingman (1961). We discuss here Kingman's result, which is a sort of Central Limit Theorem for heavy traffic and is given next.

Theorem 8.1. *Under heavy traffic, the steady-state waiting-time distribution in a $G/G/1$ queue can be approximated by an exponential distribution.*

Proof. The starting point in this study is Eq. (1.13) of Chapter 7.

$$\bar{W}(s) = \frac{\bar{W}^{-}(s)}{B^*(s)A^*(-s) - 1}$$

where $\bar{W}(s)$ is the LT of the waiting time W, while $B^*(s)$ and $A^*(s)$ are the LSTs of the service-time and interarrival-time distributions v and u, respectively, and $\bar{W}^-(s)$ is the two-sided LT of W^- defined in Section 7.1.2. The preceding is written as

$$B^*(s)A^*(-s) - 1 = \frac{\bar{W}^-(s)}{\bar{W}(s)}. \tag{1.1}$$

Maclaurin's expansion of $B^*(s)$ gives

$$B^*(s) = \sum_{k=0}^{\infty} \frac{s^k}{k!} B^{*(k)}(0) \tag{1.2}$$

where $B^{*(k)}(0)$ is the kth derivative of $B^*(s)$ at $s = 0$. Denote $E(v^k) = b_k$, then

$$B^{*(k)}(0) = (-1)^k E(v^k) = (-1)^k b_k.$$

Truncating the series on the RHS of (1.2) to three terms, we get

$$B^*(s) = 1 - b_1 s + \frac{b_2}{2} s^2 + O(s^2). \tag{1.3}$$

Similarly, writing $E(u_k) = a_k$, and truncating to three terms, we get

$$A^*(-s) = 1 + a_1 s + \frac{a_2}{2} s^2 + O(s^2). \tag{1.4}$$

Thus,

$$\begin{aligned} B^*(s)A^*(-s) - 1 &= \left(1 - b_1 s + \frac{b_2}{2} s^2\right)\left(1 + a_1 s + \frac{a_2 s^2}{2}\right) - 1 + O(s^2) \\ &= 1 + (a_1 - b_1)s + \left(\frac{a_2}{2} + \frac{b_2}{2} - a_1 b_1\right)s^2 - 1 + O(s^2) \\ &= s\left[(a_1 - b_1) + \left(\frac{a_2}{2} + \frac{b_2}{2} - a_1 b_1\right)s\right] + O(s^2). \end{aligned} \tag{1.5}$$

Clearly $s = 0$ is one of the roots of the LHS of (1.5). There is another root s_0 that satisfies

$$(a_1 - b_1) + \left(\frac{a_2}{2} + \frac{b_2}{2} - a_1 b_1\right)s_0 = 0. \tag{1.6}$$

Denote

$$a_1 = \frac{1}{\lambda}, \qquad a_2 - a_1^2 = \sigma_u^2, \qquad b_1 = \frac{1}{\mu}, \qquad b_2 - b_1^2 = \sigma_v^2.$$

Then

$$a_1 - b_1 = \left(\frac{1}{\lambda}\right)(1-\rho), \qquad \rho = \frac{\lambda}{\mu} \quad \text{and}$$

$$\frac{a_2}{2} + \frac{b_2}{2} - a_1 b_1 = \frac{1}{2}\left(\sigma_u^2 + \frac{1}{\lambda^2}\right) + \frac{1}{2}\left(\sigma_v^2 + \frac{1}{\mu^2}\right) - \frac{1}{(\lambda\mu)}$$

$$= \frac{1}{2}(\sigma_u^2 + \sigma_v^2) + \left(\frac{1}{2\lambda^2}\right)(1-\rho)^2.$$

Under the heavy-traffic condition $\rho \simeq 1$, and, thus, the second term of the preceding can be neglected, so that from (1.6), we get

$$s_0 = -\frac{\left(\frac{1}{\lambda}\right)(1-\rho)}{\left\{\frac{1}{2}(\sigma_u^2 + \sigma_v^2)\right\}}$$

$$= -\frac{2(1-\rho)}{\lambda(\sigma_u^2 + \sigma_v^2)}. \tag{1.7}$$

Thus, we can rewrite (1.5), as an approximation near $s = 0$, as

$$B^*(s)A^*(-s) - 1 = s(s - s_0)\left\{\frac{(\sigma_u^2 + \sigma_v^2)}{2}\right\}$$

$$= s(s - s_0)K \tag{1.8}$$

where

$$K = \frac{1}{2}(\sigma_u^2 + \sigma_v^2).$$

From (1.1), we have, near the origin,

$$s(s - s_0)K = \frac{\bar{W}^-(s)}{\bar{W}(s)} \quad \text{or} \tag{1.9}$$

$$\bar{W}^-(s) = s(s - s_0)K\bar{W}(s)$$

$$= (s - s_0)K[s\bar{W}(s)] \tag{1.10}$$

$$= (s - s_0)KW^*(s)$$

where $W^*(s)$ is the LST of W, i.e.,

$$\begin{aligned} W^*(s) &= \int_0^\infty e^{-st}\,dW(t) = s\int_0^\infty e^{-st}W(t)dt \\ &= s\bar{W}(s). \end{aligned}$$

Since near the origin $W^*(s) = 1$, we get

$$\bar{W}^-(s) = -s_0 K.1.$$

Thus, from (1.1) and (1.8), we have

$$\begin{aligned} \bar{W}(s) &= \frac{-s_0 K}{s(s - s_0)K} = -\frac{s_0}{s(s - s_0)} \\ &= \frac{1}{s} - \frac{1}{s - s_0}. \end{aligned} \tag{1.11}$$

Inverting the LT, we get

$$\begin{aligned} W(t) &= 1 - \exp(s_0 t) \\ &= 1 - \exp\left(-\frac{2(1 - \rho)}{\lambda(\sigma_u^2 + \sigma_v^2)}\,t\right) \end{aligned} \tag{1.12}$$

which gives the distribution function of the waiting time to an approximation. The distribution is exponential with mean given by

$$E(W) \simeq \frac{\lambda(\sigma_u^2 + \sigma_v^2)}{2(1 - \rho)}, \tag{1.13}$$

i.e., with parameter $1/E(W) = -s_0$.

Remarks.

(1) The result showing the exponential character of the waiting-time distribution of a $G/G/1$ queue may be called the *Central Limit Theorem* for queueing theory. The result is extremely robust.

(2) In Section 7.5 we discussed the upper bound for the average waiting time $E(W)$ in a $G/G/1$ system for $0 \le \rho \le 1$ (as obtained by Kingman, 1962b). He shows that

$$E(W) \le \frac{\sigma_u^2 + \sigma_v^2}{2\left(\frac{1}{\lambda}\right)(1 - \rho)}. \tag{1.14}$$

This result gives an upper bound of $E(W)$ for $0 \le \rho \le 1$. The bound is reached when $\rho \to 1$; this result has been obtained in this section. This shows that heavy-traffic mean waiting time forms the strict upper bound for the mean waiting time in a $G/G/1$ system.

(3) The distribution function (1.12) is of the form

$$\begin{aligned} W(t) &= 1 + ke^{s_0 t} \\ &= 1 + s_0 E(W) e^{s_0 t}. \end{aligned} \tag{1.12a}$$

The constant

$$k = s_0 E(W) \tag{1.12b}$$

equals -1 in the heavy-traffic situation, as previously discussed.

Fredericks (1982) proposes a class of approximations for a $G/G/1$ waiting-time–distribution function of the form (1.12a) and develops a procedure to estimate the parameters k and s_0. The dominant root and the mean delay, whenever known, can be used to determine the (constant) coefficient k, by using 1.12b).

Particular case. For an $M/G/1$ queue

$$A^*(s) = \frac{\lambda}{\lambda + s}$$

so that $A^*(-s)B^*(s) - 1 = 0$ reduces to

$$s\left[(1 - \lambda b_1) + \left(\frac{\lambda b_2}{2}\right)s\right] = 0.$$

Thus,

$$s_0 = -\frac{2(1-\rho)}{(\lambda b_2)}$$

and for exponential service time $b_2 = 2/\mu^2$ so that the mean waiting time $E(W)$ for an $M/M/1$ is given by

$$E(W) = -\frac{1}{s_0} = \frac{\rho}{\mu(1-\rho)}.$$

Note. Kingman's result holds for very heavy traffic for the general $G/G/1$ queue. The waiting-time distribution even in $M/G/1$ queues for arbitrary ρ is somewhat difficult to compute. Benes (1956) has shown that the distribution can be expressed as a geometrically decreasing weighted sum of convolutions of residual service times. The approximation is convenient for light traffic.

Marchal (1987) has given an empirical extension of Kingman's result for the Poisson input queue. Marchal extends the heavy-traffic result of Kingman to conditions of moderate traffic intensity. We now discuss the result.

8.1.2. Empirical Extension of the M/G/1 Heavy-Traffic Approximation (Marchal, 1987)

Assume that the input is Poisson, so that

$$A^*(s) = \frac{\lambda}{\lambda + s}.$$

Taking two more terms of Maclaurin's expansion, we get

$$B^*(s) = 1 - b_1 s + \frac{b_2}{2} s^2 - \frac{b_3}{6} s^3 + \frac{b_4}{24} s^4 + O(s^4). \tag{1.15}$$

Then the characteristic equation $A^*(-s)B^*(s) = 1$ reduces to

$$\left[\frac{\lambda}{(\lambda - s)}\right]\left[1 - b_1 s + \frac{b_2}{2} s^2 - \frac{b_3}{6} s^3 + \frac{b_4}{24} s^4\right] = 1$$

when all the terms above the fourth degree are ignored. The preceding can be written as $sf(s) = 0$ where

$$f(s) \equiv \left[(1 - \lambda b_1) + \left(\frac{\lambda b_2}{2}\right)s - \left(\frac{\lambda b_3}{6}\right)s^2 + \left(\frac{\lambda b_4}{24}\right)s^3\right] = 0 \tag{1.16}$$

is a cubic in s.

Now $f(-s)$ has only one variation of sign and, thus, applying Descartes' rule of signs, one finds that $f(s)$ possesses exactly one negative root, which we denote by s_0.

We consider Fredericks' approximation of the waiting-time distribution for a $G/G/1$ queue

$$W(t) = 1 + ke^{s_0 t}. \tag{1.17}$$

The mean $E(W)$ for an $M/G/1$ queue is known from the Pollaczek–Khinchine formula

$$E(W) = \frac{\lambda}{2(1 - \rho)}\left(\frac{1}{\mu^2} + \sigma_v^2\right). \tag{1.18}$$

The constant k can be estimated from

$$k = s_0 E(W).$$

Thus, we get an estimate $W(t)$. The preceding approximation has the following advantages.

Maclaurin expansion of $B^*(s)$, which involves four moments of the service-time distribution, gives a better approximation than the one involving two moments. The result has less dependence on the asymptotic effect of the heavy traffic (i.e., of traffic intensity near unity). The cubic equation has exactly one negative root, s_0. Whenever $E(W)$ is known or can be estimated, one can easily estimate k. For the $G/G/1$ queue, one can use Marchal's (1976) estimate

$$E(W) = \frac{\lambda(\sigma_u^2 + \sigma_v^2)}{2(1-\rho)} \cdot \frac{\rho^2 + \lambda^2\sigma_v^2}{1 + \lambda^2\sigma_v^2}. \tag{1.19}$$

The arguments of Marchal can be further extended to the case of a $G/G/1$ queue by starting with Maclaurin's expansions of $A^*(s)$ and $B^*(s)$.

Remarks.

(1) Marchal has considered numerical examples for some particular $M/G/1$ queue, with $G \equiv M$, $G \equiv D$, and $G \equiv E_2$. The approximations appear to be satisfactory for moderate traffic intensity, say, in the range 0.50–0.80.
(2) It can be verified that $k \to -1$ as $\rho \to 1$.

8.1.3. G/M/c Queue in Heavy Traffic

Consider a $G/M/c$ queue in steady state for which the conditional waiting time, given that an arrival has to wait, is exponential with mean $1/c\mu(1-\sigma)$ where σ is the root of the equation

$$\sigma = A^*(c\mu - c\mu\sigma), \tag{1.20}$$

$A^*(.)$ being the LST of the interarrival-time distribution. In heavy traffic where an arrival has to wait, the unconditional distribution of the waiting time approaches the conditional distribution. Thus, for $\rho \to 1$ the mean waiting time is given by

$$E(W) \simeq \frac{1}{c\mu(1-\sigma)}. \tag{1.21}$$

Change the variable in (1.20) using

$$\alpha = c\mu(1-\sigma); \tag{1.22}$$

then (1.20) becomes

$$\sigma = A^*(\alpha) = 1 - \frac{\alpha}{c\mu}. \tag{1.23}$$

Expanding $A^*(\alpha)$ as a power series in α, we get

$$\begin{aligned} A^*(\alpha) &= 1 + \alpha\left\{\frac{d}{d\alpha} A^*(\alpha)\bigg|_{\alpha=0}\right\} \\ &\quad + \frac{\alpha^2}{2}\left\{\frac{d^2}{d\alpha^2} A^*(\alpha)\bigg|_{\alpha=0}\right\} + O(\alpha^2) \qquad (1.24) \\ &= 1 - \alpha E(u) + \frac{\alpha^2}{2} E(u^2) + O(\alpha^2). \end{aligned}$$

For heavy traffic, $E(W)$ is large; from (1.21) we get $\sigma \simeq 1$ and α is small so that $O(\alpha^2)$ may be neglected. Thus, from (1.23) and (1.24), we get

$$1 - \frac{\alpha}{c\mu} = 1 - \frac{\alpha}{\lambda} + \frac{\alpha^2}{2} E(u^2).$$

Thus, we have

$$\alpha = \frac{2\left(\dfrac{1-\rho}{\lambda}\right)}{E(u^2)} = \frac{2\left(\dfrac{1}{\lambda}\right)(1-\rho)}{\sigma_u^2 + \{E(u)\}^2}. \tag{1.25}$$

For heavy traffic $\rho \to 1$, i.e.,

$$\frac{E(v)}{cE(u)} \simeq 1;$$

for exponential service time

$$[E(u)]^2 = \frac{\sigma_v^2}{c^2}$$

and so

$$\alpha = \frac{2\left(\dfrac{1}{\lambda}\right)(1-\rho)}{\sigma_u^2 + \dfrac{\sigma_v^2}{c^2}}. \tag{1.26}$$

Thus, the approximate value of the mean waiting time in a $G/M/c$ queue in heavy traffic is obtained as

$$E(W) \simeq \frac{1}{\alpha} = \frac{\sigma_u^2 + \dfrac{\sigma_v^2}{c^2}}{2\left(\dfrac{1}{\lambda}\right)(1-\rho)}. \tag{1.27}$$

Remarks.

(1) The preceding result for a $G/M/c$ system led Kingman (1964) to make the conjecture that for heavy traffic the waiting time for a $G/G/c$ queue should be exponentially distributed with mean equal to that given by (1.27). Köllerström (1974) has proved the conjecture; the approximate distribution $W(t)$ of the waiting time in a $G/G/c$ queue in heavy traffic is exponential and is given by

$$W(t) \simeq 1 - \exp\left\{-\frac{2\dfrac{(1-\rho)}{\lambda}}{\sigma_u^2 + \dfrac{\sigma_v^2}{c^2}}\, t\right\}. \tag{1.28}$$

(2) Kingman (1961) was perhaps the first to study asymptotic behavior of queues. His work, in a sense, motivated subsequent studies on asymptotic behavior, including the diffusion-process approximation considered in the next section.
(3) Borovkov (1984) (describing mainly the weak convergence of queueing processes) contains a wealth of material on asymptotic methods published by probabilists from the Soviet Union and East Germany.

8.2. Brownian Motion Process

8.2.1. Introduction

The transient-state distribution of the queue length of even the simple $M/M/1$ queue is difficult to handle; that for the $M/G/1$ queue is not known. In view of such difficulties, some methods of approximation for more general queueing models are considered. One such method that originated with the works of Iglehart (1965), Gaver (1968), and Newell (1971) involves the use of the diffusion process to approximate queue-length distribution under heavy traffic conditions. The idea is to approximate the discrete and random arrivals by a nonrandom continuum and to do the same for the departures. The analogy is of fluid flow—fluid flowing into and out of a reservoir. A moment's reflection will convince one about the appropriateness of this approach. During rush hour people coming out of a subway or an electric train from a busy station resemble a continuous flow as opposed to a discrete, random flow in a very lean hour.

There are two ways to view the asymptotic behavior of a queueing process (discrete state). One is to obtain a strong-law-of-large-numbers type of limit, and the other is to obtain a central-limit-theorem type of limit. In the former case, the limit is a deterministic (nonrandom) function of time, while in the latter case, the limit is a stochastic process (a random function of time), specifically a Brownian motion (diffusion) process.

The first type of limit is generally referred to as a FLLN (functional-law-of-large-numbers) limit, and the second as a FCLT (functional Central Limit Theorem). The corresponding models that describe the asymptotic behavior of the queueing process are called, respectively, "fluid model" and "Brownian or diffusion model"; but since both the models have continuous-state space, they are both called a "fluid type" of model. We shall confine ourselves here to the second type of model, leading to diffusion approximation.

We shall now briefly discuss a diffusion process. For details the reader may refer to a standard book on stochastic processes; see also Newell (1971), Kleinrock (1976, **vol. II**), Heyman and Sobel (1982), Gelenbe and Mitrani (1980), and Harrison (1985).

Definition. A stochastic process $\{X(t), t \geq 0\}$ satisfying the following properties is called a *Brownian motion process* with drift m and variance parameter D^2.

(i) $X(t)$ has independent increments, i.e., for every pair of disjoint time intervals, say, (s, t) and (u, v), $s < t \leq u < v$, the increments $\{X(t) - X(s)\}$ and $\{X(v) - X(u)\}$ are independent random variables;

(ii) Every increment $\{X(t) - X(s)\}$ is normally distributed with mean $m(t - s)$ and variance $D^2(t - s)$.

Property (i) implies that a Brownian motion process is a Markov process. In fact, the property of independent increments is more restrictive than the Markov property, and Property (ii) implies that it is Gaussian. Thus, we have

$$\begin{aligned} Pr\{X(t) \leq x \mid X(s) = x_0\} &= Pr\{X(t) - X(s) \leq x - x_0\} \\ &= \Phi(\alpha) \end{aligned} \tag{2.1}$$

where

$$\alpha = \frac{\{x - x_0 - m(t-s)\}}{D\sqrt{(t-s)}} \quad \text{and}$$

$$\Phi(t) = \int_{-\infty}^{t} \frac{1}{\sqrt{2\pi}} e^{-y^2/2}\, dy \tag{2.2}$$

is the DF of the standard normal variate. The quantities m and D^2 may also be interpreted as follows:

$$m = \lim_{\Delta t \to 0} \frac{E\{X(t+\Delta t) - X(t)\}}{\Delta t} \tag{2.3}$$

that is,

$$(2.3) \qquad m\Delta t = E\{X(t+\Delta t) - X(t)\} + o(\Delta t) \quad \text{and}$$

$$D^2 = \lim_{\Delta t \to 0} \frac{E\{X(t+\Delta t) - X(t)\}^2}{\Delta t}$$

$$= \lim_{\Delta t \to 0} \frac{\text{var}\{X(t+\Delta t) - X(t)\}}{\Delta t} \tag{2.4}$$

that is,

$$D^2\Delta t = \text{var}\{X(t+\Delta t) - X(t)\} + o(\Delta t).$$

The quantities m and D^2 are called the infinitesimal mean and infinitesimal variance, respectively, of the process. We assume here that m and D^2 are constants, independent of t or of x, where $X(t) = x$. By considering m and D^2 as functions of t or of x or of both t and x, we get more general processes, which we shall consider in Section 8.3.6.

Theorem 8.2. *The distribution of* $\{X(t) | X(0) = x_0\}$ for large t is exponential with parameter $-2m/D^2$.

Proof. Let

$$F(t, x; x_0)\} = Pr\{X(t) \le x | X(0) = x_0\}. \tag{2.5}$$

Then it can be shown that F satisfies the diffusion equation (or forward Kolmogorov equation)

$$\frac{\partial}{\partial t} F(t, x; x_0) = -m \frac{\partial}{\partial x} F(t, x; x_0) + \frac{D^2}{2} \frac{\partial^2}{\partial x^2} F(t, x; x_0). \tag{2.6}$$

The initial condition $X(0) = x$ gives

$$F(0, x; x_0\} = \begin{cases} 0, & x < x_0 \\ 1, & x \geq x_0 \end{cases}.$$

With a reflecting barrier placed on the x-axis, the boundary condition is

$$F(t, 0; x_0) = 0, \qquad x_0 > 0, \qquad t > 0. \tag{2.7}$$

This is called the "reflecting boundary condition." The solution of the diffusion equation (2.6) subject to the preceding conditions is given by

$$F(t, x; x_0) = \Phi\left(\frac{x - x_0 - mt}{D\sqrt{t}}\right) - e^{2xm/D^2}\Phi\left(\frac{-x - x_0 - mt}{D\sqrt{t}}\right). \tag{2.8}$$

The preceding result, which holds for both $m > 0$ and $m < 0$, gives a time-dependent solution. We are interested also in a steady-state solution.

Assume that $\lim_{t\to\infty} F(t, x; x_0) = F(x)$; then (2.6) reduces to

$$0 = -m \frac{\partial}{\partial x} F(x) + \frac{D^2}{2} F(x). \tag{2.9}$$

When $m < 0$

$$\lim_{t\to\infty} \Phi\left(\frac{x - x_0 - mt}{D\sqrt{t}}\right) = \lim_{t\to\infty} \Phi\left(\frac{-x - x_0 - mt}{D\sqrt{t}}\right) = 1$$

so that from (2.8), we have

$$F(x) = 1 - e^{2xm/D^2} = 1 - \exp\left\{(-2x)\frac{(-m)}{D^2}\right\}. \tag{2.10}$$

Thus, the distribution of $\{X(t)|X(0) = x_0\}$ for large t is exponential with mean

$$\lim_{t \to \infty} E\{X(t)|X(0) = x_0\} = \frac{D^2}{-2m}.$$

Remarks.

(1) It is evident that $F(t, x; x_0) = \Phi[(x - x_0 - mt)/D\sqrt{t}]$ is a solution of the diffusion equation (2.6). But this solution does not satisfy the boundary condition (2.7); the solution of (2.6) that satisfies the boundary condition (2.7) is given by (2.8). Though it is not difficult to solve the diffusion equation (2.6) by itself, imposition of specific boundary conditions makes its solution difficult. The boundary conditions imposed are to be meaningful. The reflecting boundary condition contained in (2.7) is meaningful in the context of the problem studied here.

For a derivation of (2.8), readers are referred to Newell (1971) and Kleinrock (1976, **vol. II**).

The diffusion equation (2.6) can also be put in terms of the PDF of $\{X(t)|X(0) = x_0\}$.

(2) We shall proceed to apply a diffusion process approach in the study of the $G/G/1$ system. The justification of such an approach is provided by an important limit theorem established by Iglehart and Whitt (1970). The limit theorem shows that the queue-length and waiting-time processes of a $G/G/c$ system can be approximated by Brownian motion processes. The importance of the approach increases because of the fact that an exact solution of even the $M/G/1$ queue is not known and, wherever known, exact solutions of still simpler queues are difficult to handle analytically.

8.2.2. Asymptotic Queue-Length Distribution

Let $A(t)$, $D(t)$, and $N(t)$ denote, respectively, the number of arrivals, number of departures, and number in the system at time t of a system. Here $\{A(t), t \geq 0\}$, $\{D(t), t \geq 0\}$, and $\{N(t), t \geq 0\}$ are stochastic processes. We assume that the system is under heavy traffic and that

$$N(t) = N(0) + A(t) - D(t).$$

The assumption that $N(t)$ does not become zero is basic in this approach. The departure process $\{D(t), t \geq 0\}$, which is otherwise dependent upon the arrival process $\{A(t), t \geq 0\}$, then becomes approximately independent of the arrival process. The number of departures increases by unity each time a service is completed and the interdeparture time will have the same distribution as service times when the system remains continually busy. Let the IID random variables t_i, $i = 1, 2, \ldots,$ denote the interarrival times and let

$$T_n = t_1 + \cdots + t_n.$$

The nth arriving customer arrives at the epoch T_n. We have the important equivalence relation

$$Pr\{A(t) \geq n\} = Pr\{T_n \leq t\}. \tag{2.11}$$

The preceding relation enables us to find the distribution of $A(t)$ from that of T_n. Since t_is are IID random variables, the Central Limit Theorem can be applied to find the asymptotic distribution of T_n. We have

$$E\{T_n\} \simeq \frac{n}{\lambda}$$

$$\text{var}\{T_n\} \simeq n\sigma_u^2$$

where $1/\lambda$ and σ_u are the mean and SD of the interarrival times, respectively.

From the Central Limit Theorem, we have

$$Pr\left\{\frac{T_n - \dfrac{n}{\lambda}}{\sigma_u\sqrt{n}} \leq x\right\} = \Phi(x). \tag{2.12}$$

To find the RHS of (2.8), we have to relate n with t. Define

$$t = x\sigma_u\sqrt{n} + \frac{n}{\lambda}.$$

For large n, the dominant term being $t \simeq n/\lambda$, we can express n in terms of t as follows:

$$n \simeq \lambda t - x\lambda\sigma_u\sqrt{t\lambda}.$$

From (2.12) we have

$$Pr\{T_n \leq t\} = \Phi(x)$$

so that

$$Pr\{A(t) \geq n\} = \Phi(x) \quad \text{or}$$

$$Pr\{A(t) \geq \lambda t - x\lambda\sigma_u\sqrt{t\lambda}\} = \Phi(x) \quad \text{or}$$

$$Pr\left\{\frac{A(t) - \lambda t}{\lambda\sigma_u\sqrt{t\lambda}} \geq -x\right\} = \Phi(x) \quad \text{or}$$

$$Pr\left\{\frac{A(t) - \lambda t}{\lambda\sigma_u\sqrt{t\lambda}} \leq x\right\} = 1 - \Phi(-x) = \Phi(x). \tag{2.13}$$

Thus, the asymptotic distribution of $A(t)$ is Gaussian, with

$$\begin{aligned} E\{A(t)\} &\simeq \lambda t \\ \text{var}\{A(t)\} &\simeq \lambda^3\sigma_u^2 t. \end{aligned} \tag{2.14}$$

Denoting the mean and SD of service-time distribution by $1/\mu$ and σ_v, it can be shown that $D(t)$ is also asymptotically normal with

$$\begin{aligned} E\{D(t)\} &\simeq \mu t \quad \text{and} \\ \text{var}\{D(t)\} &\simeq \mu^3\sigma_v^2 t \end{aligned} \tag{2.15}$$

that is,

$$D(t) \sim N(\mu t, \mu^3\sigma_v^2 t).$$

The result can be put as follows:

Theorem 8.3. *For large t and for moderate- to heavy-loaded queueing systems it can be said that*

$$N_1(t) = N(t) - N(0) = A(t) - D(t)$$

is a Gaussian process with

$$E\{N_1(t)\} \simeq \lambda t - \mu t = \mu(\rho - 1)t \quad \text{and} \tag{2.16}$$

$$\text{var}\{N_1(t)\} \simeq (\lambda^3\sigma_u^2 + \mu^3\sigma_v^2)t. \tag{2.17}$$

Remarks.

(1) It is suggested that the process $\{N(t), t \geq 0\}$, where

$$N(t) = N(0) + A(t) - D(t)$$

can be approximated by a diffusion process having infinitesimal mean m and variance D^2 given by

$$m = \lim_{\Delta t \to 0} \frac{E\{N(t + \Delta t) - N(t)\}}{\Delta t} = \lambda - \mu \tag{2.18}$$

$$D^2 = \lim_{\Delta t \to 0} \frac{\text{var}\{N(t + \Delta t) - N(t)\}}{\Delta t} = \lambda^3 \sigma_u^2 + \mu^3 \sigma_v^2. \tag{2.19}$$

Equation (2.19) follows from the fact that $\{N(t), t \geq 0\}$ has independent increments and that $\text{cov}\{N(t), N(s)\} = \text{var}\{N[\min(t, s)]\}$ holds for such a process.

(2) The expressions (2.16) and (2.18) give a reasonable approximation of the mean when $\rho > 1$; however for $\rho < 1$, (2.16) becomes negative; this defect is taken care of by considering a reflecting barrier for $N(t)$ at the origin.

(3) That (2.14) and (2.15) hold for large t also follows from a result of renewal theory (Cox 1962). For a renewal process $\{R(t), t \geq 0\}$ where the interrenewal times have mean $1/v$ and variance σ^2, we have, for large t,

$$\begin{aligned} E\{R(t)\} &\simeq vt \quad \text{and} \\ \text{var}\{R(t)\} &\simeq \sigma^2 v^3 t. \end{aligned} \tag{2.20}$$

Here $\{A(t), t \geq 0\}$ is a renewal process. For ρ close to 1, $\{D(t), t \geq 0\}$ can also be approximated as a renewal process.

(4) The results, though based on a Central Limit Theorem approach, are renewal theoretic results.

8.2.3. Diffusion Approximation for a G/G/1 Queue (Kobayashi, 1974)

Let

$$F(t, x; x_0) = Pr\{N(t) \leq x \mid N(0) = x_0\}.$$

The Brownian motion process $\{N(t), t \geq 0\}$ has the drift m and variance D^2. The assumption is that $N(t) > 0$. The lower boundary at $x = 0$ for $\{N(t), t \geq 0\}$ would act as a reflecting barrier.

Theorem 8.4. *Diffusion approximation of the steady-state queue-length distribution $\{\hat{p}_n\}$ in a G/G/1 queue is given by*

$$\begin{aligned} \hat{p}_n &= \rho(1 - \hat{\rho})(\hat{\rho})^n, \qquad n \geq 1 \\ \hat{p}_0 &= 1 - \rho \end{aligned} \tag{2.21}$$

where

$$\hat{\rho} = e^{-\gamma} = e^{2m/D^2}$$
$$= \exp\left\{-\frac{2(1-\rho)}{\mu^2(\rho^3\sigma_u^2 + \sigma_v^2)}\right\}. \qquad (2.22)$$

Proof. We find from (2.10) that the steady-state distribution

$$F(x) = \lim_{t\to\infty} F(t, x; x_0)$$
$$= 1 - \exp\left\{-\frac{2x(-m)}{D^2}\right\}, \qquad x \geq 0,$$

so that the distribution of the number in the system in steady state is exponential with

$$\text{mean} = \lim_{t\to\infty} E\{N(t)|N(0) = x_0\}$$
$$= \frac{D^2}{(-2m)}$$
$$= \frac{1}{\gamma} \quad \text{(say)}.$$

Putting the values of m and D^2 as found in (2.18) and (2.19), we get that the distribution of $N(t)$ is exponential with mean

$$\frac{1}{\gamma} = \frac{D^2}{-2m} = \frac{\mu^2(\rho^3\sigma_u^2 + \sigma_v^2)}{2(1-\rho)}. \qquad (2.23)$$

Kobayashi suggests that we may discretize the distribution. Thus, the steady-state distribution of a number N in the system is given by

$$\hat{p}_n \equiv Pr\{N = n\} = \int_n^{n+1} dF(x)$$
$$= F(n+1) - F(n)$$
$$= e^{-n\gamma}(1 - e^{-\gamma})$$
$$= (1-\hat{\rho})(\hat{\rho})^n, \qquad n = 0, 1, 2, \ldots, \qquad (2.24a)$$

where

$$\hat{\rho} = e^{-\gamma} \quad \text{(as in (2.22))}.$$

This is expected since discretization of exponential distribution leads to geometric distribution.

Kobayashi further suggests that one needs to choose a boundary condition from the simple reflecting barrier. We get $1 - \hat{p}_0 = \hat{\rho}$, whereas the exact value of server utilization is ρ. He therefore recommends that the probability that the system is empty be taken as $(1 - \rho)$ and the result (2.24a) be modified as follows:

$$\begin{aligned} \hat{p}_0 &= 1 - \rho \\ \hat{p}_n &= \rho(1 - \hat{\rho})(\hat{\rho})^{n-1}, \qquad n = 1, 2, \ldots . \end{aligned} \tag{2.24b}$$

We thus get the diffusion approximation of the steady-state distribution of the number in the system for moderately high to heavy traffic.

8.2.3.1. Particular Cases.

M/G/1 System

Here $\sigma_u^2 = 1/\lambda^2$ so that

$$\begin{aligned} \hat{L} = \lim_{t \to \infty} E\{N(t)\} &= \frac{D^2}{-2m} \\ &= \frac{\left(\dfrac{\lambda^2\sigma_v^2 + \rho^3}{\rho^2}\right)}{2(1 - \rho)} . \end{aligned} \tag{2.25}$$

From the Pollaczek–Khinchine formula, we get, using $L = \lambda W$, that

$$\begin{aligned} L = \lim_{t \to \infty} E\{N(t)\} &= \rho + \frac{\lambda^2\sigma_v^2 + \rho^2}{2(1 - \rho)} \\ &= \frac{\lambda^2\sigma_v^2 + (2\rho - \rho^2)}{2(1 - \rho)}. \end{aligned} \tag{2.26}$$

The RHS of (2.25) and (2.26) are both close when ρ is close to 1. The error in approximating the mean number in the system in equilibrium is small when ρ is close to 1.

M/M/1 System

Here $\sigma_u^2 = 1/\lambda^2$ and $\sigma_v^2 = 1/\mu^2$, so that

$$\hat{\rho} = \exp\left\{\frac{-2(1 - \rho)}{(1 + \rho)}\right\}.$$

This value of $\hat{\rho}$ is close to ρ in the neighborhood of $\rho = 1$. When $\rho \to 1$,

$\hat{\rho} \to 1$ so that (2.24) becomes

$$\hat{p}_n \simeq (1 - \hat{\rho})(\hat{\rho})^n, \; n = 0, 1, 2, \ldots;$$

and the approximation of $\hat{p}_n$ is good when ρ is close to 1 (and slightly less than 1).

8.2.4. Virtual Delay for the G/G/1 System

Let $X(t)$ denote the total workload by time t, that is, $X(t)$ is the total time required by the server to complete serving all units that arrived in the interval $(0, t]$. Suppose that $X(0) = 0$. We have

$$X(t) = v_1 + v_2 + \cdots + v_{A(t)}.$$

Let $W(t)$ denote the remaining workload (or work backlog), that is, the time required by the server to complete serving all units present at the epoch t. $W(t)$ is called the virtual waiting time, that is, the time that an imaginary customer would have to wait in the queue were he to arrive at the epoch t.

$W(t)$ is the sum of the residual service time at epoch t (of the unit being served) plus the service time of units waiting at epoch t.

Suppose that $W(0) = 0$. If the server has been busy throughout the interval $(0, t]$ and if he works continuously at a unit rate in that interval, then we have

$$\begin{aligned} W(t) &= W(0) + X(t) - t \\ &= X(t) - t \end{aligned} \tag{2.27}$$

when $W(0) = 0$. When $W(t) \geq \delta t$,

$$W(t + \delta t) - W(t) = X(t + \delta t) - X(t) - \delta t. \tag{2.28}$$

However, the simple relation given in (2.27) does not hold good in the general case. Nevertheless, it can be seen that (2.28) holds good in the general case whenever $W(t) \geq \delta t$. A look at Fig. 8.1 will convince the reader about the validity of (2.28). We have from the figure, $W(t) > \delta t$ and

$$W(t + \delta t) - W(t) = EF.$$

Again,

$$X(t + \delta t) - X(t) - \delta t = DB - CF = FH = EF$$

so that (2.28) holds. Taking expectation, we get

$$\begin{aligned} E\{X(t)\} &= E(v_i)E\{A(t)\} \\ &\simeq \frac{1}{\mu} \lambda t = \rho t. \end{aligned} \tag{2.29a}$$

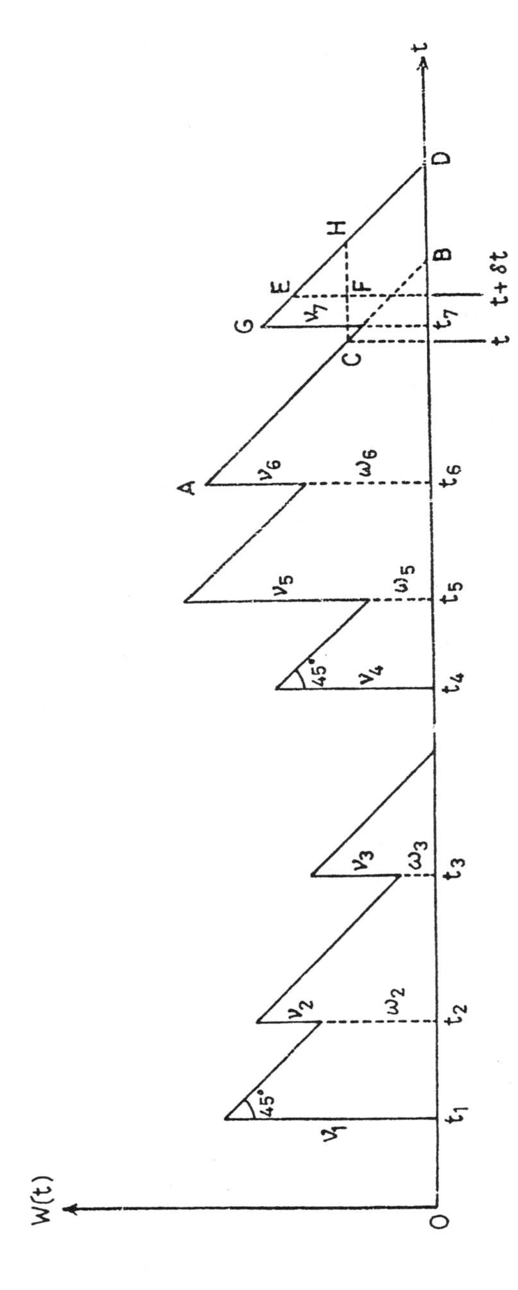

FIGURE 8.1. Graph of the virtual waiting time $W(t)$

Using the relation

$$\text{var}\{X(t)\} = \text{var}(v_i)E\{A(t)\} + \text{var}\{A(t)\}\{E(v_i)\}^2$$

we get

$$\text{var}\{X(t)\} \simeq \sigma_v^2 \lambda t + \lambda^3 \sigma_u^2 t \left(\frac{1}{\mu}\right)^2$$

$$= \lambda(\sigma_v^2 + \rho^2 \sigma_u^2)t. \tag{2.29b}$$

Assume that (2.29a) and (2.29b) hold for all t. We have

$$\lim_{\delta t \to 0} \frac{E\{X(t + \delta t) - X(t)\}}{\delta t} = \rho \quad \text{and}$$
$$\lim_{\delta t \to 0} \frac{\text{var}\{X(t + \delta t) - X(t)\}}{\delta t} = \lambda(\sigma_v^2 + \rho^2 \sigma_u^2). \tag{2.30}$$

Using (2.28), we get

$$\lim_{\delta t \to 0} \frac{E\{W(t + \delta t) - W(t)\}}{\delta t} = \rho - 1 \quad \text{and} \tag{2.31}$$

$$\lim_{\delta t \to 0} \frac{\text{var}\{W(t + \delta t) - W(t)\}}{\delta t} = \lim_{\delta t \to 0} \frac{\text{var}\{X(t + \delta t) - X(t)\}}{\delta t}$$

$$= \lambda(\sigma_v^2 + \rho^2 \sigma_u^2). \tag{2.32}$$

The mean and variance of increments in $W(\cdot)$ are proportional to δt for small δt. This is a property of the Brownian motion process. We can approximate the virtual-waiting-time process $\{W(t), t \geq 0\}$ by a Brownian motion process having a reflecting barrier at the x-axis. The parameters of this process are

$$m = \rho - 1 \quad \text{and}$$
$$D^2 = \lambda(\sigma_v^2 + \rho^2 \sigma_u^2). \tag{2.33}$$

Thus,

$$F(t, x; x_0) = Pr\{W(t) \leq x \mid W(0) = x_0\} \tag{2.34}$$

is given by (2.8) and in the limit, as $t \to \infty$

$$F(x) \simeq 1 - e^{-2x(-m)/D^2}. \tag{2.35}$$

Thus, we have the following result.

Theorem 8.5. *The diffusion approximation of the limiting distribution of the virtual waiting time $W(t)$ in a $G/G/1$ queue is exponential with mean*

$$\lim_{t\to\infty} E\{W(t)\} = \lim_{t\to\infty} E\{W(t)|W(0)=x_0\}$$

$$= \frac{D^2}{-2m} \tag{2.36}$$

$$= \frac{\lambda(\sigma_v^2 + \rho^2\sigma_u^2)}{2(1-\rho)}.$$

8.2.4.1. ***Particular Case: M/G/1 System.*** When arrivals occur in accordance with a Poisson process, $\sigma_u^2 = 1/\lambda^2$. From (2.36), we then get

$$\lim_{t\to\infty} E\{W(t)\} = \frac{\lambda\left(\sigma_v^2 + \dfrac{1}{\mu^2}\right)}{2(1-\rho)}, \tag{2.37}$$

which is the Pollaczek–Khinchin mean-value formula (for mean waiting time in the queue).

Notes.

(1) The steady-state distribution of the virtual waiting time is identical with the steady-state distribution of the actual waiting time of an arriving customer if and only if arrivals occur in accordance with a Poisson process. When the input is other than Poisson, the two steady-state distributions are different.

(2) For Poisson input, $X(t)$ is a compound Poisson process, with DF

$$Pr\{X(t) \le x\} = \sum_{n=0}^{\infty} e^{-\lambda t}\frac{(\lambda t)^n}{n!} B^{n*}(x)$$

where $B^{n*}(x)$ is the n-fold convolution of service time.

8.2.5. Approach Through an Absorbing Barrier with Instantaneous Return (Gelenbe)

In obtaining the results in Section 8.2.3 it was assumed that the lower boundary acts as a reflecting barrier for the process $\{N(t), t \ge 0\}$ where $N(t)$ is the number in the system at time t. Gelenbe (1975, 1979) has considered an alternative approach through the *instantaneous returns process*. The stochastic process $\{N(t), t \geq 0\}$ is considered to represent the position of a particle moving on a closed interval $(0, \infty]$ of the real line. When the particle reaches the lower boundary $x = 0$ of the interval, it remains there for a period of time (denoted by the RV ξ) at the end of which the particle jumps instantaneously back to the open interval $[0, \infty]$ to a random point. The origin is an absorbing barrier for a length ξ. The RV ξ corresponds to the idle period. Let $E(\xi) = 1/\lambda'$. Note that for a Poisson input $\lambda' = \lambda$. Gelenbe shows that the PDF of the steady-state distribution is given by

$$f(x) = \begin{cases} R(e^{\gamma} - 1)e^{-\gamma x}, & x \geq 1 \\ R(1 - e^{-\gamma x}) \;, & 0 \leq x \leq 1, \end{cases} \tag{2.38}$$

where γ is as given in (2.23) and

$$R = \frac{\lambda'}{\lambda' + \mu - \lambda}, \tag{2.39}$$

the condition for the existence of a steady-state solution being $\rho = \lambda/\mu < 1$.

When the input is Poisson, $R = \rho = \lambda/\mu$. Then for an $M/G/1$ queue the expected number in the steady state is approximately given by

$$\begin{aligned} E(N) &= \int_0^{\infty} x f(x) dx \\ &= \rho\left(\frac{1}{2} + \frac{1}{\gamma}\right) \\ &= \rho\left(\frac{1}{2} + \frac{\rho + \mu^2\sigma_v^2}{2(1 - \rho)}\right). \end{aligned} \tag{2.40}$$

This is close to $E(N)$ for an $M/G/1$ given by the Pollaczek–Khinchine formula.

8.2.6. *Diffusion Approximation for a G/G/c Queue: State-Dependent Diffusion Equation*

The idea of diffusion approximation for a $G/G/1$ model can be extended in principle to a multiserver $G/G/c$ model. The input process $\{A(t), t \geq 0\}$ of a multiserver system is not in anyway affected by the number of servers. Thus, (2.14) will be valid for a c-server system, i.e.,

$$E\{A(t)\} \simeq \lambda t \quad \text{and}$$

$$\text{var}\{A(t)\} = \lambda^3 \sigma_u^2 t.$$

However, Eq. (2.15) for the departure process $\{D(t), t \geq 0\}$ have to be suitably modified. When $n < c$, the departure rate will be $n\mu$ and we shall have

$$\begin{aligned} E\{D(t)\} &\simeq n\mu t \quad \text{and} \\ \operatorname{var}\{D(t)\} &\simeq n\mu^3\sigma_v^2 t \end{aligned} \tag{2.41}$$

and for $n \geq c$, the departure rate will be $c\mu$ so that

$$\begin{aligned} E\{D(t)\} &\simeq c\mu t \quad \text{and} \\ \operatorname{var}\{D(t)\} &\simeq c\mu^3\sigma_v^2 t. \end{aligned} \tag{2.42}$$

We can apply similar arguments as in the case of a single-server model and use diffusion approximation. The infinitesimal mean and variance will be state-dependent and, in place of those of (2.18) and (2.19), we shall have state-dependent m and D^2 as functions of x, where $X(t) = x$. Thus,

$$m \equiv b(x) = \lambda - [\min\{x, c\}]\mu \tag{2.43}$$

$$D^2 \equiv a(x) = \lambda^3\sigma_u^2 + [\min\{x, c)\}]\mu^3\sigma_v^2. \tag{2.44}$$

When $c = 1$, these reduce to (2.18) and (2.19), respectively.

8.2.6.1. State-Dependent Diffusion Equation.

Define the PDF of $X(t)$ by

$$f \equiv f(t, x; x_0) = Pr\{x \leq X(t) < x + dx | X(0) = x_0\}. \tag{2.45}$$

Then it can be shown that $f(t, x, x_0) \equiv f$ satisfies the diffusion equation (forward Kolmogorov equation)

$$\frac{\partial f}{\partial t} = -\frac{\partial}{\partial x}\{b(x)f\} + \frac{1}{2}\frac{\partial^2}{\partial x^2}\{a(x)f\} \tag{2.46}$$

where the diffusion parameters are given by

$$b(x) = \lim_{\Delta t \to 0} \frac{E\{X(t + \Delta t) - X(t) | X(t) = x\}}{\Delta t} \quad \text{and} \tag{2.47}$$

$$a(x) = \lim_{\Delta t \to 0} \frac{\operatorname{var}\{X(t + \Delta t) - X(t) | X(t) = x\}}{\Delta t}. \tag{2.48}$$

Note that we have put the equation in terms of the PDF instead of the DF. We now make appropriate choices of the diffusion parameters and the boundary conditions.

As discussed already for a $G/G/c$ system, $b(x)$ and $a(x)$ can be taken as given in (2.43) and (2.44).

8.2.6.2. Steady-State Solution. Let

$$\lim_{t\to\infty} f(t, x; x_0) = f(x).$$

Then Eq. (2.46) reduces to

$$\frac{d}{dx}\{b(x)f(x)\} = \frac{1}{2}\frac{d^2}{dx^2}\{a(x)f(x)\}. \tag{2.49}$$

By imposing appropriate boundary conditions, Eq. (2.49) can be solved to yield $f(x)$. Heyman and Sobel (1982) give a method of solving (2.49) with the boundary condition

$$2b(0)f(0) = \frac{d}{dx}\{a(x)f(x)\}\bigg|_{x=0}. \tag{2.50}$$

They also make a comparison between the exact values and the diffusion-approximation values (as obtained by the method) for the mean and variance of queue length for an $M/M/c$ queue for certain values of c. The agreement is quite satisfactory.

8.2.7. Diffusion Approximation for an M/G/c Model

Analytical solutions of an $M/G/c$ queue for distributions of queue length, waiting time, and busy period are not known. Thus, approximate methods assume importance. Various methods including diffusion-process approach have been put forward. Because of certain special properties of the Poisson input process, one can expect to get sharper results than for a $G/G/c$ queue through diffusion approximation.

Halachmi and Franta (1978) and Sunaga *et al.* (1978) investigated the $M/G/c$ on the basis of the heavy-traffic-limit theorem. Halachmi and Franta took as diffusion parameters the functions given in (2.43) and (2.44), and reflecting barrier as a boundary condition at the origin.

Kimura (1983) proposes as diffusion parameters

$$\begin{aligned} b(x) &= \lambda - \{\min([x], c)\}\mu \quad \text{and} \\ a(x) &= \lambda + \{\min([x], c)\}\mu^3\sigma_v^2 \end{aligned} \tag{2.51}$$

where $[x]$ is the smallest integer not smaller than x. The infinitesimal mean and variance (given in (2.51)) of the diffusion process $\{X(t), t \geq 0\}$ are piecewise continuous.

As a boundary condition he treats the origin as an "elementary return boundary," which implies (in a rough sense) that when the process

$\{X(t), t \geq 0\}$ reaches the state 0, it stays there for a random length of time ξ whose distribution is exponential and then returns to the interval $[0, \infty]$. It may be seen that for a Poisson input process, the intervals during which the system is empty (idle periods) are exponentially distributed and that from $X(\cdot) = 0$, the process jumps to $X(\cdot) = 1$ with the arrival of a fresh customer. This boundary has been used also by Gelenbe and others. Using some other properties of the Poisson input queue (such as PASTA, Wolff (1982)), Kimura obtains approximate formulas for the steady-state distributions of the queue length, the waiting time, and the busy period. It is shown that the formula of the mean waiting time obtained by Kimura through diffusion approximation agrees when $\rho \rightarrow 1$ with that obtained by Köllerström (1974) through heavy-traffic approximation.

Kimura and Ohsone (1984) extend the results to the system $M^{[X]}/G/c$ with the group arrivals. Yao (1985) suggests some refinements to the diffusion approximation of the $M/G/c$ system by Kimura (1983). Yao retains the piecewise continuity formulation of the diffusion parameters but suggests a modification of the boundary condition: replacement of the elementary return boundary by a reflecting barrier. He uses an alternative approach for the solution of the diffusion equation and incorporates some known results into the model to obtain his solution. Some numerical comparisons are also given.

8.2.8. Concluding Remarks

Though the approach of approximating a discrete-state (or jump) process by a diffusion process with continuous path is not new, application of the principle to queueing theory is of rather recent origin. The method of approximation has gained importance because of the mathematical intractability of most of the queueing models on the one hand and the satisfactory nature of the approximating solution on the other hand.

We have restricted our discussion to a steady-state type of solution of queueing models. Transient solutions, which are much more difficult to obtain, have also been attempted through diffusion approximation. (See Newell, 1971, and Kobayashi, 1974, for an account.)

Another direction in which the method of diffusion approximation has been applied and found suitable is the analysis of open and closed networks of queues. Here one has to deal with a multidimensional diffusion equation. (See Kobayashi, 1974), **part II**, and Reiser and Kobayashi, 1974.

The diffusion-process approximation has been found useful in application of queueing theory in several areas (in modeling computer and communication systems, in particular). (See Reiser, 1982, for an account.)

8.3. Queueing Systems with Vacations

8.3.1. Introduction

It may happen in several situations that the server is unavailable to the customers (primary customers) over occasional periods of time; the server may then be doing other work such as maintenance work or servicing secondary customers. The periods for which the server is unavailable are said to be server-vacation periods. Systems with server vacations can be used as models of many production, communication, and computer systems. Two examples are given next.

(i) *Production systems.* Machines producing certain items may need periodic checking and maintenance. The periods of random lengths of preventive maintenance may be considered as periods of server vacation, when the server is unavailable.

(ii) *Computer and communication systems.* A server in such a system, besides being engaged in primary functions (such as receiving, processing, and transmitting data), has to undertake secondary works such as preventive maintenance or has to scan for new work for occasional periods of time. There are several other situations when the server is unavailable to primary customers for occasional (and random) periods of time. Such systems will be called systems with (generalized) vacations.

We make the following assumptions. Each time a busy period ends and the system becomes empty, the server starts a vacation of random length of time. When the server returns from vacation and finds one or more customers waiting, he goes on serving until the system becomes empty (called *exhaustive service* discipline). If on return from a vacation (at the end of a busy period) the server finds no customer waiting, he waits for the arrival of a customer. This is called a *single-vacation system* and is denoted by V_s. On the other hand, if he finds no customer waiting, he goes on taking vacations until, on return from a vacation, he finds at least one customer waiting. This is called a *multiple-vacation system* and is denoted by V_m. We denote the nth vacation periods by v_n and assume that $\{v_n, n = 1, 2, \ldots\}$ is a sequence of IID random variables that may be independent of the arrival period or may be dependent on the arrival process. Let F_v be the DF and $\bar{f}_v$ be the LST of F_v. Consider a standard $G/G/1$ queue with interarrival time A (with mean $1/\lambda$ and SD σ_u) and service time S (with mean $1/\mu$ and SD σ_v). We shall denote a queue with general interarrival and service times with single vacation by $G/G/1\text{–}V_s$ and the corresponding queue with multiple vacation by $G/G/1\text{–}V_m$.

8.3.2. Stochastic Decomposition

It has been observed that the queues with server vacation exhibit an interesting property called the stochastic decomposition property. Under certain conditions on the sequence $\{v_n\}$, both for $G/G/1\text{–}V_s$ and $G/G/1\text{–}V_m$ models, the steady-state waiting time is the sum of two independent random variables. One is the waiting time in the same $G/G/1$ queue without vacation and the other is a random variable related to $\{v_n\}$.

For queues with Poisson input, such a decomposition property holds even for the steady-state queue-length distribution.

The decomposition property for Poisson input systems has been observed, for example, by Gaver (1962), Cooper (1970), Fuhrmann (1984), and Furhmann and Cooper (1985), while that for $G/G/1$ vacation models has been discussed by Gelenbe and Iasnogordski (1980), Servi (1984), and Doshi (1985). In a survey of queueing systems with vacations, Doshi (1986) provides a unified treatment of the topic.

8.3.3. Poisson Input Queue with Vacations: Exhaustive-Service Queue-Length Distribution

We shall now discuss the $M/G/1$ queue with single and multiple vacations under exhaustive-service discipline. Assume that the vacation sequence v_n is stationary and that the system is in steady state.

Let N^* be the number of customers present at the start of a busy period following a vacation or vacations. Clearly, $N^* \geq 1$; N^* can be deterministic or a random variable.

First, consider that N^* is a RV having PGF

$$R(z) = \sum_{n=1}^{\infty} Pr\{N^* = n\}z^n, \qquad |z| < 1.$$

Let $P(z)$ be the PGF of the number in the system at a departure epoch of a usual $M/G/1$ queue *without* vacation. Note that the distribution of the number in the system at the random epoch, at an arrival epoch, or at a departure epoch are one and the same for a Poisson input queue (PASTA). $P(z)$ is given by the Pollaczek–Khinchin formula. Let $Q(z)$ be the PGF of the number in the system at a departure epoch of an $M/G/1$ queue with vacations.

Let $V(z)$ be the PGF of the number in the system at a random point in time when the server is on vacation. For a Poisson input queue the basic decomposition result is

$$Q(z) = P(z)V(z) \tag{3.1}$$

(Fuhrmann and Cooper, 1985; Ali and Neuts, 1984).

The basic decomposition result shows that the number of customers at a departure epoch of a Poisson input queue with vacations is the sum of two random variables:

(i) the number of customers at a departure epoch at the corresponding Poisson input queue without vacation and (ii) the number of customers at a random point of time given that the server is on vacation.

While variable (i) is vacation-independent, variable (ii) is vacation-related. We now state the important decomposition result (without proof) and consider some special cases. For a proof, refer to Fuhrmann (1984) and Doshi (1986).

Theorem 8.6. *For an $M/G/1$ queue with server vacations,*

$$Q(z) = P(z)\frac{1 - R(z)}{(1 - z)E(N^*)}. \tag{3.2}$$

We consider some special cases.

(A) N^* is deterministic

(i) $Pr\{N^* = 1\} = 1$, we get the usual queue with the vacation period corresponding to the idle period of the system. Then $Q(z) = P(z)$.

(ii) N^* is a fixed number, say N, i.e., $Pr\{N^* = N\} = 1$. This corresponds to the case when the server is on vacation (or remains busy with other work or secondary customers) until the (primary) queue size builds up to a preassigned fixed number N. Known as N-policy, this was considered first by Heyman (1968), who shows that a system with such a policy possesses some optimal properties.

From (3.2) we get

$$Q(z) = P(z)\frac{1 - z^N}{(1 - z)N}. \tag{3.3}$$

In the preceding two cases under (**A**), the length of the server vacation depends on the arrival process during but not after the vacation. Under (**B**), the length of server vacation is independent of the arrival process.

(B) N^* is an RV

(a) $M/G/1\text{–}V_m$ *system*
Let A_v be the number of arrivals during a typical vacation period v. Then the PGF $\alpha(z)$ of A_v is given by

$$\alpha(z) = \sum_{n=0}^{\infty} Pr\{A_v = n\}z^n$$
$$= \bar{f}_v[\lambda(1-z)]. \tag{3.4}$$

We have

$$Pr\{A_v = 0\} = \bar{f}_v(\lambda) \tag{3.5}$$

so that

$$Pr\{A_v \geq 1\} = 1 - \bar{f}_v(\lambda). \tag{3.6}$$

Now the event $N^* = n$ is the event that the number of arrivals during the last vacation period equals n, given that this number is at least 1, i.e.,

$$Pr\{N^* = n\} = Pr\{A_v = n | A_v \geq 1\}, \qquad n = 1, 2, \ldots$$
$$= \frac{Pr\{A_v = n\}}{1 - \bar{f}_v(\lambda)}. \tag{3.7}$$

Thus,

$$\begin{aligned} R(z) &= \sum_{n=1}^{\infty} Pr\{N^* = n\}z^n \\ &= \frac{\bar{f}_v[\lambda(1-z)] - \bar{f}_v(\lambda)}{1 - \bar{f}_v(\lambda)}. \end{aligned} \tag{3.8}$$

We have

$$E(N^*) = R'(1) = \frac{-\lambda \bar{f}'_v(0)}{1 - \bar{f}_v(\lambda)} = \frac{\lambda E(v)}{1 - \bar{f}_v(\lambda)}.$$

Substituting in (3.2), we get

$$Q(z) = P(z) \frac{1 - \bar{f}_v[\lambda(1-z)]}{\lambda E(v)(1-z)}. \tag{3.9}$$

Remark 1. The second factor has an interesting interpretation. Let $Z(t)$ be the forward recurrence time of a vacation (or residual lifetime of the vacation) random variable. Then the limiting distribution Z of $Z(t)$ as $t \to \infty$ is given by

$$F_z(x) = Pr\{Z \le x\} = \frac{\int_0^x [1 - F_v(y)]dy}{E(v)} \tag{3.10}$$

where $F_v(y) = Pr(v \le y)$. (See Eq. (7.9) in Ch. 1.) Let b_n be the probability that n arrivals occur during Z and let

$$\beta(z) = \sum_{n=0}^{\infty} b_n z^n$$

be the PGF of the number of arrivals during Z. Then

$$\begin{aligned}\beta(z) &= \sum_{n=0}^{\infty} z^n \int_0^{\infty} \frac{e^{-\lambda t}(\lambda t)^n}{n!} dF_z(t) \\ &= \int_0^{\infty} \left\{\sum \frac{e^{-\lambda t}(\lambda t z)^n}{n!}\right\} \frac{1 - F_v(t)}{E(v)} dt \\ &= \int_0^{\infty} \frac{e^{-\lambda t(1-z)}\{1 - F_v(t)\}}{E(v)} dt \\ &= \frac{1 - \bar{f}_v[\lambda(1-z)]}{\lambda E(v)(1-z)},\end{aligned} \tag{3.11}$$

which is equal to the second factor on the RHS of (3.9). Thus, while the first factor is the PGF of the number at departure epoch in the standard $M/G/1$ queue without vacation, the second factor is the PGF of the number of arrivals during the limiting forward recurrence time of the vacation period.

Remark 2. We have $\alpha'(1) = \lambda E(v)$, so that the second factor on the RHS of (3.9) can be written as

$$\frac{1 - \alpha(z)}{(1-z)\alpha'(1)}.$$

Thus, for $M/G/1\text{–}V_m$, (3.2) can be put as

$$Q(z) = P(z)\frac{1 - \alpha(z)}{(1-z)\alpha'(1)}, \tag{3.12}$$

$\alpha(z)$ being the PGF of the number of arrivals during the vacation.

(b) *$M/G/1-V_s$ model*
Here there is only one vacation and there may be no arrivals or one arrival or more than one arrival during the server-vacation period. If there is no arrival, the server waits for an arrival to occur and then $N^* = 1$. If there is an arrival during the vacation, then N^* is equal to the number of arrivals during the vacation. Thus,

$$Pr\{N^* = 1\} = Pr\{A_v = 0\} + Pr\{A_v = 1\}$$

$$Pr\{N^* = n\} = Pr\{A_v = n\}, n = 2, 3, \ldots .$$

Thus, using (3.4) and (3.5) we get

$$\begin{aligned} R(z) &= \sum_{n=1}^{\infty} Pr\{N^* = n\}z^n \\ &= Pr\{A_v = 0\}z + \sum_{n=1}^{\infty} Pr\{A_v = n\}z^n \\ &= z\bar{f}_v(\lambda) + \bar{f}_v[\lambda(1-z)] - \bar{f}_v(\lambda) \\ &= \bar{f}_v[\lambda(1-z)] - (1-z)\bar{f}_v(\lambda) \quad \text{and} \end{aligned} \tag{3.13}$$

$$\begin{aligned} E(N^*) = R'(1) &= -\lambda\bar{f}'_v(0) + \bar{f}_v(\lambda) \\ &= \lambda E(v) + \bar{f}_v(\lambda). \end{aligned} \tag{3.14}$$

Substitution in (3.2) gives

$$Q(z) = P(z)\,\frac{1 - \bar{f}_v[\lambda(1-z)] + (1-z)\bar{f}_v(\lambda)}{(1-z)[\lambda E(v) + \bar{f}_v(\lambda)]}. \tag{3.15}$$

8.3.4. Poisson Input Queue with Vacations: Waiting-Time Distribution

By *sojourn time* or *waiting time* of a customer, we shall here mean his queueing time plus his service time. We shall consider steady-state waiting time of an arbitrary customer in a vacation system with Poisson input. We assume that:

(i) the queue discipline is FIFO (for while queue-length distribution is not affected by queue discipline, waiting-time distribution is affected) and

(ii) the waiting time of a customer is independent of the input process that occurs after the epoch of arrival of the customer considered. Note that this condition is not satisfied in case of a queue under N-policy; our result will not hold good for such a queue.

Let $W_1(.)$ denote the DF of the waiting time of an arbitrary customer in a standard $M/G/1$ queue (without vacation) and let $W_1^*(\cdot)$ denote its LST. Let $W(.)$ and $W^*(.)$ denote the corresponding functions in an $M/G/1$ queue with vacation under the assumption previously stated. We have the following result due to Fuhrmann and Cooper (1985).

Theorem 8.7. *For an $M/G/1$ queue with vacations,*

$$W^*(s) = W_1^*(s)V\left(1 - \frac{s}{\lambda}\right). \tag{3.16}$$

Proof. Under FIFO discipline, the customers left behind by an (arbitrary) departing customer are precisely those customers that arrived during the waiting time of the departing customer. It follows that

$$\begin{aligned} Q(z) &= \int_0^\infty \exp\{-\lambda(1-z)t\}dW(t) \\ &= W^*[\lambda(1-z)] \end{aligned}$$

so that, putting $\lambda(1 - z) = s$, we get

$$W^*(s) = Q\left(1 - \frac{s}{\lambda}\right) \tag{3.17}$$

for an $M/G/1$ queue with vacation. Similarly for the queue without vacation

$$W_1^*(s) = P\left(1 - \frac{s}{\lambda}\right). \tag{3.18}$$

Using the basic decomposition result (3.1) one gets

$$W^*(s) = W_1^*(s)V\left(1 - \frac{s}{\lambda}\right).$$

Case i. $M/G/1\text{–}V_m$

Putting $s = \lambda(1 - z)$ in (3.9) we get

$$Q\left(1 - \frac{s}{\lambda}\right) = P\left(1 - \frac{s}{\lambda}\right)\frac{1 - \bar{f}_v(s)}{sE(v)}$$

so that for such a system

$$W^*(s) = W_1^*(s)\,\frac{1 - \bar{f}_v(s)}{sE(v)}. \tag{3.19}$$

Note that

$$V\left(1-\frac{s}{\lambda}\right)=\frac{1-f_v(s)}{sE(v)}$$

is the LST of the forward-recurrence time of a vacation (as can be observed from an earlier discussion). In this vacation model, the waiting-time distribution decomposes into two independent components. One is the waiting-time distribution in the corresponding model without vacation; the other is the forward-recurrence time of the vacation. When the vacation distribution is given, the decomposition result reduces the problem to a convolution problem.

We have followed Fuhrmann's approach. Alternative methods of derivation of decomposition result have been given by Shanthikumar (1988) through the level-crossing argument and by Doshi (1985) through the sample-path argument.

Case ii. $M/G/1\text{–}V_s$ *model*
Putting $s = \lambda(1 - z)$ in (3.15) we get (using (3.17) and (3.18))

$$W^*(s) = W_1^*(s)\,\frac{\bar{f}_v(\lambda)+\left(\dfrac{\lambda}{s}\right)[1-\bar{f}_v(s)]}{\lambda E(v)+\bar{f}_v(\lambda)}. \tag{3.20}$$

8.3.5. M/G/1 System with Vacations: Nonexhaustive Service

So far we assumed that once the server starts service, he serves all the customers one by one until none is left. As against this exhaustive service discipline there could be situations giving rise to nonexhaustive service discipline, under which the server vacation may start even when some customers are present in the system (left in the queue without being served). Two cases may arise.

(i) The preemptive case where vacations may preempt an ongoing service, e.g., as in the case of a breakdown of the service mechanism, and

(ii) nonpreemptive case where the vacations may commence only at epochs of service completion or vacation termination, e.g., as in the case of scheduled preventive maintenance.

We shall restrict ourselves here to the nonpreemptive case. Assume that service discipline is LIFO. Consider an $M/G/1$ system with vacations. Let

R denote the number of customers already present when a typical vacation began and let $\zeta(z)$ be its PGF. Consider an epoch of vacation commencement when the queue length is R. The vacation and the customers who arrive during this vacation will start a cycle that will not be affected by the R customers present at its commencement.

Suppose that we suppress all vacations during this cycle. Then the distribution of the queue length (excluding the R customers initially present) during this period is the same as that of an $M/G/1$ queue with exhaustive service discipline. Thus, the number of customers Q_1 at a departure epoch of an $M/G/1$ vacation queue with nonexhaustive service discipline will be equal to $R + Q$ where Q is the number of customers at a departure epoch of the corresponding $M/G/1$ vacation queue with exhaustive service discipline. Now Z and Q are independent. Denoting the PGF of Q_1 by $Q_1(z)$, we shall get the following three-way decomposition result.

Theorem 8.7. *For an $M/G/1$–V_m queue with nonexhaustive service, the PGF $Q_1(z)$ of the number at a departure epoch is given by*

$$\begin{aligned} Q_1(z) &= \zeta(z)Q(z) \\ &= \zeta(z)P(z)\frac{1-\alpha(z)}{(1-z)\alpha'(1)}. \end{aligned} \tag{3.21}$$

For a rigorous proof and observations on this interesting decomposition result, refer to Fuhrmann and Cooper (1985) *and Doshi* (1986).

Notes.

(1) The preceding three-way decomposition result will not hold if the number present at the commencement of a vacation and the number of arrivals during a vacation are not independent. However, the basic two-way decomposition result (3.1) will hold good.

(2) For exhaustive service, $Pr(R = 0) = 1$ and $\zeta(z) = 1$; then (3.21) reduces to (3.12).

8.3.6. Concluding Remarks

8.3.6.1. G/G/1 Model. An analogous type of decomposition result involving waiting time holds for a $G/G/1$–V_m model. This has been demonstrated, for example, by Gelenbe and Iasnogorodski (1980), Doshi (1985, 1986), and Fricker (1986). The sample-path arguments put forward by Doshi (1985) could be extended to cover $G/G/1$–V_m and $G/G/1$–V_s models.

8.3.6.2. Variations of the Vacation Model. There are various related models, for example, a model with start-up time. Here, when a customer arrives to start a busy period, the server goes through a set-up or start-up time (SUT) of random length U before starting actual service. The server is unavailable to primary customers during this start-up time.

This problem and its variations have been considered by several researchers, for example, Pakes (1972), Lemoine (1975), and Doshi (1985).

For a Poisson input queue with SUT U, with DF $F_u(.)$, it is shown that the decomposition can be put as

$$Q(z) = P(z)\frac{1 - R(z)}{(1 - z)\lambda E(U)} \tag{3.22}$$

where

$$\begin{aligned} R(z) &= 1 \times z + \sum_{n=0}^{\infty} z^n \int_0^{\infty} \frac{e^{-\lambda t}(\lambda t)^n}{n!}\, dF_u(t) \\ &= z + \bar{f}_u[\lambda(1 - z)] \end{aligned} \tag{3.23}$$

$\bar{f}_u(.)$ being the LST of $F_u(.)$.

The decomposition property has been used by Medhi and Templeton (1990) in the study of an $M/G/1$ queue under N-policy, with general start-up time. Fuhrmann and Cooper (1985) have considered application of decomposition property in an $M/G/1$ queue with vacations to two continuum cyclic queueing models.

A finite-capacity vacation-type queue has been considered, for example, by Teghem (1987) and Loris-Teghem (1988), for an $M/G/1$ model and by Jacob and Madhusoodanan (1987) for an $M/G\ (a, b)/1$ model. Introduction of a vacation component makes many queueing systems more realistic. It can also be visualized that many queueing problems can be simplified by considering them as vacation-type problems, and their solution too can be simplified.

8.3.6.3. Multiserver Models. It appears that relatively little attention has been paid to multiserver queues with vacations. Multiserver queues $M/M/c$ with vacations have been considered, for example, by Mitrani and Avi-Itzhak (1967), Levy and Yechiali (1976), and Neuts and Lucotani (1979). The queues with general service time do not seem to have been studied as of 1989.

Remark. It may be observed in passing that there are similar, though not analogous, product-form results of interest in other areas. For example, we may mention the product-form or factorization result of sufficient statistics in the theory of statistical estimation. Another interesting result is the "product-form" representation of the conditional hazard function $h(t|\mathbf{X}, \theta)$ on a vector of covariates $\mathbf{X}$ and unobserved heterogeneity component θ; it is as follows:

$$h(t|\mathbf{X}, \theta) = h_0(t)\psi(\mathbf{X})\phi(\theta)$$

where $h_0(t)$ is the baseline hazard function (Cox (1972)). This representation has been fruitfully employed in several types of studies, such as in econometrics (for example, Lancaster, (1979) Heckman and Singer, (1984) and in business in the study of household-brand-switching behavior (Jain (1990, to appear).)

8.4. Design and Control of Queues

Our main concern so far has been to describe probabilistically the behavior of a system and to find the various associated performance measures, given the arrival and service patterns, queue discipline and other configurations. Practical questions that arise in applications of queueing theory relate to determination of parameters, patterns, and/or policies for which a model will be optimal in some specified sense. In general, one may have little or no control over the arrival patterns, though sometimes arrivals can be controlled through truncation or other means. On the other hand, it may be possible to exercise control over the service mechanism (for example, through the service pattern or number of service channels to be operated), the service policy, and such other configurations in a queueing system in order that the system performance may attain optimal value in a certain sense. For example, control may be through adjustment of service rate (Mitchell, 1973; Doshi, 1978) or by turning on and off a service mechanism according to a policy involving the state of the system (Heyman, 1968). Objective functions as well as constraints can be formulated in terms of such measures as average cost, average waiting time, and other performance measures of interest. Solutions need optimization techniques. Problems that arise in such situations are called problems of design and control of queues. It is, however, not always easy to make a clear distinction between design and control problems and their corresponding models.

Classical optimization methods are generally used in designs; problems, formulations and analyses of control problems involve application of theory and techniques of renewal reward process, Markov decision processes, martingale, dynamic programming, and so on.

Optimization models have been increasingly used in the design problems arising in several applied areas, such as production processes and telecommunication networks. Wherever queueing phenomena arise in such problems, performance measures of associated queueing systems often occur in the objective functions and/or in the constraints of the optimization models. The solutions of certain optimization problems depend partly on the *concavity* and *convexity properties* of these measures. As such investigation of such properties has been receiving increasing attention in recent times, for example, see Rolfe (1971), Dyer and Proll (1977), Tu and Kumun (1983), Weber (1983), Grassman (1983) Lee and Cohen (1983), and Harel and Zipkin (1987).

The design problem associated with server allocation has several applications in such areas as multiple center manufacturing systems (see, for example, Shanthikumar and Yao (1986, 1987, 1988)) as well as allocation of vehicles and fleets in transportation, and police patrol and ambulances, etc. (for example, Parikh (1977), Green (1984), Berman *et al.* (1985), Chaiken and Dermont (1978), and so on).

A detailed discussion of the topic (design and control of queues) is beyond the scope of this book. We would however try to indicate briefly the nature of the problems that arise. For a general survey of the topics, see, for example, Stidham and Prabhu (1974), Crabill *et al.* (1977), Sobel (1974), and Serfozo (1981). Hillier and Lieberman (1967) described a number of interesting design problems. One simple design model considered is as follows.

Consider a single-server model with known λ. Assume that the cost per server per unit time is A and that the cost of waiting per customer per unit time is B. The design problem consists of finding μ that will minimize the expected cost per unit time.

$$E(C) = A\mu + BL$$

where L is the average number of customers in the system.

For an $M/M/1$ queue

$$E(C) = A\mu + \frac{B\lambda}{(\mu - \lambda)}$$

so that the value of μ for which $E(C)$ is minimum is the value μ^*, if any, that satisfies

$$\frac{d}{dC} E(C) = 0 \quad \text{and}$$

$$\frac{d^2}{dC^2} E(C) < 0;$$

we get

$$\mu^* = \lambda + \sqrt{\frac{\lambda B}{A}}.$$

It may be noted that even for an $M/M/C$ model, $E(C)$ is at a minimum for $c = 1$ and so for the same μ^*.

Consider now a control problem. One control policy that consists of turning on and off the server is the N-policy formulated by Heyman (1968) (and discussed in the previous section and Section 6.4.5). The facility is shut down (turned off) as soon as all present are served and reactivated (turned on) as soon as the queue size, when the server is idle, builds up to N. N is called the control parameter. Heyman considers (i) a start-up cost, (ii) a shut-down cost, (iii) a server running cost per unit time (when the server is active), and (iv) a customer-holding (-waiting) cost per customer per unit time. He shows that the optimal type of policy is the N-policy.

Similarly, T-policy is another control policy, according to which the service facility is turned off for a fixed period of time T, from the instant of each service completion leaving the system empty, (see Heyman (1977)). Another policy is D-policy, according to which the service facility reopens as soon as the total work load (after each service completion leaving the system empty) exceeds a critical level D, (see Sivazlian (1977)). (Refer to Tijms (1986) for analyses of control models under these policies by renewal reward processes.)

Studies relate to determination of optimal policies according to specified objectives, for example, minimization of average waiting time and minimization of average cost as per specific cost structure.

As did Heyman (1968) and Bell (1971), Sobel (1969) also considers cost structure along with start-up and shut-down costs for a single-server system. McGill (1969) considers a general switching-cost model for a system with a variable number of exponential servers. Bell (1975) considers the average cost criterion, while discounted cost over a finite horizon for a c-server Markovian model is considered by Huang *et al.* (1977). Assuming that the cost structure includes customer-holding (waiting) cost and service-channel-holding (server-running) cost as well as linear-switching cost, Szarkowicz and

Knowles (1985) consider optimal control of an $M/M/c$ system and show that a control-limit policy is optimal under less restrictive conditions. They use also dynamic programming formulation.

Remark. Control problems for bulk service queues have been considered, for example, by Kosten, Deb and Serfozo (1973), Ignall and Kolesar (1974), Weiss (1979), and Powell and Humblet (1986). Consider a Poisson input system operated under the policy: When the server is available and there are fewer than Q customers waiting, the service does not begin; if there are Q or more waiting and the server is available, service begins with all those waiting. This is an infinite-capacity bulk-service queue of the type $M/G(Q, \infty)/1$. Assume that a fixed start-up cost K is incurred each time service is initiated and the waiting cost per customer is h per unit of time. Deb and Serfozo (1973) show that the optimal type of policy is the control-limit policy, which requires that service begins if and only if the number of waiting customers is as large as Q. Weiss (1979) finds expressions for average long-run cost per unit time and for the optimal Q. He applies renewal theory on the assumption that the epoch at which service begins each time (with none left in the queue, because of infinite server capacity), are regeneration points. The situation will not be the same for a general-bulk-service queue with *finite* server capacity.

Since mass-transit vehicles are sort of natural batch servers, and since a shuttle between two destinations can be considered as a single-server system, this kind of bulk-service model can be used as a model for such systems. Similarly, multiserver models could also fit as realistic models for mass-transportation systems. In view of this, control problems for batch-service systems have assumed importance.

For a brief survey, see, for example, Medhi (1984a,b). Powell and Humblet (1986) consider unified treatment for a number of vehicle-dispatch strategies, such as vehicle-holding strategy (general-bulk-service rule), vehicle-cancelation strategy as well as a combination of these two strategies. A Markov chain approach is considered. They define Q_n as the steady-state queue length at the nth dispatch instant (instant of nth service completion) and show that, under certain conditions, the Markov chain $\{Q_n, n \geq 0\}$ is ergodic, and they obtain the PGF of the steady-state queue length at dispatch instants in terms of the PGFs of associated variables arising out of the control strategy. They also obtain a decomposition type of result.

$$Q(z) = Q_1(z)B(z)$$

where $Q(z)$ and $Q_1(z)$ are the PGFs of the queue length at dispatch instants of corresponding queues with and without control strategy, respectively, and $B(z)$ is a function connected with the control strategy.

As computer communication networks are modeled as queueing networks, the question of optimal flow control in queueing networks has been receiving attention. (See, for example, Lazar, 1983, and Sauer *et al.* 1981.)

All these studies show the importance and use of control and design of queues in practical applications.

Problems and Complements

8.1. Diffusion approximation for the busy period

Consider an $M/G/1$ system. Let $W(t)$ be the virtual delay and let a busy period start at t_0 with the arrival of a customer whose service time v has DF $B(\,)$ with LST $B^*(\,)$ and moments b_k. The delay process $\{W(t), t \geq 0\}$ can be approximated as a diffusion process with infinitesimal mean and variance given by

$$c = \rho - 1$$
$$D^2 = \lambda b_2.$$

Let $f(t, x)$ be the first passage time from 0 to x of the Wiener process with infinitesimal mean $-c$ and variance D^2 and $f(t)$ be the PDF of the busy period for the approximate process. Show that

$$f(t, x) = x(2\pi D^2 t^3)^{-1/2} \exp\left\{\frac{-(x + ct)^2}{2D^2 t}\right\} \quad \text{and}$$

$$f(t) = (2\pi D^2 t^3)^{-1/2} \int_0^\infty x \exp\left\{-\frac{(x + ct)^2}{2D^2 t}\right\} dB(x).$$

For an $M/M/1$ queue

$$f(t) = \{\mu\rho(\pi^2 \rho t)^{-1/2}\} \exp\left\{-\frac{(1 - \rho)\mu t}{4\rho}\right\}$$
$$- \left\{\frac{(3\rho - 1)}{2}\right\}\mu \exp\{(2\rho - 1)\mu t\}\Phi_c\left[\left\{\frac{(3\rho - 2)}{2}\right\}\left(\frac{\mu t}{\rho}\right)^{1/2}\right]$$

where Φ_c is the complementary DF of standard normal variate,

$$\Phi_c(x) = (2\pi)^{-1/2} \int_x^{\infty} e^{-y^2/2}\, dy.$$

For an $M/D/1$ queue with constant service time d

$$f(t) = (2\pi\lambda t^3)^{-1/2} \exp\left[\frac{-\{d + (\rho - 1)t\}^2}{2\rho\, dt}\right]$$

(Heyman, 1974).

8.2. $M/G/1$ under N-policy with zero start-up time

Suppose that demand for service arises in accordance with a Poisson process with rate λ. The service times are IID random variables with DF B having first *two* finite moments $b_1 = \mu$ and b_2. The facility starts (instantaneously) only when N units are present after a busy period with exhaustive service terminates (i.e., the server becomes idle).

A fixed set-up cost $K > 0$ is incurred every time the facility is reopened, and a holding cost $h(>0)$ per unit time per unit present is also incurred for every unit present in the system. This is an $M/G/1$ system under N-policy with zero start-up time. N is the control parameter.

(a) Show that the long-run fraction of time the service facility is busy equals ρ independently of the control parameter N.

(b) Show that, with probability 1, the long-run average cost per unit time is given by

$$\frac{\lambda(1-\rho)K}{N} + h\left\{\rho + \frac{\lambda^2 b_2}{2(1-\rho)} + \frac{N-1}{2}\right\}$$

and that the average cost is minimal for one of the two integers nearest to

$$N = \sqrt{\left\{\frac{2\lambda(1-\rho)K}{h}\right\}}$$

(Tijms, 1986)

8.3. $M/G/1$ queue under T-policy with zero startup time

Here the facility is controlled in a different manner. Every time the server becomes idle after exhaustive service, the service facility is utilized for other work for fixed length of time T, and then the facility is reactivated only when there is at least one unit (i.e., multiple-server vacation with fixed vacation time T). Suppose that K is the fixed set-up cost, h is the holding cost per unit time per unit present, and that startup time is zero.

(a) Defining a cycle C as the interval between two consecutive epochs at which a vacation period starts, show that

$$E(C) = \frac{T}{1-\rho}.$$

(b) Show that with probability 1, the long-run average cost per unit time is given by

$$\frac{K(1-\rho)}{T} + h\left\{\rho + \frac{\lambda^2 b_2}{2(1-\rho)} + \frac{\lambda T}{2}\right\}$$

and that the average cost per unit time is mimimal for

$$T = \sqrt{\frac{2(1-\rho)K}{h\lambda}}$$

(Tijms, 1986).

8.4. $M/G/1$ system under N-policy and general start-up time
Suppose that instead of zero start-up time, the start-up times are IID random variables with common DF $D(.)$ with mean u and LST $D^*(.)$ The system can be in three different types of states: I (turned off), S (turned-on with server doing preservice work) (as soon as queue size builds up to N after a turned-off period I), and B (server rendering service to customers that starts as soon as turned-on period is over and terminates when all present are served, leaving none in the queue) (exhaustive service). Let $p_{n,i}$ denote the steady-state probability that there are n in the system given that the state is i, $i \in \{I, S, B\}$. Let $P_I(z)$, $P_s(z)$, and $P_B(z)$ be the corresponding PGFs. Show that

$$p_{n,I} = p_{0,I} = k(1-\rho), \qquad 0 \le n \le N-1, \quad \text{and}$$

$$P_s(z) = p_{0,I} z^N \frac{\{D^*(\lambda - \lambda z) - 1\}}{(z-1)}.$$

Using the decomposition property, show further that (with $k = 1/(N + \lambda u)$ and $B^*(\,)$ as the LST of service-time distribution)

$$P_B(z) = \frac{k\rho(1-\rho)\{z^N D^*(\lambda - \lambda z) - 1\} B^*(\lambda - \lambda z)}{z - B^*(\lambda - \lambda z)}.$$

For busy-period distribution, see Section 6.4.5 (Medhi and Templeton, 1990).

8.5. $M/G(Q, \infty)/1$: average long run cost. Assume that the input is Poisson with rate λ, and the service times are IID random variables, denoted by v, with moments $b_k = E(v^k)$. The service rule is general bulk service with infinite capacity having a minimum of Q in a batch. Let K be the startup cost and h be the waiting cost per customer per unit time. Let R be the interval between two renewals, i.e., between the epochs of two successive service initiations, and let Y be the cost between two renewals. Show that

$$E(R) = b_1 + (1/\lambda) \sum_{n=0}^{Q-1} (Q - n)P_n,$$

where P_n is the probability that exactly n customers arrive during a service period, and

$$E(Y) = K + h\lambda b_2 + (h/2\lambda) \sum_{n=0}^{Q-1} (Q^2 - Q - n^2 + n)P_n.$$

Hence, find the mean waiting time in an $M/G(Q, \infty)/1$ queue. (Weiss, 1979).

References

Ali, O. M. E., and Neuts, M. F. (1984). A service system with two stages of waiting and feedback of customers. *J. Appl. Prob.* **21**, 404–423.

Altiok, T. (1987). Queues with group arrivals and exhaustive service discipline. *Queueing Systems* **2**, 307–320.

Bell, C. E. (1971). Characterization and computation of optimal policies for operating on $M/G/1$ queueing system with removable server. *Opns. Res.* **19**, 208–218.

Bell, C. E. (1975). Turning off a server with customer present. Is this any way to run on $M/M/c$ queue with removable servers? *Opns. Res.* **23**, 571–574.

Benes, V. E. (1956). On queueing with Poisson arrivals. *Ann. Math. Stat.* **28**, 670–677.

Berman, O., Larson, R. C., and Shiu, S. S. (1985). Optimal server location in a network operating as an $M/G/1$ queue. *Opns. Res.* **33**, 746–771.

Borovkov, A. A. (1984). *Asymptotic Methods in Queueing Theory.* Wiley, New York.

Chaiken, J. M., and Larson, R. C. (1972). Methods of allocating urban emergency units: a survey. *Mgmt. Sci.* **19**, 110–130.

Chaiken, J. M., and Dermont, N. P. (1978). A patrol car allocation model: background, capabilities, and algorithms. *Mgmt. Sci.* **24**, 1280–1300.

Cooper, R. B. (1970). Queues served in cyclic order: waiting times. *Bell Sys. Tech. J.* **49**, 399–413.

Cox, D. R. (1962). *Renewal Theory.* Methuen, London.

Cox, D. R. (1972). Regression models and life tables. *J.R.S.S.* **B34**, 187–200.

Cox, D. R., and Smith, W. L. (1961). *Queues.* Methuen, London.

Crabill, T. B., Gross, D., and Magazine, M. J. (1977). A classified bibliography of research on optimal design and control of queues. *Opns. Res.* **25**, 219–232.

Deb, R. K., and Serfozo, R. F. (1973). Optimal control for batch service queues. *Adv. Appl. Prob.* **5**, 340–361.

Doshi, B. T. (1978). Optimal control of the service rate in an $M/G/1$ queueing system. *Adv. Appl. Prob.* **10**, 682–701.

Doshi, B. T. (1979). Generalized semi-Markov decision processes. *J. Appl. Prob.* **16**, 618–620.

Doshi, B. T. (1985). A note on stochastic decomposition in a $GI/G/1$ queue with vacation or set-up times. *J. Appl. Prob.* **22**, 419–428.

Doshi, B. T. (1986). Queueing systems with vacations—a survey. *Queueing Systems* **1**, 29–66. (Contains a list of 50 references.)

Dyer, M. E., and Proll, L. G. (1977). On the validity of marginal analysis for allocating servers in $M/M/c$ queues. *Mgmt. Sci.* **23**, 1019–1022.

Federgruen, A., and Green, L. (1986). Queueing systems with service interruptions. *Opns. Res.* **34**, 752–768.

Fredericks, A. A. (1982). A class of approximations for the waiting time distribution in $GI/G/1$ queueing system. *Bell. System Tech. J.* **61**, 295–325.

Fricker, C. (1986). Etude d'une file $GI/G/1$ à service autonome (avec vacances du serveur). *Adv. Appl. Prob.* **18**, 283–286.

Fuhrmann, S. W. (1984). A note on the $M/G/1$ queue with server vacations. *Opns. Res.* **32**, 1368–1373.

Fuhrmann, S. W., and Copper, R. B. (1985). Stochastic decompositions in an $M/G/1$ queue with generalized vacations. *Opns. Res.* **33**, 1117–1129.

Gaver, D. P., Jr. (1962). A waiting line with interrupted service, including priorities. *J.R.S.S.* **B24**, 73–90.

Gaver, D. P., Jr. (1968). Diffusion approximations and models for certain congestion problems. *J. Appl. Prob.* **5**, 607–623.

Gelenbe, E. (1975). On approximate computer system models. *J. Ass. Comp. Mach.* **22**, 261–269.

Gelenbe, E. (1979). Probabilistic models of computer systems. Part II: Diffusion approximations, waiting times and batch arrivals. *Acta Informatica* **12**, 285–303.

Gelenbe, E., and Mitrani, I. (1980). *Analysis and Synthesis of Computer Systems.* Academic Press, London.

Gelenbe, E., and Iasnogorodski, R. (1980). A queue with server of walking type (autonomous service). *Ann. Inst. Henri Poincaré.* **XVI**, 63–73.

Grassman, W. (1983). The convexity of the mean queue size of the $M/M/c$ queue with respect to the traffic intensity. *J. Appl. Prob.* **20**, 916–919.

Green, L. (1984). A multiple dispatch queueing model of police patrol operations. *Mgmt. Sci.* **30**, 653–664.

Halachmi, B., and Franta, W. R. (1978). A diffusion approximation to the multiserver queue. *Mgmt. Sci.* **24**, 522–529; Erratum, 1448.

Hall, P. (1985). Heavy traffic approximations for busy period in an $M/G/\infty$ queue. *Stoch. Proc. & App.*, **19**, 259–269.

Harel, A. (1987). Sharp bounds and simple approximations for the Erlang delay and loss formulas. *Mgmt. Sci.* (to appear).

Harel, A., and Zipkin, P. (1987). Strong convexity results for queueing systems. *Opns. Res.* **35**, 405–418.

Harel, A. and Zipkin, P. (1989). The convexity of a general performance measure for the $M/G/c$ queue. *J. Appl. Prob.* (to appear).

Harris, C. M., and Marchal, W. G. (1988). State dependence in $M/G/1$ server-vacation models. *Opns. Res.* **36**, 560–565.

Harrison, M. J. (1985). *Brownian Motion and Stochastic Flow Systems.* Wiley, New York.

Heckman, J., and Singer, B. (1984). A method of minimizing the impact of distributional assumptions in econometric models for duration data. *Econometrica* **52**, 271–320.

Heyman, D. P. (1968). Optimal control policies for $M/G/1$ queueing systems. *Opns. Res.* **16**, 362–382.

Heyman, D. P. (1974). An approximation for the busy period of the $M/G/1$ queue using a diffusion model. *J. Appl. Prob.* **11**, 159–169.

Heyman, D. P. (1975). A diffusion model approximation for the $GI/G/s$ queue in heavy traffic. *Bell Syst. Tech. J.* **54**, 1637–1646.

Heyman, D. P. (1977). The T policy for the $M/G/1$ queue. *Mgmt. Sci.* **23**, 775–778.

Heyman, D. P., and Sobel, M. J. (1982). *Stochastic Models in Operations Research.* McGraw Hill, New York.

Hillier, F. S., and Lieberman, G. J. (1967). *Introduction to Operations Research.* Holden-Day, San Francisco.

Huang, C. C., Brumelle, S. L., and Sawaki, K. (1977). Optimal control for multiserver queueing systems under periodic reviews. *NRLQ*, **24**, 127–135.

Iglehart, D. L. (1965). Limit diffusion approximations for the many server queue and the repairmen problem. *J. Appl. Prob.* **2**, 429–441.

Iglehart, D. L., and Whitt, W. (1970). Multiple channel queues in heavy traffic. I & II. *J. Appl. Prob.* **2**, 150–177, 355–369.

Ignall, E., and Kolesar, P. (1974). Optimal dispatching of an infinite capacity shuttle: control at a single terminal. *Opns. Res.* **22**, 1003–1024.

Jacob, M. J., and Madhusoodanan, T. P. (1987). Transient solution for a finite capacity $M/G(a, b)/1$ queueing system with vacations to the server. *Queueing Systems* **2**, 381–386.

Keilson, J., and Servi, L. D. (1987). The dynamics of an $M/G/1$ vacation model. *Opns. Res.*, **35**, 575–582.

Keilson, J., and Servi, L. D. (1989). Blocking probability for $M/G/1$ vacation systems with occupancy level dependent schedules. *Opns. Res.* **37**, 134–140.

Kingman, J. F. C. (1961). The single server queue in heavy traffic. *Proc. Camb. Phil. Soc.* **57**, 902–904.

Kingman, J. F. C. (1962a). On queues in heavy traffic. *J.R.S.S.* **B24**, 383–392.

Kingman, J. F. C. (1962b). Some inequalities for the queue $GI/G/1$. *Biometrika* **49**, 315–324.

Kingman, J. F. C. (1964). The heavy traffic approximation in the theory of queues, in: *Proceedings of Symposia on Congestion Theory*, (Smith, W. L. and Wilkinson, W. E. eds.), 137–169, University of North Carolina Press, Chapel Hill, NC.

Kingman, J. F. C. (1970). Inequalities in the theory of queues. *J.R.S.S.* **B32**, 102–110.

Kingman, J. F. C. (1980). Queue disciplines in heavy traffic. *Math. Opns. Res.* **7**, 262–271.

Kimura, T. (1983). Diffusion approximation for an $M/G/m$ queue. *Opns. Res.* **31**, 304–321.

Kimura, T., and Ohsone, T. (1984). A diffusion approximation for an $M/G/m$ queue with group arrivals. *Mgmt. Sci.* **30**, 381–388.

Kleinrock, L. (1976). *Queueing Systems*, Vol. II, *Computer Applications.* Wiley, New York.

Kobayashi, H. (1974). Application of the diffusion approximation to queueing networks: Part I: Equilibrium queue distributions. Part II: Non equilibrium distributions and applications to computer modeling. *J. Assoc. Comput. Mach.* **21**, 316–328, 459–469.

Kobayashi, H. (1978). *Modeling and Analysis: An Introduction to System Performance Evaluation Methodology.* Addison-Wesley, Reading, MA.

Köllerström, J. (1974). Heavy traffic theory for queues with several servers I, *J. Appl. Prob.* **11**, 544–552.

Lancaster, T. (1979). Econometric methods for the duration of unemployment. *Econometrica* **47**, 939–956.

Lee, H. L., and Cohen, M. A. (1983). A note on the convexity of performance measures of $M/M/c$ queueing systems. *J. Appl. Prob.* **20**, 920–923.

Lee, T. T. (1984). $M/G/1/N$ queue with vacation time and exhaustive service discipline. *Opns. Res.* **32**, 774–785.

Lemoine, A. (1975). Limit theorems for generalized single server queues: the exceptional system. *SIAM J. Appl. Math.* **28**, 596–606.

Levy, H., and Kleinrock, L. (1986). A queue with starter and a queue with vacations: Delay analysis by decomposition. *Opns. Res.* **34**, 426–436.

Levy, Y., and Yechiali, U. (1975). Utilization of idle time in an $M/G/1$ queueing system. *Mgmt. Sci.* **22**, 202–211.

Levy, Y., and Yechiali, U. (1976). $M/M/s$ queues with server vacations. *INFOR*, **14**, 153–163.

Lippman, S. A. (1975). Applying a new device in the optimization of exponential queueing systems. *Opns. Res.* **23**, 687–710.

Loris-Teghem, J. (1988). Vacation policies for an $M/G/1$ type queueing system with finite capacity. *Queueing Systems* **3**, 41–52.

Marchal, W. G. (1976). An approximate formula for waiting time in single server queues. *AIIE Trans.* **8**, 473.

Marchal, W. (1978). Some simpler bounds on the mean queueing time. *Opns. Res.* **26**, 1083–1088.

Marchal, W. G. (1987). An empirical extension of the $M/G/1$ heavy traffic approximation. *Annals O.R.* **8**, 93–101.

Medhi, J. (1984a). Bulk service queueing models and associated control problems. In: *Statistics: Applications and New Directions*, 369–377; Indian Statistical Institute, Calcutta, India.

Medhi, J. (1984b). *Recent Developments in Bulk Queueing Models.* Wiley Eastern, New Delhi, India.

Medhi, J., and Templeton, J. G. C. (1990). A Poisson input queue under N-policy and with general start-up time. *Comp. & Opns. Res.* (to appear).

Miller, B. I. (1968). Finite state continuous time Markov decision processes with finite planning horizon. *SIAM J. Control.* **6**, 266–280.

Mitchell, B. (1973). Optimal service rate selection in an $M/G/1$ queue. *Siam J. Appl. Math.* **24**, 19–30.

Mitrani, I. L., and Avi-Itzhak, B. (1968). A many server queue with service interruptions. *Opns. Res.* **16**, 628–638.

Miyazawa, M. (1986). Approximation of the queue length distribution of an $M/G/s$ queue by the basic equations. *J. Appl. Prob.* **23**, 443–448.

Neuts, M. F., and Lucotani, D. (1979). A Markovian queue with N servers subject to breakdowns and repairs. *Mgmt. Sci.* **25**, 849–861.

Neuts, M. F., and Ramalhoto, M. F. (1984). A service model in which the server is required to search for customers. *J. Appl. Prob.* **21**, 157–166.

Newell, G. F. (1968). Queues with time-dependent arrival rates I: the transition through saturation. *J. Appl. Prob.* **5**, 436–451.

Newell, G. F. (1971). *Applications of Queueing Theory*, (2nd ed., 1982). Chapman & Hall, London.

Ohno, K., and Ichiki, K. (1987). Computing optimal policies for controlled tandem queueing systems. *Opns. Res.* **35**, 121–126.

Pakes, A. G. (1972). A $GI/M/1$ queue with a modified service mechanism. *Ann. Inst. Stat. Math.* **24**, 589.

Parlar, M. (1984). Optimal dynamic service rate control in time dependent $M/M/S/N$ queues. *Int. J. Systems Sci.* **15**, 107–118.

Parikh, S. C. (1977). On a fleet sizing and allocation problem. *Mgmt. Sci.* **23**, 972–977.

Pourbabai, B., and Sonderman, D. (1986). Server utilization factors in queueing loss systems with ordered entry and heterogeneous servers. *J. Appl. Prob.* **23**, 236–242.

Powell, W. B. (1981). *Stochastic Delays in Transportation Terminals: New Results in the Theory and Application of Bulk Queues.* Ph.D Dissertation, MIT, Cambridge, MA.

Powell, W. B. (1985). Analysis of vehicle holding and cancellation strategies in bulk arrival, bulk service queues. *Trans. Sci.* **19**, 352–377.

Powell, W. B. (1986). Iterative algorithms for bulk arrival, bulk service queues with Poisson and non-Poisson arrivals. *Trans. Sci.* **20**, 65–80.

Powell, W. B., and Humblet, P. (1986). The bulk service queue with a general control strategy: theoretical analysis and new computational procedure. *Opns. Res.* **34**, 267–275.

Prabhu, N. U. (1974). Stochastic control of queueing systems. *NRLQ*, **21**, 411–418.

Reiser, M. (1982). Performance evaluation of data communication systems. *Proc. of IEEE*, **70**, 171–196. (Includes a list of 142 references.)

Reiser, M., and Kobayashi, H. (1974). Accuracy of the diffusion approximation for some queueing systems. *IBM J. Res. Devel.*, **18**, 110–124.

Rolfe, A. J. (1971). A note on the marginal allocation in multi-server facilities. *Mgmt. Sci.* **17**, 656–658.

Ross, S. M. (1970a). Average cost semi-Markov decision processes. *J. Appl. Prob.* **7**, 649–656.

Ross, S. M. (1970b). *Applied Probability Models with Optimization Applications.* Holden Day, San Francisco, CA.

Rue, R. C., and Rosenshine, M. (1981). Some properties of optimal control policies for entries to an $M/M/1$ queue. *NRLQ* **28**, 225–232.

Serfozo, R. (1981). Optimal control of random walks, birth and death processes and queues. *Adv. Appl. Prob.* **13**, 61–83.

Shanthikumar, J. G. (1988). On stochastic decomposition in $M/G/1$ queues with generalized server vacations. *Opns. Res.* **36**, 566–569.

Shanthikumar, J. G., and Yao, D. D. (1986). The effects of increasing service rates in a closed queueing network. *J. Appl. Prob.* **23**, 474–483.

Shanthikumar, J. G., and Yao, D. D. (1987). Optimal server allocation in a system of multi-server stations. *Mgmt. Sci.* **33**, 1173–1191.

Shanthikumar, J. G., and Yao, D. D. (1988). On server allocation in multiple center manufacturing systems. *Opns. Res.* **36**, 333–342.

Sivazlian, B. D. (1979). Approximate optimal solution for a D-policy in an $M/G/1$ queueing system. *AIIE Trans.* **11**, 341–343.

Smith, J. G., and Daskalaki, S. (1988). Buffer space allocation in automated assembly lines. *Opns. Res.* **36**, 343–358.

Smith, W. L. (1953). On the distribution of queueing times. *Proc. Cam. Phil. Soc.* **49**, 449–461.

Sobel, M. J. (1969). Optimal average cost policy for a queue with start-up and shut down costs. *Opns. Res.* **17**, 145–162.

Sobel, M. J. (1974). Optimal operation of queues in *Mathematical Methods of Queueing Theory*, (Clarke, A. B. ed.), pp. 231–261, Springer-Verlag, Berlin, New York.

Stidham, S. Jr., and Prabhu, N. U. (1974). Optimal control in queueing systems. In: *Mathematical Methods of Queueing Theory*, (Clarke, A. B. ed.). pp. 263–294. Springer-Verlag, Berlin, New York.

Stidham, S. (1985). Optimal control of admissions to a queueing system. *IEEE Trans. Automat. Control* **AC-30**, 705–713.

Sunaga, T., Kondo, E., and Biswas, S. K. (1978). An approximation method using continuous models for queueing problems. *J. Opns. Res. Soc. Japan* **21**, 29–44.

Szarkowicz, D. S., and Knowles, T. W. (1985). Optimal control of an $M/M/s$ queueing system. *Opns. Res.* **33**, 644–660; **Err. 34**, 184.

Teghem, J., Jr. (1987). Optimal control of a removable server in an $M/G/1$ queue with finite capacity. *Euro. J. Opnl. Res.* **31**, 358–367.

Teghem, J., Jr. (1986). Control of the service process in a queueing system. *Euro. J. Opnl. Res.* **30**, 141–158.

Tijms, H. C. (1986). *Stochastic Modelling and Analysis: A Computational Aspect.* Wiley, New York.

Tu, H. Y., and Kumin, H. (1983). A convexity result for a class of $GI/G/1$ queueing systems. *Opns. Res.* **31**, 948–950.

Van-Nunen, J. A. E. E., and Puterman, M. L. (1983). Computing optimal control limits of $GI/M/s$ queueing systems with controlled arrivals. *Mgmt. Sci.* **29**, 725–734.

Weber, R. R. (1980). On the marginal benefit of adding servers to $G/GI/m$ queues. *Mgmt. Sci.* **26**, 946–951.

Weber, R. R. (1983). A note on waiting times in single server queues. *Opns. Res.* **31**, 950–951.

Weiss, H. J. (1979). The computation of optimal control limits for a queue with batch services. *Mgmt. Sci.* **25**, 320–328.

Weiss, H. J. (1981). Further results on an infinite capacity shuttle with control at a single terminal. *Opns. Res.* **29**, 1212–1217.

Whitt, W. (1974). Heavy traffic limit theorems for queues: A Survey. In: *Mathematical Methods of Queueing Theory* (Clarke, A. B., ed.), 307–350. Springer-Verlag, Berlin, New York.

Whitt, W. (1985). The best order for queues in series. *Mgmt. Sci.* **31**, 475–487.

Wolff, R. W. (1982). Poisson arrivals see time averages. *Opns. Res.* **30**, 223–231.

Wolff, R. W. (1987). Upper bounds on work in system for multi-channel queues. *J. Appl. Prob.* **24**, 547–551.

Wolff, R. W. (1989). *Stochastic Modeling and the Theory of Queues.* Prentice-Hall.

Yao, D. D. (1985). Refining the diffusion approximation for the $M/G/m$ queue. *Opns. Res.* **33**, 1266–1277.

Yao, D. D. (1986). Convexity properties of the over-flow in an ordered-entry system with heterogeneous servers. *Opns. Res. Lett.* **5**, 145–147.

Appendix 1 Abbreviations and Symbols

BCMP	Baskett, Chandy, Muntz, Palacios
C–K, C.K.	Chapman–Kolmogorov
DF, CDF	distribution function
DFR	decreasing failure rate
e	column vector with each of its elements equal to unity
$E(X)$	expectation of the random variable X
FCFS	first-come/first-served
FCLT	functional Central Limit Theorem
FIFO	first-in/first-out
FLLN	functional-law-of-large-numbers
$F * G$	convolution of two independent distributions with DFs F and G
F^{n*}, $F^{(n)*}$	convolution of n IID random variables with common distribution F

$F^{*(k)}(s)$	kth-derivative of $F^*(s)$ (with $F^{*(1)}(s) \equiv F^{*\prime}(s)$)
IID	identically and independently distributed
iff	if and only if
IMLR	increasing mean residual life
LAA	lack of anticipation assumption
LCFS	last-come/first-served
LHS	Left-hand side
LST	Laplace–Stieltjes Transform
LT	Laplace Transform
MPBA	Multiple Poisson bulk-arrival
$\mathbf{0}$	column vector with each of its elements equal to zero (also matrix with all zero elements)
$\mathbf{P} = (p_{ij})$	a matrix with elements p_{ij}
PASTA	Poisson arrivals see time averages
PDF	probability density function
PGF	probability generating function
P–K, PK	Pollaczek–Khinchin
$Pr(A)$, $P(A)$, $Pr\{A\}$, $P\{A\}$	probability of the event A
RHS	right-hand side
RV	random variable
SD	standard deviation
SMP	semi-Markov process
SUT	start-up time or set-up time
TPM	transition probability matrix
var(X)	variance of the random variable X
WRT	with respect to
$\boldsymbol{\alpha} = (a, \ldots, a_n)$	a vector with elements, $a_1, \ldots, a_n$

Appendix 2 Properties of Laplace Transforms

$$L\{f(t)\} = \bar{f}(s) = \int_0^\infty e^{-st} f(t)dt \quad (t \geq 0)$$

one-to-one correspondence exists between $f(t)$ and $\bar{f}(s)$.

1. Linearity property

$$L\{a_1 f_1(t) + \cdots + a_k f_k(t)\} = a_1 \bar{f}_1(s) + \cdots + a_k \bar{f}_k(s)$$

2. Translation property

$$\text{(i)} \quad L\{e^{-at} f(t)\} = \bar{f}(s + a)$$

$$\text{(ii)} \quad L\begin{Bmatrix} f(t - a), & t > a \\ 0, & t < a \end{Bmatrix} = e^{-as} \bar{f}(s)$$

3. Change-of-scale property

$$L\{f(at)\} = \left(\frac{1}{a}\right) \bar{f}\left(\frac{s}{a}\right)$$

4. LT of derivatives

$$L\{f'(t)\} = s\bar{f}(s) - f(0)$$

and in general

$$L\{f^{(n)}(t)\} = s^n\bar{f}(s) - \sum_{i=1}^{n} s^{n-i} f^{(i-1)}(0)$$

for derivative $f^{(n)}(t)$ of order n, $n = 1, 2, \ldots$.

5. LT of integrals

$$\text{(i)} \quad L\left\{\int_0^t f(x)dx\right\} = \frac{\bar{f}(s)}{s}$$

$$\text{(ii)} \quad L\left\{\int_0^t \int_0^u f(x)dx\,du\right\} = \frac{\bar{f}(s)}{s^2}$$

6. Convolution property

$$L\left\{\int_0^t g(t-y)f(y)dy\right\} = \bar{f}(s)\bar{g}(s)$$

7. Limit property: Initial value property

$$\lim_{t\to\infty} f(t) = \lim_{s\to 0} s\bar{f}(s)$$

8. Limit property: Final value property

$$\lim_{t\to 0} f(t) = \lim_{s\to\infty} s\bar{f}(s)$$

9. Multiplication by power of t

$$L\{t^n f(t)\} = (-1)^n \bar{f}^{(n)}(s)$$

Appendix 3 Table of Laplace Transforms

$$\bar{f}(s) = \int_0^\infty e^{-st} f(t)\,dt$$

	$\bar{f}(s)$	$f(t)$
1.	$\frac{1}{s}$	1
2.	$\frac{1}{s^n}$	$\frac{t^{n-1}}{(n-1)!}, \; n = 2, 3, \ldots .$
3.	$\frac{1}{s^a}$	$\frac{t^{a-1}}{\Gamma(a)}, \; a > 0$
4.	$\frac{1}{\sqrt{s}}$	$\frac{1}{\sqrt{\pi t}}$
5.	$\frac{1}{s-a}$	e^{at}

6.	$\dfrac{1}{(s-a)(s-b)}$	$\dfrac{1}{a-b}(e^{at}-e^{bt}),\ a \neq b$
7.	$\dfrac{1}{(s-a)^n}$	$\dfrac{t^{n-1}e^{at}}{(n-1)!},\ n = 2, 3, \ldots$
8.	$\dfrac{a}{s+a}$	$ae^{-at},\ a > 0$
9.	$\dfrac{a}{s^2+a^2}$	$\sin at$
10.	$\dfrac{s}{s^2+a^2}$	$\cos at$
11.	$\left(\dfrac{a}{s+a}\right)^k$	$\dfrac{a^k t^{k-1} e^{-at}}{\Gamma(k)},\ k > 0,\ a > 0$
12.	$\dfrac{2as}{(s^2+a^2)^2}$	$t \sin at$
13.	$\dfrac{s^2-a^2}{(s^2+a^2)^2}$	$t \cos at$
14.	$\ln\left(1+\dfrac{1}{s}\right)$	$\dfrac{1-e^{-t}}{t}$
15.	e^{-as}	$f(t) = \delta(t-a) = 1,\ t = a$ $= 0,\ t \neq a$ (Dirac δ function located at a)
16.	$\dfrac{e^{-as}}{s}$	$f(t) = 0,\ 0 < t < a$ $= 1,\ a < t$ (unit step function)
17.	$\dfrac{e^{-as}}{s^2}$	$f(t) = 0,\ 0 < t < a$ $= t - a,\ a < t$
18.	$\dfrac{e^{1/s}}{s^{n+1}}$	$t^{n/2} I_n(2\sqrt{t})$

19.	$\dfrac{\{s-(s^2-a^2)\}^n}{\sqrt{s^2-a^2}}$	$a^n I_n(at),\ n > -1$
20.	$\{s-(s^2-a^2)\}^n$	$\dfrac{n}{t}\, I_n(at),\ n = 1, 2, \ldots$

Subject Index

C

D

E

F

G

N

O

P

Q

R

S

T

U

V

W

Y

Z

Author Index

A

B

C

D

E

F

G

H

I

J

K

L

M

N

O

P

R

S

T

U

V

W

Y

Z